Lecture Notes in Mathematics

Edited by A. Dold, Heidelberg and B. Eckmann, Zürich

362

Proceedings of the Conference on the Numerical Solution of Ordinary Differential Equations

19,20 October 1972, The University of Texas at Austin

Proceedings 1972

Edited by Dale G. Bettis, The University of Texas at Austin/USA

Springer-Verlag
Berlin · Heidelberg · New York 1974

AMS Subject Classifications (1970): 65-02, 65Lxx, 70F10

ISBN 3-540-06602-0 Springer-Verlag Berlin · Heidelberg · New York
ISBN 0-387-06602-0 Springer-Verlag New York · Heidelberg · Berlin

Offsetdruck: Julius Beltz, Hemsbach/Bergstr.

PREFACE

Recently there have been many advances in the development of methods for solving ordinary differential equations. These advances consist of both refinements and extensions of previous methods and new methods, for example, variable-order variable-step and generalized multi-step methods; higher order Runge-Kutta methods; and extrapolation methods. These methods are normally tested and compared by solving problems of small systems of ordinary differential equations that have known solutions. These test problems are normally selected because they reveal particular limitations of the methods. In science and engineering the difficulties which the user often encounters are characterized by complexities which are not inherent to the test problems, i.e., limitations associated with computer storage and run time, practical estimates of the error of the machine solution, programming difficulties, etc.

The primary purpose of the Austin Conference was two-fold: (i) the presentation and the comparison of recent advances in the methods for solving ordinary differential equations; (ii) the discussion of the application of these methods to the complex problems encountered by the researcher.

Particular emphasis was devoted to methods and techniques for solving the differential equations of the N-Body problem, which consist of determining the motion and the subsequent evolution of a system of N particles.

There were approximately 180 participants at the Austin Conference and 44 speakers. The abstracts of the presentations appeared in Volume 4 Number 4 of the Bulletin of the American Astronomical Society. This present volume consists of papers given at the Conference.

The Austin Conference was sponsored jointly by the Society for Industrial and Applied Mathematics and by the Division of Dynamical Astronomy of the American Astronomical Society. The Institute for Advanced Studies in Orbital Mechanics, at the University of Texas at Austin was responsible for the overall organization of the Conference.

The support of the Conference by the National Science Foundation was greatly appreciated.

The organizing committee included Guy Janin (Geneva Observatory and The University of Texas at Austin), M. Lecar (Harvard University) and G. Scheifele (Swiss Federal Institute of Technology, Zurich and The University of Texas at Austin).

The planning and organization of the Austin Conference was the result of contributions from many individuals, including representatives of the sponsoring organizations and the speakers and the participants. Special recognition is extended to Dr. John R. Pasta and Dr. Val G. Tareski of the National Science Foundation, Dr. J. Derral Mulholland of the American Astronomical Society, Dr. W.J. Jameson, Jr. from the Society of Industrial and Applied Mathematics, and Dr. Robert T. Gregory, Dr. David M. Young, Jr., Dr. Earnest F. Gloyna, and Dr. Byron D. Tapley from The University of Texas at Austin.

Most of the manuscripts were typed at Austin after the conference. Special appreciation is given to Mrs. Pam Rossano for her careful and patient typing.

D. G. Bettis

Department of Aerospace Engineering
and Engineering Mechanics

The University of Texas at Austin

TABLE OF CONTENTS

LIST OF CONTRIBUTORS

Afaq Ahmad	Hunter College of the City University of New York, New York
J. Baumgarte	Swiss Federal Institute of Technology, Zurich, Switzerland
Paul R. Beaudet	Computer Sciences Corporation, Falls Church, Virginia
Dale G. Bettis	The University of Texas, Austin, Texas
R. Broucke	Jet Propulsion Laboratory, Pasadena, California, and University of California, Los Angeles, California
J.C. Butcher	Dept. of Mathematics, University of Auckland, Auckland, New Zealand
R.P. Canale	The University of Michigan, Ann Arbor, Michigan
P.J. Cefola	Computer Sciences Corporation, Silver Spring, Maryland
Jerome R. Cherniack	Smithsonian Institution, Astrophysical Observatory, Cambridge, Massachusetts
Leon Cohen	Hunter College of the City University of New York, New York
Roy Danchick	Aerojet ElectroSystems Corporation, Azusa, California
P.J. Firnett	Informatics, Inc., Los Angeles, California
D.B. Frazho	The University of Michigan, Ann Arbor, Michigan
L.J. Gallaher	Georgia Institute of Technology, Rich Electronic Computer Center, Atlanta, Georgia
Otis Graf	The University of Texas, Austin, Texas
Avram Hayli	Observatoire de Besancon, Besancon, France
Bernard E. Howard	University of Miami, Coral Gables, Florida
Guy Janin	The University of Texas, Austin, Texas
Fred T. Krogh	California Institute of Technology, Jet Propulsion Laboratory, Pasadena, California
Myron Lecar	Smithsonian Institution, Astrophysical Observatory, Cambridge, Massachusetts
Rudolph Loeser	Smithsonian Institution, Astrophysical Observatory, Cambridge, Massachusetts

A.C. Long	Computer Sciences Corporation, Silver Spring, Maryland
R.H. Miller	University of Chicago, Chicago, Illinois
Hays Moore	Computer Sciences Corporation, Silver Spring, Maryland
K.S. Nimitz	Computer Sciences Corporation, Silver Spring, Maryland
I.E. Perlin	Georgia Institute of Technology, Rich Electronic Computer Center, Atlanta, Georgia
W.F. Powers	The University of Michigan, Ann Arbor, Michigan
Haywood Smith, Jr.	University of South Florida, Tampa, Florida
Frank Stenger	University of Utah, Salt Lake City, Utah
E. Stiefel	Swiss Federal Institute of Technology, Zurich, Switzerland
Joseph Stoer	Institut fur Angewandte Mathematik, Wurzburg, West Germany
B.A. Troesch	University of Southern California, Los Angeles, California
C.E. Velez	NASA/Goddard Space Flight Center, Greenbelt, Maryland
Roland Wielen	Astronomisches Rechen-Insitut, Heidelberg, West Germany

EXTRAPOLATION METHODS FOR THE SOLUTION OF INITIAL VALUE PROBLEMS AND THEIR PRACTICAL REALIZATION

by

Josef Stoer

Institut für Angewandte Mathematik

Würzburg, West Germany

1. Introduction

During the past years, extrapolation methods have been widely used for solving initial value problems for ordinary differential equations. In particular, after the publication of an Algol-program 1966 (see [2]), which was based on the fundamental results of Gragg [6,7] on the existence of asymptotic expansions for discretization methods, this method got a certain popularity and was soon tested and compared with other methods for solving initial value problems. In 1968 N. Clark [3] carried out extensive tests at Argonne and, based on empirical evidence, worked out an improved version of the algorithm and translated it into Fortran. This improved program was then carefully compared with other methods by Crane and Fox [4] at Bell Laboratories in 1969. Still more systematic comparisons were made by Hull et al. [8] in 1971. All these tests were very favorable for extrapolation methods. They confirmed their overall reliability, speed, and accuracy, but also showed certain weak points of the algorithm published in [2] (for later reference, we denote the algorithm (code) of [2] by "Algorithm A", the version of Clark [3] by "Algorithm B"). In particular, algorithms A and B were relatively inefficient for low accuracies and if the initial step-size chosen by the user was grossly inadequate. The reason for this is the crude (but safe) mechanism for changing stepsizes adopted in algorithm A and B which also caused some inefficiencies when these algorithms were tried on problems which require drastic stepsize changes very frequently. (But, surprisingly, Hull et al. found out in [8] that even though in these cases the best predictor/corrector methods of variable order needed much fewer function calls of the right hand side of the differential equation than extrapolation methods, the total computing time needed was less for extrapolation methods, which shows the simplicity and robustness of these methods!)

It is the purpose of this paper to discuss the pertinent aspects of extrapolation algorithms and to propose some remedies for the deficiencies mentioned. According to extensive tests with a preliminary version of a new extrapolation algorithm ("Algorithm C") carried out at Würzburg and by R. Bulirsch at Cologne, the efficiency of Algorithm C is improved some 30%-60% with problems which need many step-size changes.

2. In this section, we will describe some known results on extrapolation algorithms which will be used later on. The theoretical foundation of extrapolation methods was laid by W. Gragg in his thesis [6] and in [7]. He considered the modified midpoint rule for solving the initial value problem

$$(1) \qquad \begin{aligned} y' &= f(x,y) \\ y(x_0) &= y \end{aligned}$$

Without loss of generality we may assume that (1) is a differential equation for one unknown function y(x) only; the following results remain true, if (1) is a system of differential equations, y and f vectors of functions and if all results, relations, etc. are interpreted componentwise.

For a given x and each stepsize h of the form $h = \frac{H}{n}$, where $H := x - x_0$ is the basic stepsize and n is a natural number, the modified midpoint rule yields an approximation $\eta(x,h)$ for y(x) in the following way

$$
\begin{aligned}
&\eta_0 := y_0\\
&\eta_1 := \eta_0 + \frac{h}{2}\, f(x_0,\eta_0)\\
(2)\qquad &\text{for } i := 1,2,\ldots,\ 2n-1\\
&\qquad \eta_{i+1} := \eta_{i-1} + h\, f(x_i,\eta_i)\\
&\eta(x,h) := \tfrac{1}{2}[\eta_{2n} + \eta_{2n-1} + \frac{h}{2}\, f(x_{2n},\ \eta_{2n})]\ ,
\end{aligned}
$$

where $x_i := x_0 + i\cdot\frac{h}{2}$, $i=0,1,\ldots$. Gragg showed that $\eta(x,h)$ has an asymptotic expansion in h^2 of the following form

$$(3)\qquad \eta(x,h) \sim y(x) + e_1(x)h^2 + e_2(x)h^4 + \ldots ,$$

provided f is sufficiently often differentiable in a neighborhood of the exact solution. Here, the coefficients $e_i(x)$ are differentiable functions with $e_i(x_0) = 0$ so that

$$(4)\qquad e_i(x) = H\cdot e_i'(x_0) + O(H^2)\ .$$

Now, in an extrapolation algorithm a sequence $F = \{n_i\}$ of natural numbers with $1 = n_0 < n_1 < \ldots$ is chosen (we choose (see e.g. [2]) the sequence

$$(5)\qquad F = \{\ 1,2,3,4,6,8,12,\ \ldots\ \}$$

of Bulirsch) defining a sequence of step-sizes $h_i := \frac{H}{n_i}$, $i = 0,1,2,\ \ldots$

Then a tableau of approximation, $T_{i,k}$ for $y(x)$ of the following form is computed:

$$
\text{(T)} \qquad
\begin{array}{l|ccccc}
 & 0 & 1 & 2 & & \ell \\
\hline
\eta(x,h_0)=: & T_{00} & & & & \\
 & & T_{11} & & & \\
\eta(x,h_1)=: & T_{10} & & T_{22} & & \\
 & & T_{21} & & & \\
\eta(x,h_2)=: & T_{20} & \vdots & & & T_{\ell\ell} \\
\vdots & & \vdots & & & T_{\ell+1,\ell} \\
 & & T_{\ell 1} & & & \\
\eta(x,h_\ell)=: & T_{\ell 0} & \dashrightarrow & T_{\ell+1,2} & & \\
 & & T_{\ell+1,1} & & & \\
\eta(x,h_{\ell+1}) = & T_{\ell+1,0} & & & &
\end{array}
$$

In the case of rational extrapolation, T_{ik} is defined by

$$T_{ik} = \hat{T}_{ik}(0) ,$$

where $\hat{T}_{ik}(h)$ is that interpolating rational function in h^2 of the form

$$\hat{T}_{ik}(h) = \frac{c_0 + c_1 h^2 + + c_\mu h^{2\mu}}{d_0 + d_1 h^2 + + d_\nu h^{2\nu}} \qquad \mu := [\tfrac{k}{2}] , \; \nu = k - \mu$$

satisfying

$$\hat{T}_{ik}(h_j) = \eta(x,h_j) \quad \text{for } j = i, i-1, \dots, i-k$$

As is well known (see [1,2]), there are simple recursion formulae for T_{ik} permitting computation of T_{ik} from its left neighbors $T_{i,k-1}$, $T_{i-1,k-1}$, $T_{i-1,k-2}$

$$T_{i-1,k-2} \; \xrightarrow[T_{i,k-1}]{T_{i-1,k-1}} \; T_{ik}$$

It is further known [7] that the error $e_{ik} := T_{ik} - y(x)$ has the following form

(6) $$e_{ik} = h_i^2\, h_{i-1}^2 \ldots h_{i-k}^2 \cdot (\sigma_k(x) + O(h_{i-k}^2)) \;, \quad i \to \infty$$

where $\sigma_{k+1}(x)$ is the quotient of Hankel determinants:

$$\sigma_{2q-1}(x) := \frac{-H_0^{(q+1)}(x)}{H_0^{(q)}(x)} \;, \quad \sigma_{2q}(x) := \frac{H_1^{(q+1)}(x)}{H_1^{(q)}(x)}$$

$$H_p^{(q)}(x) := \det\,(\mathcal{H}_q^{(p)}(x)),\ \mathcal{H}_q^{(p)}(x) := \begin{pmatrix} e_p(x), & e_{p+1}(x), \ldots, & e_{p+q-1}(x) \\ e_{p+1}(x), & & \vdots \\ \vdots & & \vdots \\ e_{p+q-1}(x) & , \ldots, & e_{p+2q-2}(x) \end{pmatrix}$$

Because of (4) we have asymptotically for $H \to 0$

$$\mathcal{H}_q^{(p)}(x) = H^q\, E_q^{(p)} + O(H^{q+1}) \;,\ E_q^{(p)} := \frac{d}{dx}\, \mathcal{H}_q^{(p)}(x) \Big|_{x=0}$$

so that for some constant τ_k

(7) $$\sigma_k(x) = \tau_k \cdot H + O(H^2) \;,$$

provided of course $\det(E_0^{(q)})$, $\det(E_1^{(q)}) \neq 0$.

It follows from (6), (7) asymptotically for $H \to 0$

(8) $$e_{ik} = \tau_k H^{2k+3} \cdot \frac{1}{f_{ik}} + O(H^{2k+4})$$

where

$$f_{ik} := (n_i n_{i-1} \ldots n_{i-k})^2 \;.$$

Of course, the construction of (T) should be stopped with an element

T_{ik}, as soon as the relative error $\varepsilon_{ik} = \frac{e_{ik}}{s}$, $s \approx \max_{z\in[x_0,x]} |y(z)|$,

of T_{ik} is equal to the relative accuracy eps wanted for the solution:

$$\text{eps} \le |\varepsilon_{ik}| = \left|\frac{\tau_k H^{2k+3}}{s}\right| \frac{1}{f_{ik}} + O(H^{2k+4}) .$$

To make this stopping criterion feasible, we need an estimate for ε_{ik}: It follows from (8) for $i > k$

$$e_{ik} - e_{i-1,k} = T_{ik} - T_{i-1,k} = -\tau_k H^{2k+3} f_{i-1,k}^{-1} [1-(\frac{n_{i-k-1}}{n_i})^2] + O(H^{2k+4}) ,$$

so that because of $(\frac{n_{i-k-1}}{n_i})^2 \ll 1$ with reasonable accuracy

$$\text{(9)}\quad \begin{aligned} &\text{a)} \quad \left|\frac{\tau_k H^{2k+3}}{s}\right| \approx \left|\frac{T_{ik} - T_{i-1,k}}{s}\right| \cdot f_{i-1,k} \\ &\text{b)} \quad |\varepsilon_{ik}| \approx \left|\frac{T_{ik} - T_{i-1,k}}{s}\right| \cdot \frac{n_{i-k-1}^2}{n_i^2} \end{aligned} \qquad , i > k .$$

These estimates can be computed along with (T).

3. In this section, the problem of choosing or changing the step-size H for extrapolation algorithms is considered. This problem has two practically important aspects for the construction of efficient programs:

1) It should be detected as early as possible, if the currently used step-size (proposed by the user of the program) is too large, and if it is efficient to do so.

2) At the end of each integration step the program should provide the user with a reasonable proposal for the step-size of the next integration step.

To solve these problems we first have to find out what an "optimal step-size" is for extrapolation algorithms. We start with the observation that for given eps and for each index pair (i,k) with $i \geq k \geq 0$ there is a largest step-size H_{ik} such that (T) stops at T_{ik}, $|\varepsilon_{ik}| \approx$ eps, $|e_{ik}| \approx$ eps·s. This step size H_{ik} is approximately given by (compare (8)

(10) $$H_{ik} := \sqrt[2k+3]{\frac{\text{eps}\cdot s}{|\tau_k|} f_{ik}} .$$

The strategy S_{ik} of choosing step-sizes then consists in always choosing the step-size $H = H_{ik}$ for the construction of (T).

In order to find the optimal strategy S_{ik}, we have to compare the costs of S_{ik}. A measure for the cost of S_{ik} is given by the number

$$W_{ik} := \frac{A_i}{H_{ik}} ,$$

where A_i is the number of function calls of the right hand side f of (1) necessary for the construction of the Tableau (T) up to the element T_{ik}, that is the number of function evaluations of f needed for computing $\eta(x,h_j)$, $j=0,1,\ldots,i$. Thus W_{ik} gives the amount of labor involved per step length.

It is easily seen that for the sequence (5) the numbers A_i are given by:

$$\{A_i\}_{i=0,1\ldots} = \{3, 7, 13, 21, 33, 49, 73, \ldots\}$$

where

$$A_0 := 3, \quad A_{i+1} := A_i + 2\cdot n_{i+1} \quad \text{for } i > 0 .$$

Using the definition (10) of H_{ik} we get

(11) $$W_{ik} = \sqrt[2k+3]{\left|\frac{\tau_k}{s}\right|} \cdot A_i \cdot \sqrt[2k+3]{\frac{1}{f_{ik}\cdot\text{eps}}} .$$

A comparison of W_{ik} for fixed k and $i = k, k+1, \ldots$ gives for the sequence (5)

$$W_{kk} < W_{k+1,k} < W_{k+2,k} < \ldots$$

so that S_{kk} is best among S_{ik} for $i \geq k$.

A similar comparison of the strategies S_{kk}, $k = 0,1,\dots$ is not possible since W_{kk} also depends on the problem to be solved, namely via the numbers τ_k in (11). A comparison is only possible under additional assumptions on the growth properties of τ_k: For the sake of simplicity we make the following assumption:

(A) The numbers τ_k, are such that the coefficients

$$\sqrt[2k+3]{\left|\frac{\tau_k}{s}\right|} = c, \text{ for } k = 0,1,\dots$$

are independent of k.

For problems of this type, W_{kk} is given by

$$W_{kk} = c\cdot\alpha_k \;, \quad \alpha_k = \alpha_k(\text{eps}) := A_k \cdot \sqrt[2k+3]{\frac{1}{f_{kk}\cdot\text{eps}}} \qquad (12)$$

For the sequence (5), typical values of α_k are given in the following table:

Table of α_k(eps)

eps \ k	0	1	2	3	4	5	6	7
10^{-1}	6.5	8.4	10.8	13.3	$\longrightarrow$			
10^{-2}	13.9	13.3	15.1	17.2	20.3	$\longrightarrow$		
10^{-3}	30.0	21.1	20.9	22.2	25.1	$\longrightarrow$		
10^{-4}	65.0	33.4	29	28.7	31	33.5	$\longrightarrow$	
10^{-5}	$\longleftarrow$	53.2	40.3	37.1	38.2	40	$\longrightarrow$	
10^{-6}	$\longleftarrow$	84.0	56.2	48	47.1	47.8	51.3	$\longrightarrow$
10^{-7}		$\longleftarrow$	78.1	62	58	57.2	59.8	$\longrightarrow$
10^{-8}			$\longleftarrow$	80.2	71.5	68.4	69.8	73
10^{-9}			$\longleftarrow$	103.5	88.2	81.6	81.4	83.6
10^{-10}				$\longleftarrow$	108.7	97.6	95	95.7
10^{-11}				$\longleftarrow$	134	116.6	110.8	109.5
10^{-12}					$\longleftarrow$	139.2	129	125.3

The numbers α_k are increasing in the direction of the arrows. Each line contains a minimal element $\alpha_\ell(\text{eps}) = \min_k \alpha_k(\text{eps})$ (underlined) showing that for each eps there is an <u>optimal strategy</u> $S_{\ell\ell}$ (under assumption (A)). This minimum is not very marked and is approximately given by the rule of thumb

$$\ell = \ell(\text{eps}) \approx [(m+2.5)/2] \text{ , if eps } = 10^{-m} \text{ .} \tag{13}$$

Of course, this determination of the optimal strategy $S_{\ell\ell}$ is only valid, if (A) holds and it may be questioned whether this is at least approximately true for many practical problems or not. A strong indication for the reasonableness of (A) and the rule (13) are the results of Clark [3] who found out empirically by many tests that to each eps belongs an optimal Tableausize ℓ which is about the same as the one suggested by (13). Moreover, the (2k+3)- rd root in (11) strongly damps an abnormal growth or decrease of τ_k , a further indication that assumption (A) is not unreasonable.

Denote by

$$(14) \qquad \tilde{H} := H_{\ell\ell} = \sqrt[2\ell+3]{\frac{\text{eps}\cdot s}{|\tau_\ell|} f_{\ell\ell}}$$

the step-size belonging to the optimal strategy $S_{\ell\ell}$ and let

$$(15) \qquad \alpha := \sqrt[2\ell+3]{\text{eps}\cdot f_{\ell\ell}} = \sqrt[2\ell+3]{\text{eps}\cdot n_0^2 n_1^2 \cdots n_\ell^2} .$$

Then under assumption (A) it is true that

$$(16) \qquad \left|\frac{\tau_k \tilde{H}^{2k+3}}{s}\right| = \alpha^{2k+3} \quad \text{for all } k = 0,1,2,\ldots,\ell .$$

<u>Proof:</u> The definition of $\tilde{H}$ (14) gives

$$\tilde{H} \cdot \sqrt[2\ell+3]{\left|\frac{\tau_\ell}{s}\right|} = \alpha$$

and assumption (A)

$$\tilde{H} \cdot \sqrt[2k+3]{\left|\frac{\tau_k}{s}\right|} = \alpha \quad \text{for all } k = 0,1,\ldots,\ell ,$$

which is equivalent to (16) .

Note that relation (16) can be used to compute the optimal step-size $\tilde{H}$ from the elements of (T) in the following way:

Assume that eps is given, ℓ has been determined, say by (13), and the corresponding α by (15) and suppose that for some $j > 0$ the partial Tableau (T_j) consisting of the elements T_{ik} of (T) satisfying $i \le j$, $k \le \min(j,\ell)$ has already been computed using the basis step-size H:

$$(T_j)\qquad \begin{matrix} T_{oo} & & & \\ & T_{11} & & \\ T_{10} & \cdot & \cdot & \\ \cdot & \cdot & \cdot & \\ \cdot & \cdot & & T_{jm} \\ \cdot & \cdot & \cdot & \\ \cdot & & \cdot & \\ \cdot & T_{j1} & & \\ T_{jo} & & & \end{matrix}\qquad ,\quad m := m(j) = \min(j,\ell)$$

The optimal step-size $\tilde{H}$ has the form $\tilde{H} = H/u$. Then under assumption (A) we get from (16) and (9)

$$\alpha^{2k+3} = \left|\frac{\tau_k H^{2k+3}}{s}\right| \frac{1}{u^{2k+3}} \approx \left|\frac{T_{ik} - T_{i-1,u}}{s}\right| \frac{f_{i-1,k}}{u^{2k+3}} \quad \text{for } i > k.$$

Using this relation for $k := j-1$, $i := j$ if $j \leq \ell$, and for $k := \ell$, $i := j$ if $j > \ell$, we get the following estimates for the step-size reduction factor u

$$(18)\qquad u \approx \begin{cases} \dfrac{1}{\alpha} \sqrt[2j+1]{\left|\dfrac{T_{j,j-1} - T_{j-1,j-1}}{s}\right| f_{j-1,j-1}}, & \text{if } j \leq \ell \\[2ex] \dfrac{1}{\alpha} \sqrt[2\ell+3]{\left|\dfrac{T_{j,\ell} - T_{j-1,\ell}}{s}\right| \cdot f_{j-1,\ell}}, & \text{if } j > \ell. \end{cases}$$

With the partial tableau (T_j), two cases are possible, $u \leq 1$ or $u > 1$ for the reduction factor computed by (18), which need a separate discussion.

a) If $u \leq 1$ then the current step-size H is not too large for a problem satisfying assumption (A). So (T_{j+1}) is formed, if the latest element T_{jm}, $m = \min(j,\ell)$, of T_j is not yet accurate enough, otherwise the algorithm is stopped and T_{jm} is accepted as approximation to y(x). In the later case, $\tilde{H} := H/u$ is proposed as the step-size for the <u>next</u> integration step.

T_{jm} is accurate enough f if, for example

$$\text{(19)}\quad \begin{aligned} &\text{a)}\quad \left|\frac{T_{jj} - T_{j-1,j-1}}{s}\right| \cdot \beta_{jm} \le \text{eps}\ , \quad \text{if } j \le \ell \ (m = j) \\ &\text{b)}\quad \left|\frac{T_{j,\ell} - T_{j-1,\ell}}{s}\right| \cdot \beta_{jm} \le \text{eps}\ , \quad \text{if } j > \ell \ (m = \ell), \end{aligned}$$

where β_{jm} is a factor with

$$\text{(20)}\qquad \beta_{jm} \approx \begin{cases} 1/(n_j^2 - 2) & \text{if } j \le \ell \qquad (j \ge 1) \\ n_{j-\ell-1}^2/n_j^2 & \text{if } j > \ell\ , \end{cases}$$

but may be chosen larger depending on one's cautiousness. The test (19b) is motivated by the asymptotic error estimate (9b), the test (19a) is based on the assumption that (for small e_{jj}) the error $|e_{jj}|$ of T_{jj} is smaller than the error $|e_{j,j-1}|$ of $T_{j,j-1}$, so that

$$\left|\frac{T_{jj} - T_{j-1,j-1}}{s}\right| \frac{1}{n_{j-2}^2} \le \text{eps}$$

implies

$$\begin{aligned} (n_j^2-2)\text{eps} \ge |T_{jj} - T_{j-1,j-1}| &= |T_{j,j-1} - T_{j-1,j-1} + y(x) - T_{j,j-1} + T_{jj} - y(x)| \\ &\ge |T_{j,j-1} - T_{j-1,j-1}| - |e_{j,j-1}| - |e_{jj}| \\ &\ge |T_{j,j-1} - T_{j-1,j-1}| - 2|e_{j,j-1}| \\ &\approx |e_{j,j-1}|(n_j^2 - 2) \end{aligned}$$

because of (9) and therefore

$$|e_{jj}| \le |e_{j,j-1}| \le \text{eps}.$$

b) If $u > 1$ holds for the partial tableau (T_j), then the current step-size H is too large: For problems satisfying (A) the element $T_{\ell\ell}$ will not be accurate enough. The first Tableau (T_{i_o}) for which the leading element $T_{i_o\ell}$ is accurate enough, $|e_{i_o,\ell}| \le \text{eps}\cdot s$, will have an index $i_o > \max(j,\ell)$ ... .

For problems satisfying assumption (A) this index $i_o = i_o(u)$ can be computed from u: Because of (8), (15), and (16) we have

$$\left|\frac{e_{i_o,\ell}}{s}\right| = \left|\frac{\tau_\ell H^{2\ell+3}}{s}\right| \cdot \frac{1}{f_{i_o,\ell}} = \left|\frac{\tau_\ell \tilde{H}^{2\ell+3}}{s}\right| \frac{u^{2\ell+3}}{f_{i_o,\ell}} = \frac{\alpha^{2\ell+3}u^{2\ell+3}}{f_{i_o,\ell}}$$

$$= \text{eps}\cdot u^{2\ell+3} \cdot \frac{f_{\ell\ell}}{f_{i_o,\ell}} \overset{!}{\lesssim} \text{eps}$$

so that

(21) $$i_o(u) := \min\{i > \ell \mid \sqrt[2\ell+3]{\frac{f_{i\ell}}{f_{\ell\ell}}} \geq u\}$$

The following table of numbers $\beta_{i,\ell} := \sqrt[2\ell+3]{\frac{f_{i\ell}}{f_{\ell\ell}}}$ for the sequence (5)

$i \setminus \ell$	2	3	4	5	6	7
ℓ	1	1	1	1	1	1
$\ell+1$	1.486	1.488	1.46	1.465	1.448	1.453
$\ell+2$	2.025	2.02	2.02	2.02	2.02	2.015
$\ell+3$	2.69	2.75	2.74	2.78	2.74	2.79
$\ell+4$	3.68	3.75	3.80	3.82	3.82	3.86

Table of $\beta_{i\ell}$

shows that for a reasonable range of (i,ℓ), $\beta_{i,\ell}$ is approximately

$$\beta_{i\ell} \approx 1.46\cdot(1.374)^{i-\ell-1} \quad \text{for } i \geq \ell+1 ,$$

giving the following simple approximation to $i_o(u)$:

(22) $$i_o(u) := \begin{cases} \ell+1 & \text{if } 1 \leq u \leq 1.063 \\ \ell+2 + \text{entier}(3.14\cdot \ln(u) - 1.19), & \text{if } u \geq 1.063 \end{cases}$$

If u is very large, it is certainly worthwhile to replace the current step-size H by the reduced step-size $\tilde{H} = H/u$ and to start the construction of (T) again. In this case the work done so far is lost, and therefore it would be unwise to restart the tableau with $\tilde{H}$ if u is close to 1. A rational decision of when to restart can be found by looking at the costs involved: If the old step-size H is retained then for problems satisfying assumption (A) we need $A_{i_o(u)}$ function calls of f for the integration of (1) from x_o to $x_o + H$. If the old tableau is stopped at stage T_j (see (17)) and H is replaced by $\tilde{H} = H/u$ and the tableau is restarted with the reduced step-size, then A_j function calls are discarded. Then, as $\tilde{H}$ is optimal, A_ℓ function calls are needed to integrate (1) from x_o to $x_o + \tilde{H}$. Since $\tilde{H} = H/u < H$, about $\frac{H}{\tilde{H}} = u$ integration steps of size $\tilde{H}$ are needed to cover the distance from x_o to $x_o + H$. So, a restart costs $A_j + u \cdot A_\ell$ function calls. Therefore, in case $u > 1$, one should restart with the reduced step-size $\tilde{H} := H/u$ only if

$$A_j + u \cdot A_\ell < A_{i_o(u)} \quad . \tag{23}$$

Summing up, an extrapolation algorithm with a reasonable mechanism for changing step-sizes looks like this

1) Input x_o, y_o, H, eps.
2) Compute ℓ from (13), α from (15) .
3) Compute (T_o)
4) For j=1,2,...
   a) Compute (T_j) from (T_{j-1}); set $m := \min(j,\ell)$;
   b) Determine u by (18);
   c) Check by (19), whether T_{jm} is accurate enough; if yes, set $x_o := x$, $y_o := T_{jm}$, $H := H/u$ and stop; otherwise go to d) .
   d) If $u \le 1$, set $j := j+1$ and go to 4a); otherwise go to e) .
   e) If $u > 1$ compute $i_o(u)$ by (22). If $A_j + u \cdot A_\ell < A_{i_o(u)}$, set $H := H/u$ and go to 3) ; otherwise set $j := j+1$ and go to 4a) .

At the end of this algorithm, x_o is replaced by x, y_o by y(x) and H by another presumably better step-size for the next integration step.

4. A test version of the method described above (= method C) has been coded and tested at the University of Würzburg on the EL X8 computer (machine precision 10^{-12}) and in a somewhat simplified form also at the University of Cologne (machine precision 10^{-15}) by R. Bulirsch. Both versions gave about the same improvements over algorithm A and B. A final version will be published in Num. Math. in the Handbook Series.

The following typical examples show the gain of efficiency:

1) Restricted three-body problem (see Durham et al. [5])

$$\ddot{x} = x + 2\dot{y} - \mu' \frac{x+\mu}{[(x+\mu)^2 + y^2]^{3/2}} - \mu \frac{x-\mu'}{[(x-\mu')^2 + y^2]^{3/2}}$$

$$\ddot{y} = y - 2\dot{x} - \mu' \frac{y}{[(x+\mu)^2 + y^2]^{3/2}} - \mu \frac{y}{[(x-\mu')^2 + y^2]^{3/2}}, \mu' = 1 - \mu$$

initial values:

a) orbit 1: $t_o = 0$, $x_o = 1.2$, $\dot{x}_o = 0$, $y_o = 0$, $\dot{y}_o = -1.04935\ 75098\ 3$

period T = 6.19216 93313 96... $\mu = 0.01212\ 85627\ 65312$

b) orbit 2: $t_o = 0$, $x_o = 0.994$, $\dot{x}_o = 0$, $y_o = 0$, $\dot{y}_o = -2.03173\ 26295\ 573$

period T = 11.12434 03372 66085.. $\mu = 0.01227\ 7471$

The following figures show the number of function calls N needed for one period versus the accuracy eps. The following extrapolation algorithms have been compared

1) Method A; the old algorithm of [2].
2) Method B: the improved Fortran version of A by Clark [3].
   The corresponding results are reproduced from [3].
3) Method C: the new algorithm of this paper.

Two versions of Method C were tried differing by the choice of β_{jm} in the error test

(19): In Method C_1, the conservative choice $\beta_{jm} = 1$ was made, in Method C_2, β_{jm} was chosen according to (20), which is not so stringent. Of course, C_1 needed more function calls but produced more accurate results than C_2. Nevertheless, in (almost) all cases also the results of C_2 were as accurate as could be expected considering the condition of the problem and the accuracy eps required. So, roughly speaking, C_1 gives unnecessarily accurate results: With C_1 the error per step is much less than eps, with C_2 it is comparable to eps. Since methods A and B were also based on the choice $\beta_{jm}= 1$, the results of A and B should be compared with C_1 rather than with C_2. (The results for A and C_1 were computed at Cologne, the results for C_2 at Würzburg).

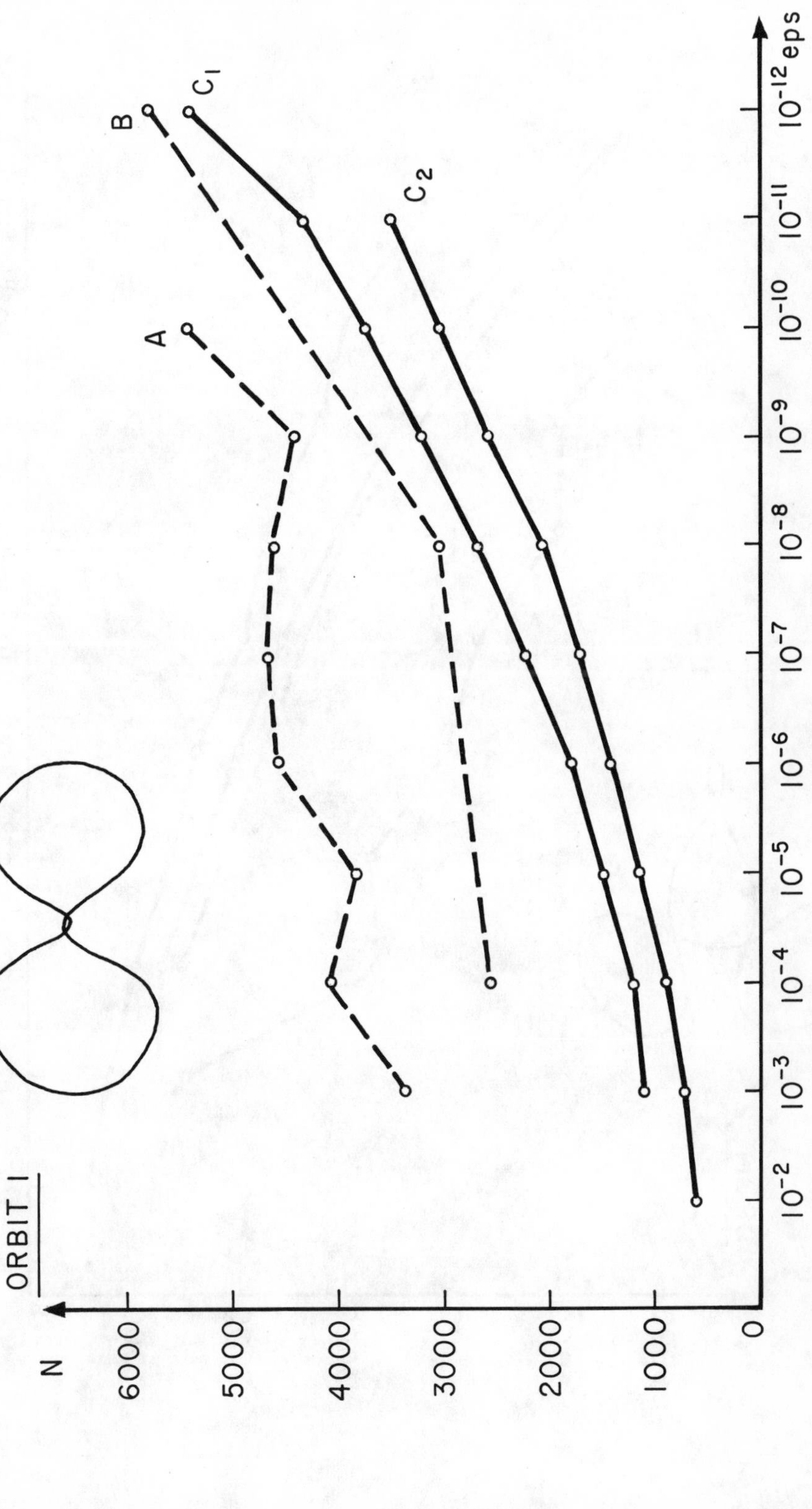
ORBIT I
N
6000
5000
4000
3000
2000
1000
0
A
B
C_1
C_2
10^{-2}
10^{-3}
10^{-4}
10^{-5}
10^{-6}
10^{-7}
10^{-8}
10^{-9}
10^{-10}
10^{-11}
10^{-12} eps

FIG. 1

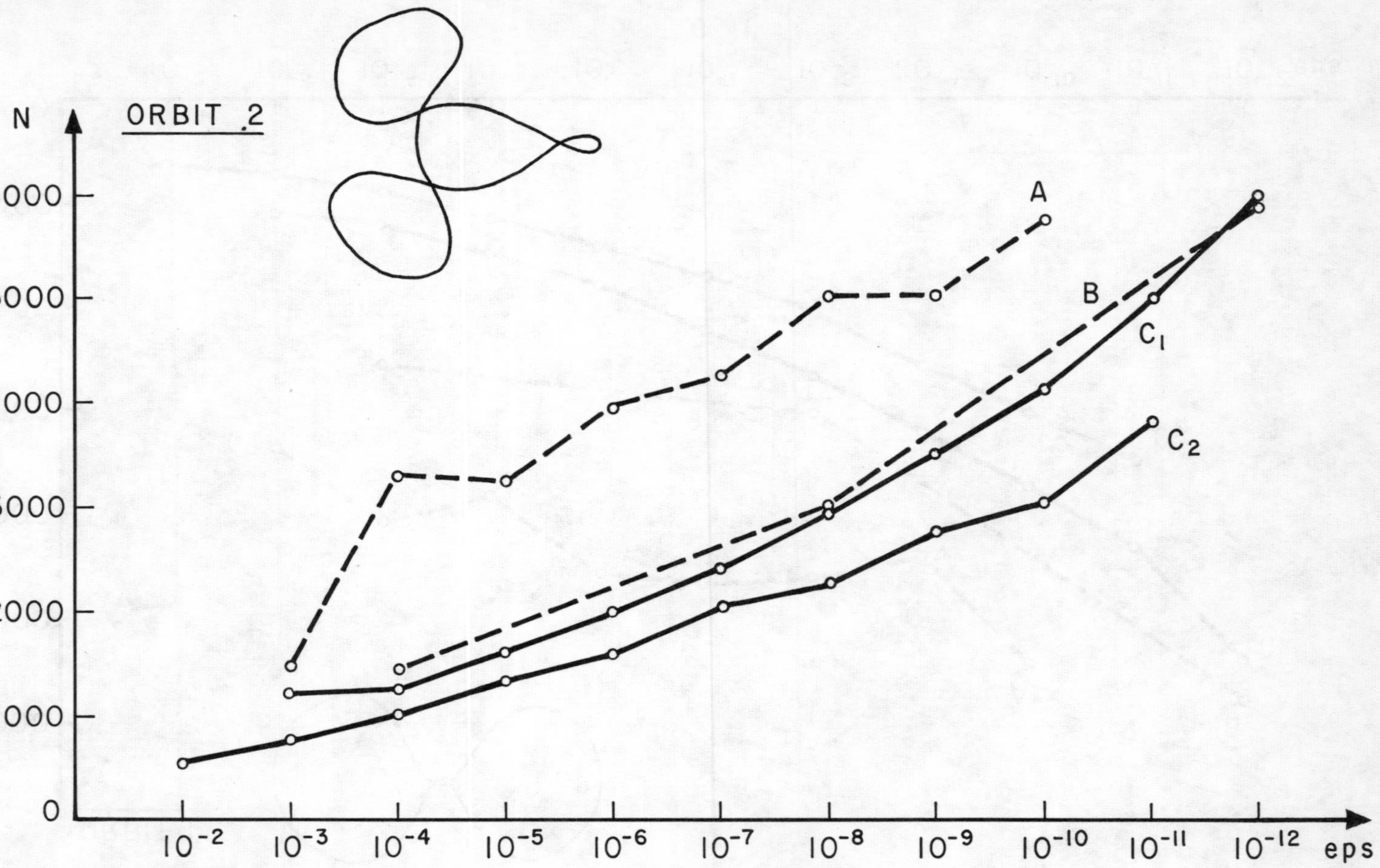
ORBIT 2
N
6000
5000
4000
3000
2000
1000
0
A
B
C1
C2
10-2
10-3
10-4
10-5
10-6
10-7
10-8
10-9
10-10
10-11
10-12
eps

FIG. 2

The following table shows that, other than method A, method C_1 (and likewise C_2) is almost insensitive against a bad choice of the initial step-size H. The results refer to orbit 1 and eps = 10^{-7}:

H	N (Method C_1)	N (Method A)
0.001	2301	5496
0.01	2225	4639
0.1	2229	4682
1.0	2328	4473
3.0	2295	4863
6.1921.. (= period)	2272	5207

2) Computation of an ellipse of eccentricity ε (see Hull et al. [8]).

$$\ddot{x} = \frac{-x}{r^3}, \quad r^2 := x^2 + y^2$$

$$\ddot{y} = \frac{-y}{r^3}$$

initial values: $t_o := 0$, $x_o := 1 - \varepsilon$, $\dot{x}_o := 0$

$y_o := 0$ $\quad \dot{y}_o := \frac{1+\varepsilon}{1-\varepsilon}$

Integration interval: $0 \le t \le 20$.

The following table contains the number N of function calls needed for method B, method C_2 and the multistep method of Adams-Krogh of variable order ("Method D"), which needed the least number of function calls among the methods tested by Hull et al. [8]. The results for methods B and D are taken from [8]. It should be noted however that [8]used a different error criterium for B and D: They run their calculating in such a way that for each integration step the absolute error/step length,

e_{ik}/H , was at most eps (and not the relative error ε_{ik} as with C_2).

	$\varepsilon = 0.5$			$\varepsilon = 0.7$			$\varepsilon = 0.9$		
eps	B	C_2	D	B	C_2	D	B	C_2	D
10^{-3}	841	632	425	1664	782	556	3581	1302	961
10^{-6}	1750	1288	863	3031	1437	1149	4810	2002	1813
10^{-9}	3378	2288	1286	4283	2626	1906	6713	3695	2831

The results show that the improvement of C_2 over B increases with ε , that is with problems needing drastic step-size changes. For problems of this type, C_2 becomes comparable with D even with respect to the number of function calls. Now, one of the observations of [8] was that the computing time for extrapolation methods was less than for the methods needing the fewest function calls, as for example method D. So, Hull et al. concluded that there is no uniformly best (with respect to computing time) integration algorithm, but that it depends on the complexity of f (1) whether, say B is better than D or not. Summing up, our test results seem to show that this break-even point has shifted in favor of the extrapolation algorithms with an improved step-size mechanism.

Acknowledgement:

The author wishes to thank R. Bulirsch for many discussions and for the contribution of several test results obtained by him. Also, the assistance of G. Schuller and B. Kiekebusch-Müller of the Computing Center of the University of Würzburg in coding and testing several versions of extrapolation algorithms was very valuable.

References

1. R. Bulirsch and J. Stoer, "Fehlerabschätzungen und Extrapolation mit rationalen Funktionen bei Verfahren von Richardson-Typus," Num. Math.(1964), 6, 413-427

2. R. Bulirsch and J. Stoer, "Numerical Treatment of Ordinary Differential Equations by Extrapolation Methods," Num. Math.(1966), 8, 1-13

3. N.W. Clark, "A Study of Some Numerical Methods for the Integration of Systems of First Order Ordinary Differential Equations," Report ANL-7428, March 1968, Argonne National Laboratory

4. P.C Crane and P.A. Fox, "A Comparative Study of Computer Programs for Integrating Differential Equations," Numerical Mathematics Computer Program Library One-Basic routines for general use, 1969, Vol. 2, issue 2, Bell Telephone Laboratories Inc., Murray Hill, New Jersey

5. H.L. Durham, O.B. Francis, L.J. Gallaher, H.G. Hale, and I.E. Perlin, "Study of Methods for the Numerical Solution of Ordinary Differential Equations," NASA-CR-57430, 1964, Huntsville, Alabama

6. W.B. Gragg, "Repeated Extrapolation to the Limit in the Numerical Solution of Ordinary Differential Equations," Thesis UCLA (1963)

7. W.B. Gragg, "On Extrapolation Algorithms for Ordinary Initial Value Problems," J. SIAM Numer. Anal.(1965), Ser. B 2, 384-403

8. T.E. Hull, W.H. Enright, B.M. Fellen, and A.E. Sedgwick, "Comparing Numerical Methods for Ordinary Differential Equations," J. SIAM Numer. Anal. (1972), 9, 603-637

CHANGING STEPSIZE IN THE INTEGRATION OF DIFFERENTIAL EQUATIONS USING MODIFIED DIVIDED DIFFERENCES*

by

Fred T. Krogh

California Institute of Technology
Jet Propulsion Laboratory
4800 Oak Grove Drive
Pasadena, Calif. 91103

Abstract

Multistep methods for solving differential equations based on numerical integration formulas or numerical differentiation formulas (for stiff equations) require special provision for changing the stepsize. New algorithms are given which make the use of modified divided differences an attractive way to carry out the change in stepsize for such methods. Error estimation and some of the important factors in stepsize selection and the selection of integration order are also considered.

1. Introduction

We have considered a number of methods for changing the stepsize of multistep methods in [1], and the use of modified divided differences in particular in [2]. The algorithm proposed here is different than that given in [2] in that the stepsize is not changed on every step, and the procedure for computing integration coefficients is designed to take advantage of this situation. Even when the stepsize is changed on every step, the new algorithm computes the required coefficients more efficiently than the algorithms in [2] - [7].

*This paper presents the results of one phase of research carried out at the Jet Propulsion Laboratory, California Institute of Technology, under Contract NAS7-100, sponsored by the National Aeronautics and Space Administration.

However, if there are more than approximately 2 first order stiff equations and the stepsize is changed on every step, then the overall algorithm is slightly less efficient than a carefully organized Lagrangian formulation. (See [7] for example.) Reference [7], which gives the only algorithm of the type considered here for stiff equations, reactivated our interest in this area by giving an algorithm with a computational cost that only goes up linearly with the integration order, as opposed to quadratically for methods based on numerical integration formulas.

The use of a completely variable stepsize gives more flexibility than what we propose here, and other methods considered in [1] require less computation. We believe the new method is a good compromise between the conflicting goals of flexibility, computational economy, and stability and reliability.

The following section gives algorithms for computing integration, interpolation, and differentiation coefficients in a framework useful for the step-by-step integration of ordinary differential equations of arbitrary order. The interpolation and differentiation formulas are useful for the case of stiff equations.

In section 3, details connected with the implementation of these algorithms in a program for integrating differential equations are considered. Computational details associated with implementing the algorithms in an efficient way are considered in section 4.

Section 5 considers the problem of interpolating to points which do not coincide with the end of an integration step. The paper concludes with a discussion of some details such as selection of integration order and stepsize, which are difficult to make rigorous statements about, but which are very important in making an integration program efficient.

2. Algorithms for Computing Coefficients and Updating Differences

Let $w(t)$ be a function given at discrete points, t_i, with $t_{i+1} > t_i$ for all i, and consider the polynomial interpolating w at the points $t_n, t_{n-1}, \ldots, t_{n-q+1}$ given by the Newton divided difference formula

$$P_{q-1,n}(t) = w[t_n] + (t-t_n)w[t_n, t_{n-1}] + \cdots$$
$$+ (t-t_n)(t-t_{n-1})\cdots(t-t_{n-q+2})w[t_n, t_{n-1}, \ldots, t_{n-q+1}] \tag{2.1}$$

where

$$w[t_n, t_{n-1}, \ldots, t_{n-i}] = \begin{cases} w(t_n) & i=0 \\ \dfrac{w[t_n, \ldots, t_{n-i+1}] - w[t_{n-1}, \ldots, t_{n-i}]}{t_n - t_{n-i}} & i=1, 2, \ldots \end{cases} \tag{2.2}$$

Also consider the polynomial

$$P^*_{q,n}(t) = P_{q-1,n}(t) + (t-t_n)(t-t_{n-1})\cdots(t-t_{n-q+1})w[t_{n+1}, t_n, \ldots, t_{n-q+1}] \tag{2.3}$$

which interpolates w at $t_{n+1}, t_n, \ldots, t_{n-q+1}$. Although apparently $P^*_{q,n} \equiv P_{q,n+1}$, later P is associated with a predictor formula and P^* with a corrector formula, and the divided difference $w[t_{n+1}, \ldots, t_{n-q+1}]$ in equation (2.3) is computed using a predicted value of $w(t_{n+1})$.

We are interested in the problem of efficiently integrating, interpolating, and differentiating the polynomials $P_{q-1,n}$ and $P^*_{q,n}$ at $t = t_{n+1}$ and then obtaining the polynomials $P_{q-1,n+1}$ and $P^*_{q,n+1}$.

As in [2] we make a change of variable and introduce some additional notation. Let

$$
\left.
\begin{aligned}
&h_i = t_i - t_{i-1} \\
&\tau = (t-t_n)/h_{n+1} \\
&\xi_i(n+1) = h_{n+1} + h_n + \cdots + h_{n+1-i} \\
&\alpha_i(n+1) = h_{n+1}/\xi_i(n+1) \\
&\beta_0(n+1) = 1 \\
&\beta_i(n+1) = [\xi_0(n+1)\xi_1(n+1)\cdots\xi_{i-1}(n+1)]/[\xi_0(n)\cdots\xi_{i-1}(n)] \\
&\varphi_0(n) = w(t_n) \\
&\varphi_i(n) = \xi_0(n)\xi_1(n)\cdots\xi_{i-1}(n)w[t_n, t_{n-1}, \ldots, t_{n-i}] \\
&H_k(n+1) = h_{n+1}/k, \; H_0(n+1) = h_{n+1} \\
&s_k(n+1) = \begin{cases} H_0(n+1)H_1(n+1)\cdots H_{k-1}(n+1) & k > 0 \\ 1 & k = 0 \\ [H_1(n+1)H_2(n+1)\cdots H_{-k}(n+1)]^{-1} & k < 0 \end{cases}
\end{aligned}
\right\} \tag{2.4}
$$

The (n) or (n+1) following a variable name may be dropped if the value to be assigned to n is obvious. From equations (2.2) and (2.4) it is easy to obtain

$$\varphi_{i+1}(n+1) = \varphi_i(n+1) - \beta_i(n+1)\varphi_i(n) \tag{2.5}$$

from which it is apparent that if $h_{n+1} = h_n = \cdots = h_{n+2-i}$ then the modified divided difference $\varphi_i(n+1)$ is equivalent to the i-th backward difference of w at $t = t_{n+1}$. Later we shall find it useful to estimate $\varphi_i(n+1)$ from the values of $\varphi_i(n)$. These estimates are obtained using equation (2.5) and estimating that $\varphi_q \equiv 0$. Thus

$$
\begin{aligned}
&\varphi_q^{(e)}(n+1) = 0 \\
&\varphi_i^{(e)}(n+1) = \varphi_{i+1}^{(e)}(n+1) + \beta_i(n+1)\varphi_i(n), \quad i=q-1, q-2, \ldots, 0.
\end{aligned}
\tag{2.6}
$$

Clearly

$$\varphi_i(n+1) = \varphi_i^{(e)}(n+1) + [w(t_{n+1}) - \varphi_0^{(e)}(n+1)], \quad i=0, 1, \ldots, q. \tag{2.7}$$

The procedure used to update the modified divided differences from t_n to t_{n+1} is given by equations (2.6) and (2.7). The expression $\beta_i(n+1)\varphi_i(n)$ which is required in equation (2.6) is also used in the formulas for integration, interpolation and differentiation, since its use permits a more economical evaluation of the required coefficients. Thus we introduce

$$\varphi_i^*(n) = \beta_i(n+1)\varphi_i(n) \tag{2.8}$$

From equations (2.1), (2.3), (2.4), (2.6), (2.7), and (2.8), simple substitution gives

$$P_{q-1,n}(t) = P_{q-1,n}(t_n+h_{n+1}\tau) = \sum_{i=0}^{q-1} c_{i,n}(\tau)\varphi_i^*(n) \tag{2.9}$$

and

$$P_{q,n}^*(t) = P_{q-1,n}(t) + c_{q,n}(\tau)[w(t_{n+1})-\varphi_0^{(e)}(n+1)] \tag{2.10}$$

where

$$c_{i,n}(\tau) = \begin{cases} 1 & i = 0 \\ \left(\dfrac{h_{n+1}\tau}{\xi_0(n+1)}\right)\left(\dfrac{h_{n+1}\tau+\xi_0(n)}{\xi_1(n+1)}\right) \cdots \left(\dfrac{h_{n+1}\tau+\xi_{i-2}(n)}{\xi_{i-1}(n+1)}\right) & i \geq 1 \end{cases} \tag{2.11}$$

$$c_{i,n}(\tau) = \begin{cases} 1 & i = 0 \\ \alpha_0(n+1)\tau = \tau & i = 1 \\ [\alpha_{i-1}(n+1)\tau + (\xi_{i-2}(n)/\xi_{i-1}(n+1)]c_{i-1,n} & i \geq 2 \end{cases} \qquad (2.12)$$

To treat the case of extrapolating to $t = t_{n+1}$, set $\tau = 1$ and observe that

$$\alpha_{i-1}(n+1) + (\xi_{i-2}(n)/\xi_{i-1}(n+1)) = 1 \quad (i \geq 2) \qquad (2.13)$$

and hence $c_{i,n}(1) \equiv 1$.

The formula for the k-th derivative of P with respect to t is obtained by computing the k-th derivative of $c_{i,n}(\tau)$ with respect to τ. Thus from equation (2.12)

$$c'_{i,n}(\tau) = \alpha_{i-1}(n+1)c_{i-1,n}(\tau) + [\alpha_{i-1}(n+1)\tau + \frac{\xi_{i-2}(n)}{\xi_{i-1}(n+1)}]c'_{i-1,n}(\tau)$$

$$\vdots$$

$$c^{(k)}_{i,n}(\tau) = k\alpha_{i-1}(n+1)c^{(k-1)}_{i-1,n}(\tau) + [\alpha_{i-1}(n+1)\tau + \frac{\xi_{i-2}(n)}{\xi_{i-1}(n+1)}]c^{(k)}_{i-1,n}(\tau)$$

Setting $\tau = 1$ and using equation (2.13)

$$c^{(k)}_{i,n}(1) = k\alpha_{i-1}(n+1)c^{(k-1)}_{i-1,n}(1) + c^{(k)}_{i-1,n}(1) \qquad (2.14)$$

Define

$$d_{i,k} = \frac{1}{k!}\, c^{(k)}_{i,n}(1) \qquad (2.15)$$

Then from the above ($d_{i,k} = 0$ for $i < k$)

$$d_{i,k} = \begin{cases} 1 & k=0,\ i=0, 1, \ldots, q \\ \alpha_{i-1}(n+1)+d_{i-1,1} & k=1,\ i=1, 2, \ldots, q \\ \alpha_{i-1}(n+1)d_{i-1,k-1}+d_{i-1,k} & k=2, 3, \ldots;\ i=k, k+1, \ldots, q \end{cases} \tag{2.16}$$

From equations (2.4), (2.9), (2.10), and (2.15) there follows

$$\left[\frac{d^k}{dt^k} P_{q-1,n}(t)\right]_{t=t_{n+1}} = s_{-k}(n+1) \sum_{i=k}^{q-1} d_{i,k}\varphi_i^*(n) \tag{2.17}$$

$$\left[\frac{d^k}{dt^k} P^*_{q,n}(t)\right]_{t=t_{n+1}} = \left[\frac{d^k}{dt^k} P_{q-1,n}(t)\right]_{t=t_{n+1}} + s_{-k}(n+1)d_{q,k}[w(t_{n+1})-\varphi_0^{(e)}(n+1)] \tag{2.18}$$

where the $d_{i,k}$ can be determined as indicated in equation (2.16).

For the case of integration we proceed much as we did for differentiation, except that now integration by parts is used. Let $c_{i,n}^{(-k)}(\tau)$ denote the k-fold integral $\int_0^\tau \int_0^\tau \cdots \int_0^\tau c_{i,n}(\tau)d\tau$. Then (starting with equation (2.12))

$$c_{i,n}^{(-1)}(\tau) = (\alpha_{i-1}(n+1)\tau + \frac{\xi_{i-2}(n)}{\xi_{i-1}(n+1)})c_{i-1,n}^{(-1)}(\tau)-\alpha_{i-1}(n+1)c_{i-1,n}^{(-2)}(\tau)$$

$$\vdots$$

$$c_{i,n}^{(-k)}(\tau) = (\alpha_{i-1}(n+1)\tau + \frac{\xi_{i-2}(n)}{\xi_{i-1}(n+1)})\, c_{i-1,n}^{(-k)}(\tau)-k\alpha_{i-1}(n+1)c_{i-1,n}^{(-k-1)}(\tau)$$

$$c_{i,n}^{(-k)}(1) = c_{i-1,n}^{(-k)}(1) - k\alpha_{i-1}(n+1)c_{i-1,n}^{(-k-1)}(1)$$

Define

$$g_{i,k} = (k-1)!\,c_{i,n}^{(-k)}(1) \tag{2.19}$$

From equation (2.11) and the above

$$g_{i,k} = \begin{cases} 1/k & i = 0 \\ 1/[k\cdot(k+1)] & i = 1 \\ g_{i-1,k} - \alpha_{i-1}(n+1)g_{i-1,k+1} & i = 2, 3, \ldots, q;\ k = q+d-i, \ldots, 1 \end{cases} \tag{2.20}$$

where d is the largest value of k for which one desires $g_{q,k}$. From equations (2.4), (2.9), (2.10), (2.19), and the definition of $c_{i,n}^{(-k)}$ it follows that the k-fold integrals of P and P^* are given by

$$\int_{t_n}^{t_{n+1}} \int_{t_n}^{t} \cdots \int_{t_n}^{t} P_{q-1,n}(t)dt = s_k(n+1) \sum_{i=0}^{q-1} g_{i,k}\varphi_i^*(n) \tag{2.21}$$

$$\int_{t_n}^{t_{n+1}} \int_{t_n}^{t} \cdots \int_{t_n}^{t} P_{q,n}^*(t)dt = \int_{t_n}^{t_{n+1}} \int_{t_n}^{t} \cdots \int_{t_n}^{t} P_{q-1,n}(t)dt$$
$$+ s_k(n+1)g_{q,k}[w(t_{n+1}) - \varphi_0^{(e)}(n+1)] \tag{2.22}$$

where the $g_{i,k}$ can be determined as indicated in equation (2.20).

3. Implementation of the Algorithms

Consider the single d-th order differential equation

$$Y^{(d)} = f(t, Y, Y', \ldots, Y^{(d-1)}), \quad Y^{(k)}(t_0) = Y_0^{(k)}, \quad k=0, 1, \ldots, d-1 \tag{3.1}$$

and let

$$\begin{aligned} y_n &= y(t_n) = \text{computed approximation to } Y(t_n) \\ p_n &= \text{predicted value of } y_n \end{aligned} \tag{3.2}$$

Except where noted to the contrary, the extension to systems of differential equations of what is done in this section is a simple matter of applying what is done for the single equation to each equation in the system. Permitting different values for d and/or q (the integration order) for different equations in the system can also be done in an obvious way.

The formulas of the previous section permit a numerical solution of equation (3.1) with d+1 different choices of w. Let

$$w = y^{(d-j)} \tag{3.3}$$

where j takes one of the values $0, 1, \ldots, d$. The smaller the value of j, the more accurate the formulas for solving equation (3.1). But for some problems (stiff equations), too small a value for j will unduly restrict the stepsize because of stability problems. If $j > 0$, then in the general case an approximate solution to a system of nonlinear equations must be obtained on every step. Thus the case $j = 0$ is to be preferred if the stepsize is not thereby unduly restricted. This choice gives the well known Adams-Bashforth-Moulton method when $d = 1$ and the stepsize is constant, and to methods which we shall simply call Adams methods for all d. The case $j = 1$, $d = 1$ gives a class of formulas first suggested by Curtiss and Hirschfelder [8] and recently popularized by Gear [9], [10]. If the $Y^{(d)}$ in equation (3.1) is replaced by 0, then equation (3.1) is an algebraic equation if $d = 1$, and an implicit differential equation of order $d-1$ for $d > 1$. Including such equations, which we shall call implicit for all d, is a simple matter although one is restricted to $j \geq 1$ for this case.

Although equation (3.1) could be broken up into d first order equations, thereby simplifying much of what follows, the extra complexity of dealing with a d-th order equation directly is well worth while in many cases. The most obvious advantage is that only one set of differences is required, thus saving on both storage and the computation of differences. In the case of the Adams methods we have found that integrating 2-nd order equations directly sometimes permits a larger stepsize or enables one to integrate efficiently with a PEC (Predict-Evaluate derivatives-Correct) method instead of the more usual PECE method. The best choice depends on both the differential equations and the initial conditions. For example, in terms of derivative evaluations to obtain a given accuracy on $x'' = -x/r$, $y'' = -y/r$, $r = (x^2+y^2)^{1/2}$ (a simple 2-body problem), if the motion is circular $(PECE)_2$ is approximately twice as efficient as either $(PECE)_1$ or $(PEC)_2$, where the subscript 2 indicates the integration of the above two 2-nd order equations, and 1 the integration of the equivalent first order equations $x' = u$, $u' = -x/r^3$, $y' = v$, $v' = -y/r^3$. On the other hand if the motion is elliptic with eccentricity .6, then $(PEC)_2$ is approximately twice as efficient as $(PECE)_2$ or $(PECE)_1$. Examples can also be given where reduction to a system of first order equations is best; see e.g. [2] or [11]. The advantages or disadvantages of integrating d-th order stiff equations directly is as far as we know an open question. One advantage as we shall show below is that the direct integration permits a reduction in the effective number of equations which must be solved at each step. The reader who is not interested in the general case may find it advantageous to substitute specific values for d and j in the following text. If this is done, any equation with a final index less than the starting index should be deleted.

For predictor formulas, equations (3.3), (2.21), (2.17), (2.9), and (2.4) give

$$\left.\begin{aligned} p_{n+1}^{(d-j-1)} &= y_n^{(d-j-1)} + h \sum_{i=0}^{q-1} g_{i,1}\varphi_i^*(n), \\ p_{n+1}^{(d-j-k)} &= y_n^{(d-j-k)} + H_1[y_n^{(d-j-k+1)} + H_2[y_n^{(d-j-k+2)} + \cdots \\ &\quad +H_{k-1}[y^{(d-j-1)} + h \sum_{i=0}^{q-1} g_{i,k}\varphi_i^*(n)]]\cdots], \quad k=2,3,\ldots,d-j \end{aligned}\right\} \tag{3.4}$$

$$p_{n+1}^{(d-j+k)} = s_{-k} \sum_{i=k}^{q-1} d_{i,k}\varphi_i^*(n), \quad k=1,2,\ldots,j, \tag{3.5}$$

and, of course,

$$p_{n+1}^{(d-j)} = \varphi_0^{(e)}(n+1) = \sum_{i=0}^{q-1} \varphi_i^*(n) \tag{3.6}$$

For the case j = 0, the corrector formulas are

$$y_{n+1}^{(d-k)} = p_{n+1}^{(d-k)} + s_k g_{q,k}[f(t_{n+1}, p_{n+1} p'_{n+1}, \ldots, p_{n+1}^{(d-1)}) - \varphi_0^{(e)}(n+1)], \quad k = 1, 2, \ldots, d. \tag{3.7}$$

$$y_{n+1}^{(d)} = f(t_{n+1}, y_{n+1}, y'_{n+1}, \ldots, y_{n+1}^{(d-1)}) \tag{3.8}$$

The case j > 0 requires the (approximate) solution of the following system of equations.

$$y_{n+1}^{(d-j-k)} = p_{n+1}^{(d-j-k)} + s_k g_{q,k}[y_{n+1}^{(d-j)} - \varphi_0^{(e)}(n+1)], \quad k=1,2,\ldots,d-j \tag{3.9}$$

$$y_{n+1}^{(d-j+k)} = p_{n+1}^{(d-j+k)} + s_{-k} d_{q,k}[y_{n+1}^{(d-j)} - \varphi_0^{(e)}(n+1)], \quad k=1,2,\ldots,j-1 \tag{3.10}$$

$$f(t_{n+1}, y_{n+1}, y'_{n+1}, \ldots, y_{n+1}^{(d-1)}) = p_{n+1}^{(d)} + s_{-j} d_{q,j}[y_{n+1}^{(d-j)} - \varphi_0^{(e)}(n+1)] \tag{3.11}$$

The expression

$$e = [y_{n+1}^{(d-j)} - \varphi_0^{(e)}(n+1)] = \varphi_q(n+1) \tag{3.12}$$

which appears in equations (3.9) and (3.10) is also required for updating the difference table. For the control of round-off error <u>it is essential that e be solved for directly</u>, rather than computing e from equation (3.12) after obtaining $y_{n+1}^{(d-j)}$. Substituting equations (3.9) and (3.10) into equation (3.11), the problem of solving equations (3.9)-(3.11) for e is reduced to

$$\begin{aligned} f(t_{n+1}, p_{n+1} + s_{d-j} g_{q,d-j} e, \ldots, p_{n+1}^{(d-j-1)} + s_1 g_{q,1} e, p_{n+1}^{(d-j)} + e, p_{n+1}^{(d-j+1)} \\ + s_{-1} d_{q,1} e, \ldots, p_{n+1}^{(d-1)} + s_{-j+1} d_{q,j-1} e) = p_{n+1}^{(d)} + s_{-j} d_{q,j} e \end{aligned} \tag{3.13}$$

Of course, in the general case equation (3.13) is a system of equations with order equal to the total number of differential equations in the system. There are many ways that one might obtain an approximate solution to this system; see e.g. [12]. For general purpose use a two iteration constant slope Newton method appears to be a good choice; see [13]. The (1 x 1) matrix required to apply the Newton method to equation (3.13) is given by

$$A = \left\{ (\frac{\partial f}{\partial y}) s_{d-j} g_{q,d-j} + \cdots + (\frac{\partial f}{\partial y^{(d-j-1)}}) s_1 g_{q,1} + (\frac{\partial f}{\partial y^{(d-j)}}) \right. \tag{3.14}$$

$$\left. + (\frac{\partial f}{\partial y^{(d-j+1)}}) s_{-1} d_{q,1} + \cdots + (\frac{\partial f}{\partial y^{(d-1)}}) s_{-j+1} d_{q,j-1} \right\} - s_{-j} d_{q,j}$$

Although it is a little awkward notationally, it is completely straightforward to extend A to the case of a system of equations, even if different values of d, j, and/or q are used for different equations. The same matrix can and should be used over several steps. This is not important for a single equation, but for large systems a significant gain in efficiency results from using the same factorization of A over as many steps as possible. Of course, the partial derivatives of f need not be computed on any step that A is left unchanged.

The two stage iteration process proceeds as follows. Solve

$$Ae_1 = p_{n+1}^{(d)} - f(t_{n+1}, p_{n+1}, p'_{n+1}, \ldots, p_{n+1}^{(d-1)}) \tag{3.15}$$

for e_1. Compute $c_{n+1}, c'_{n+1}, \ldots, c_{n+1}^{(d-1)}$ using equations (3.9) and (3.10) with y_{n+1} replaced by c_{n+1} and $[y_{n+1}^{(d-j)} - \varphi_0^{(e)}(n+1)]$ replaced by e_1. $(c_{n+1}^{(d-j)} = p_{n+1}^{(d-j)} + e_1)$ Then solve

$$Ae_2 = p_{n+1}^{(d)} + s_{-j} d_{q,j} e_1 - f(t_{n+1}, c_{n+1}, c'_{n+1}, \ldots, c_{n+1}^{(d-1)}) \tag{3.16}$$

for e_2. Substituting c_{n+1} for p_{n+1} and e_2 for $[y_{n+1}^{(d-1)} - \varphi_0^{(e)}(n+1)]$ in equations (3.9) and (3.10) the final values of $y_{n+1}, \ldots, y_{n+1}^{(d-1)}$ are obtained. The difference table is updated using $e = e_1 + e_2$, and the ratio $\|e_2\| / \|e_1\|$ gives an indication of the convergence of the iteration and thus is useful in deciding how frequently a new matrix A should be introduced for the iteration. Note that two evaluations of f are required, the same as is required in equations (3.7) and (3.8). For implicit equations, simply replace $p_{n+1}^{(d)}$ and $s_{-j}d_{q,j}$ with 0 in equations (3.11), (3.13)-(3.16).

For the purpose of obtaining error estimates, observe that equations (2.6) and (2.8) give

$$\sum_{i=0}^{q-1} g_{i,k}\varphi_i^*(n) = \sum_{i=0}^{q-1} g_{i,k}[\varphi_i^{(e)}(n+1) - \varphi_{i+1}^{(e)}(n+1)] = \sum_{i=0}^{q-1} g_{i,k}^*\varphi_i^{(e)}(n+1) \qquad (3.17)$$

where

$$g_{i,k}^* = \begin{cases} g_{i,k} & i = 0 \\ g_{i,k} - g_{i-1,k} & i > 0 \end{cases} \qquad (3.18)$$

Since $g_{q,k} = g_{0,k}^* + g_{1,k}^* + \cdots + g_{q,k}^*$, equations (2.7) and (3.17) permit us to write

$$\sum_{i=0}^{q-1} g_{i,k}\varphi_i^* + g_{q,k}[w(t_{n+1}) - \varphi_0^{(e)}(n+1)] = \sum_{i=0}^{q} g_{i,k}^*\varphi_i(n+1) \qquad (3.19)$$

Clearly the same type formulas can be obtained with $d_{i,k}$ substituted for $g_{i,k}$, and

$$d^*_{i,k} = \begin{cases} d_{i,k} & i = 0 \\ d_{i,k} - d_{i-1,k} & i > 0 \end{cases} \qquad (3.20)$$

It is also clear that replacing $g_{q,k}$ in equation (3.19) with $g_{q-1,k}$ will simply change the upper limit of the sum on the right side of equation (3.19) from q to q-1. The corrector formulas as given in equations (3.7), (3.9)-(3.11) have an order one greater than the predictor formulas in equations (3.4) and (3.5). If $g_{q,k}$, $g_{q,k}$, $d_{q,k}$ and $d_{q,j}$ in equations (3.7), (3.9), (3.10), and (3.11) respectively were replaced by $g_{q-1,k}$, $g_{q-1,k}$, $d_{q-1,k}$, $d_{q-1,j}$ then the correctors would have the same order as the predictors. We have given our algorithm with correctors of higher order than the predictors because in the case of a constant stepsize and d = 1, the Adams methods for q = 1, 2, . . . , 12 have significantly better stability characteristics when the corrector has an order one greater than the predictor. (For q = 13, . . . , 19, as far as we have checked, the opposite is true.) For the case j = 1, d = 1, Klopfenstein [13] has shown that the method using a corrector with order one greater than the order of the predictor has the same region of asymptotic absolute stability as the method using the same order corrector. (This means that for h sufficiently large, the methods are equally sensitive to error in the matrix A of equation (3.14).) In both cases one also has the advantage of using a corrector which will give better accuracy on most problems. (As far as we know, no effort has been made to compare the algorithm which uses a corrector with order one greater than that of the predictor to the more usual one for the case d > 1 or for the case when the stepsize is not constant.)

It is frequently stated that for the purpose of error estimation the predictor and corrector should be of the same order. If the usual error estimate is added to the corrector with the same order as the predictor, one increases the order of the corrector by one, obtaining the type of algorithm we recommend. We suggest using the same error estimate for the case when the corrector presumably is more accurate. Also see Shampine and Gordon [14] where this matter is considered in more detail. For a given stepsize, if the integration order is selected to minimize the local error, then by necessity any error estimate will be quite crude. (Despite the crude error estimates one gets with this policy of selecting the integration order, we believe it is the best policy since it tends to reduce global errors for a given amount of work, and since reasonable estimates of the global error are usually difficult to obtain from local errors even if they are known with high precision.)

Let

$$E_{n+1}^{(d-k)} = y_{n+1}^{(d-k)} - \hat{y}_{n+1}^{(d-k)} \tag{3.21}$$

where $\hat{y}_{n+1}^{(d-k)}$ is the result of using a corrector with the same order as the predictor to compute $y_{n+1}^{(d-k)}$, and $E_{n+1}^{(d-k)}$ is to serve as an indication of the error in $y_{n+1}^{(d-k)}$.

Clearly for the case j=0, equations (3.7), (3.18), and (3.21) give

$$E_{n+1}^{(d-k)} = s_k g^*_{q,k}[f(t_{n+1}, p_{n+1}, p'_{n+1}, \ldots, p_{n+1}^{(d-1)}) - \varphi_0^{(e)}(n+1)] \tag{3.22}$$

Although the local error in $y_{n+1}^{(d-k)}$ is of higher order in h_{n+1} the larger the value of k, this is not true for the global error; see [11]. The global error in all cases has order one less than the order of the local error in computing $y^{(d-1)}$.

Computing good theoretical error bounds for the case $j > 0$ is more work than can be justified. Thus we suggest estimating the change in e due to using correctors of different orders by the change in $e_1 (= e_1 - \hat{e}_1)$. This gives a good approximation if $\|e_1\|$ is considerably larger than $\|e_2\|$, which should be the case since ordinarily one will want to recompute A if $\|e_2\| > \alpha \|e_1\|$, where $\alpha \approx 1/8$. With $\hat{A}$ defined as is A in equation (3.14), except with every q replaced by q-1, it is easy to obtain from equations (3.15), (3.15) with Ae_1 replaced by $\hat{A}\hat{e}_1$, (3.20), and (3.21)

$$\hat{A}\, E_{n+1}^{(d-j)} = s_{-j} d_{q,j}^{*} e_1 \tag{3.23}$$

By appropriately bounding the stepsize (to prevent A and $\hat{A}$ from becoming too poorly conditioned), one can guarantee that $E_{n+1}^{(d-j)}$ will be changed very little if $\hat{A}$ is replaced by A in equation (3.23). This is what we recommend to the cautious user. Those who regard the error estimate primarily as a means to reasonable stepsize control may want to simplify equation (3.23) by assuming the partial derivatives are all zero (always a reasonable assumption for sufficiently small stepsize), obtaining

$$E_{n+1}^{(d-j)} \approx - \frac{d_{q,j}^{*}}{d_{q-1,j}} e_1 = - \frac{d_{q,j}^{*}}{d_{q-1,j}} (c_{n+1}^{(d-j)} - p_{n+1}^{(d-j)}) \tag{3.24}$$

The use of equation (3.24) for large stepsizes can be justified for some types of problems. For example, it provides safe error bounds if A is diagonally dominant with all negative elements on the diagonal. Given $E_{n+1}^{(d-j)}$, equations (3.9), (3.10), (3.12), (3.18), (3.20) and (3.21) clearly yield

$$E_{n+1}^{(d-j-k)} = s_k(g_{q-1,k} E_{n+1}^{(d-j)} + g_{q,k}^{*} e_1), \quad k=1,2,\ldots,d-j \tag{3.25}$$

$$E_{n+1}^{(d-j+k)} = s_{-k}(d_{q-1,k} E_{n+1}^{(d-j)} + d_{q,k}^{*} e_1), \quad k=1,2,\ldots,j-1 \tag{3.26}$$

Almost as important as obtaining an estimate of the local error is estimating the effect that various strategies of selecting the stepsize will have on future estimated errors. As a first step consider the case when the divided difference (and hence the corresponding derivative) which is in the error estimate is constant, and $h_{n+k} = h_n$, $k \geq 1$. If there has been a recent change in stepsize, then local error estimates will change from step to step until q steps without a stepsize change have occurred. This is due to the factors $\xi_0 \xi_1 \cdots \xi_{q-1}$ which multiply $w[t_n, t_{n-1}, \ldots, t_{n-q}]$ to form $\varphi_q(n)$; see equation (2.4). For this simple case, after q steps without a change in the stepsize, the error estimate will have changed from its current value by a factor of (approximately)

$$\sigma_q = \frac{h_n(2h_n)\cdots(qh_n)}{h_n(h_n+h_{n-1})\cdots(h_n+h_{n-1}+\cdots+h_{n-q+1})} \tag{3.27}$$

In order to limit the frequency with which the stepsize is changed and to reduce the work required to decide how much to change the stepsize, we suggest giving the user the option of specifying the two parameters

$$\begin{aligned} &\rho_i(\rho_i > 1) = \text{the basic factor by which the stepsize is to be increased.} \\ &\rho_d(\rho_d < 1) = \text{the basic factor by which the stepsize is to be decreased.} \end{aligned} \tag{3.28}$$

The closer to one these parameters are selected, the more frequent changes in the stepsize will be, and thus the more overhead that is required for computing integration coefficients and difference tables. (See the next section.) At the same time, increased flexibility in selecting the stepsize (within reason) enables the solution to be computed to a given accuracy with fewer derivative evaluations. A reasonable choice for these parameters in most applications is $\rho_i = 2$, $\rho_d = 1/2$; but for problems with extremely expensive derivative evaluations, values as close to one as $\rho_i = 1.1$, $\rho_d = .9$ may prove useful.

We propose the following strategy for selecting the stepsize.

1. After computing error estimates, but before the second derivative evaluation of the step, check to see if the estimated error is "too" big. If so, go back to the beginning of the current step and try again with the stepsize reduced by a factor of $\min\{1/2, \rho_d\}$.
2. At the completion of the step, estimate what the error would be on the next step if the stepsize were held constant. If this estimated error is "too" big, reduce the stepsize by a factor of ρ_d before starting the next step.

The "too" big in the test for redoing a step should be at least twice as large as the "too" big in the test for simply reducing the step. With such a policy a step will require being repeated only rarely, thus saving the derivative evaluation that is wasted in such cases and also some of the overhead associated with backing up.

If no reduction in the stepsize is required, then increase the stepsize by a factor of ρ_i^k, where k is the smallest integer for which (any estimated error)$\cdot$(max $\{\sigma_q, 1\})\cdot(\rho_i^q)^{k+1}$ is "too" big. The "too" big used here should be no bigger than one tenth the "too" big used in the test for deciding if the stepsize should be reduced at the end of the step unless there has been a fairly long and consistent history of the error decreasing from one step to the next, in which case it pays to gradually increase the tolerance used here until it is the same size as that used for decreasing the stepsize. We have found it prudent to restrict k (in ρ_i^k) so that $\rho_i^k \leq \max(2, \rho_i)\rho_i^j$, where h increased on the previous step by ρ_i^j. The factors ρ_i^q used in estimating the growth in the error should be stored during the initialization procedure for all values of q which may be used.

4. Computational Details

In presenting the algorithms below, the following notation is used.

n_e number of differential equations.

$d(\ell)$ order of the ℓ-th equation.

$q(\ell)$ integration order used on the ℓ-th equation.

$j(\ell)$ the $(d(\ell)-j(\ell))$-th derivative of $y(\ell)$ is used in forming the differences $\varphi(i, \ell)$, $i=0, 1, \ldots$

$y^{(k)}(\ell)$ current value of $y^{(k)}$ for the ℓ-th equation, $k=0, 1, \ldots, d(\ell)-1$.

$\hat{y}^{(k)}(\ell)$ value of $y^{(k)}(\ell)$ from the previous step.

$f(\ell)$ current value of $f(t, y, y', \ldots, y^{(d-1)})$ for the ℓ-th equation.

$\varphi(i, \ell)$ i-th modified divided difference for the ℓ-th equation.

$e(\ell)$ for stiff equations, the value of e for the ℓ-th equation. (See equations (3.12), (3.15), (3.16).)

h current value of the stepsize.

$\xi(k), \alpha(k), \beta(k), H(k), s(k)$ current values of $\xi_k, \alpha_k, \beta_k, H_k$, and s_k. (See equation (2.4).)

$g(i, k), d(i, k)$ coefficients for integration and differentiation formulas. Same as $g_{i,k}$, $d_{i,k}$. (See equations (2.16), (2.20), (3.4), and (3.5).)

$g^*(i, k)$, $d^*(i, k)$ coefficients required for error estimation. (See equations (3.18), (3.20), (3.22), and (3.23).)

$\sigma(k)$ Same as σ_k in equation (3.27).

q_I $\max\limits_{\{\ell: j(\ell)<d(\ell)\}} \{q(\ell)\}$ = maximum order integration formula used.

q_D $\max\limits_{\{\ell: j(\ell)>0\}} \{q(\ell)\}$ = maximum order differentiation formula used.

m_I $\max\{d(\ell)-j(\ell)\}$ = maximum number of repeated integrations.

m_D $\max\{j(\ell)\}$ = maximum number of repeated differentiations.

n_h number of steps that h has been constant (not counting the current step).

(4.1)

We have found that permitting different values for d, q, and j for the different equations in a system, and permitting any q to change from one step to the next, provides a useful flexibility. Different values for d can give a more efficient integration if equations of different orders are being integrated; different values for q can make for more efficient integrations and makes available valuable information for diagnostic purposes; and different values for j can significantly reduce the size of the system of nonlinear equations which must be solved on every step if only a few equations in a large system cause the stiffness. When implemented as described here, this flexibility is obtained with little cost over what is required when the same values are used for each equation. Where it is not obvious, we indicate the simplifications that can be obtained when one or another of d, q, or j is fixed. In order that the implementation be as efficient as possible, we require $d(\ell) \le 4$. The extension to larger values of d is trivial, but in practice larger values of d are rarely used. (Such equations can always be broken up into lower order equations.)

The description of the algorithm for computing integration, differentiation, and related coefficients, assumes that the following initial values have been assigned as indicated. These coefficients are never changed by the algorithm.

$$\left.\begin{aligned}
&\alpha(0) = \beta(0) = s(0) = \sigma(1) = 1.\\
&d(1,1) = d^*(1,1) = 1.\\
&d(i,k) = d^*(i,k) = 0,\ i = 1, 2, \ldots, k-1;\ k = 2, 3, \text{ and } 4.\\
&g(0,k) = 1/k,\ g(1,k) = 1/[k(k+1)],\ k = 1, 2, 3, \text{ and } 4.
\end{aligned}\right\} \qquad (4.2)$$

The following variables are used internal to the algorithm.

q_s	$\max\{q_I, q_D, 2\}$ = step number of the method.
n_s	number of steps for which α, β, σ, and ξ coefficients determined by a constant stepsize have been computed.
$n_I(n_D)$	number of steps for which integration (differentiation) coefficients determined by a constant stepsize have been computed.
τ_1, τ_2	locations used for temporary storage when computing ξ.

(4.3)

(4.3 cont'd.)

q_I^* value of q_I the last time integration coefficients were computed. (Initially $q_I^* = 0$.)

$\hat{q}_I$ $= q_I^* + m_I$.

B(k) $= 1/k(k+1), k=1, 2, \ldots, \max\{q_I+m_I-1\}$ (used to initialize V(k)).

V(k) $= g(n_I-1, k)$ if $k \le \hat{q}_I-n_I$; $= g(\hat{q}_I-k, k)$, for $k = \hat{q}_I-n_I+1, \ldots, \hat{q}_I-1$ (used to initialize W(k)).

W(k) $= g(n, k)$ (at C20 in the algorithm below) $k=1, 2, \ldots, \hat{q}_I-n$). (4.3)

For use in selecting the order, it is useful to carry along one more difference than is required by the integration (or differentiation) formula. Since the value of β used in forming this last difference is not very critical, a simple extrapolation formula is used to obtain the last β (see C13). In order to make good decisions on when to increase the order we have found it necessary to examine differences of at least 4 different orders. In order to have enough differences for the order selection then it is necessary to restrict q_s to be ≥ 2. With such a policy it is necessary to set $\xi(0)$ equal to the starting stepsize when starting an integration. (We also use a different method for selecting the order on the first few steps.) For best efficiency the place to go at statements C20 and C28 should be set initially (once per integration if m_I and m_D are constant) based on the values of m_I and m_D. (In FORTRAN this is best done with the assigned GO TO statement.) It is assumed that if $m_I=0$ then $q_I=0$, if $m_D=0$ then $q_D=0$, and that neither q_I nor q_D can be increased by more than one on any step. We have left out the calculation of $g^*(n, k)$ for $k \ge 2$ at C21-C23 since we never estimate errors in anything but $y^{(d-1)}$. Error estimates for $y^{(d-j)} (j > 1)$ should not be used for stepsize selection since the error estimates tend to be much too small when starting. (Due to the small stepsize required by the low order.)

Algorithm for Computing Coefficients

C1.	[Set q_s=step number of method.]	$q_s \leftarrow \max\{q_I, q_D, 2\}$.
C2.	[Test if stepsize did not change.]	if $n_h \neq 0$, go to C5.
C3.	[Set new values for H and s.]	$H(0) \leftarrow h$; $H(k) \leftarrow h/k$, $s(k) \leftarrow H(k-1)s(k-1)$, $k=1,2,\ldots,m_I-1$; $s(-k) \leftarrow s(1-k)/H(k)$, $k=1,2,\ldots,m_D$.
C4.	[Set variables indicating a step change.]	$n_s \leftarrow 1$; $n \leftarrow 1$; $n_I \leftarrow 1$; $n_D \leftarrow 1$; $\tau_1 \leftarrow h$; go to C11.
C5.	[Test if integration order did not increase.]	if $q_I \leq q_I^*$, go to C7.
C6.	[Compute new V's required by the increase in q_I.]	$V(\hat{q}_I) \leftarrow B(\hat{q}_I)$; if $n_I=2$, go to C7; $V(k) \leftarrow V(k) - \alpha(\hat{q}_I - k)V(k+1)$, $k = \hat{q}_I - 1, \hat{q}_I - 2, \ldots, \hat{q}_I - n_I + 2$.
C7.	[Test if stepsize has been constant long enough.]	if $q_s \leq n_s$, go to C14.
C8.	[Update n_s and set the index n.]	$n_s \leftarrow n_s + 1$; $n \leftarrow n_s$.
C9.	[Compute precisely, those coefficients which remain fixed if h is held constant.]	$\beta(n_s - 1) \leftarrow 1$; $\sigma(n_s) \leftarrow 1$; $\alpha(n_s - 1) \leftarrow 1/n_s$; $\tau_1 \leftarrow n_s h$.
C10.	[Test if step has been constant for q_s steps.]	if $n \geq q_s$, go to C13.
C11.	[Compute coefficients which will change on next step, even if h is held constant.]	$\tau_2 \leftarrow \xi(n-1)$; $\xi(n-1) \leftarrow \tau_1$; $\beta(n) \leftarrow \beta(n-1)\tau_1/\tau_2$; $\tau_1 \leftarrow \tau_2 + h$; $\alpha(n) \leftarrow h/\tau_1$; $\sigma(n+1) \leftarrow (n+1)\sigma(n)\alpha(n)$.
C12.	[Test if more coefficients need be computed.]	$n \leftarrow n+1$; if $n < q_s$, go to C11.
C13.	[Set $\xi(q_s - 1)$ and approximate $\beta(q_s)$.]	$\xi(q_s - 1) \leftarrow \tau_1$; $\beta(q_s) \leftarrow \beta^2(q_s - 1)/\beta(q_s - 2)$
C14.	[Test if no integ. coeff. are required.]	if $n_I > q_I$, go to C26.
C15.	[Set indices for computing integ. coefficient.]	$n \leftarrow n_I$; $n_I \leftarrow n_I + 1$; $q_I^* \leftarrow q_I$; $\hat{q}_I \leftarrow q_I^* + m_I$; $j \leftarrow \hat{q}_I - n$.

C16.	[Test if stepsize did not change.]	if $n > 1$, go to C18.
C17.	[Initialize V(k) and W(k).]	$V(k) \leftarrow B(k)$, $W(k) \leftarrow V(k)$, $k=1, 2, \ldots, j$; go to C25.
C18.	[Update V(k) (and initialize W(k)).]	$V(k) \leftarrow V(k) - \alpha(n-1)V(k+1)$, $W(k) \leftarrow V(k)$, $k=1, 2, \ldots, j$; go to C20.
C19.	[Inner loop for computing integration coefficients.]	$j \leftarrow j-1$; $W(k) \leftarrow W(k) - \alpha(n-1)W(k+1)$, $k=1, 2, \ldots, j$.
C20.	[Go store integration coefficients.]	go to C25-m_I.
C21.	[C21 ≡ C25-4.]	$g(n, 4) \leftarrow W(4)$.
C22.	[C22 ≡ C25-3.]	$g(n, 3) \leftarrow W(3)$.
C23.	[C23 ≡ C25-2.]	$g(n, 2) \leftarrow W(2)$.
C24.	[C24 ≡ C25-1.]	$g(n, 1) \leftarrow W(1)$; $g^*(n, 1) \leftarrow g(n, 1) - g(n-1, 1)$
C25.	[Test if more integ. coeff. required.]	$n \leftarrow n+1$; if $n \leq q_I$, go to C19.
C26.	[Test if no differentiation coeff. required.]	if $n_D > q_D$, go to C34.
C27.	[Set indices for computing diff. coeff.]	$n \leftarrow n_D$; $n_D \leftarrow n_D+1$; if $n=1$, go to C33.
C28.	[Go compute and store diff. coeff.]	go to C33-m_D
C29.	[C29 ≡ C33-4.]	$d^*(n, 4) \leftarrow \alpha(n-1)d(n-1, 3)$; $d(n, 4 \leftarrow d^*(n, 4) + d(n-1, 4)$.
C30.	[C30 ≡ C33-3.]	$d^*(n, 3) \leftarrow \alpha(n-1)d(n-1, 2)$; $d(n, 3) \leftarrow d^*(n, 3) + d(n-1, 3)$.
C31.	[C31 ≡ C33-2.]	$d^*(n, 2) \leftarrow \alpha(n-1)d(n-1, 1)$; $d(n, 2) \leftarrow d^*(n, 2) + d(n-1, 2)$.
C32.	[C32 ≡ C33-1.]	$d^*(n, 1) \leftarrow \alpha(n-1)$ $d(n, 1) \leftarrow d^*(n, 1) + d(n-1, 1)$.
C33.	[Test if more diff. coeff. required.]	$n \leftarrow n+1$; if $n \leq q_D$ go to C28.
C34.	[End of computing coefficients.]	Exit.

A striaghtforward implementation of the difference formulation of the Adams method involves accessing the difference tables in three different loops: to form the predicted values, to form the differences from predicted derivative values in order to obtain corrected values, and to form the difference tables at the end of the step from the final corrected derivative values. Each of these loops involves overhead associated with initializing indices and with the loop itself, and each must be passed through for each equation. An algorithm is given below which requires accessing the difference tables in only one loop. (The difference tables also must be accessed when correcting, estimating errors, and selecting orders, but the entire difference table is not required for these operations.) The capability of treating equations with different orders is obtained using preassigned transfers (similar to what was done in the algorithm for computing integration and differentiation formula coefficients) rather than a loop which would require additional overhead. The variable order Adams program DVDQ [15] uses three loops for operations on the difference tables, and loops on the order (for both predicting and correction) to permit equations of different orders. Thus a similar program based on the approach used here should require significantly less overhead than is reported for DVDQ in [16] and [17]. (Note, DVDQ uses a different method for changing stepsize, see [1] and it has no provision for stiff equations.)

The algorithm given below should be executed just after computing integration coefficients, which in turn is the first thing done on a step. The algorithm includes among the jobs it does:

1. An updating of the difference table based on $[y_n^{(d-j)} - \varphi_0^{(e)}(n)]$ from the previous step.
2. The calculation of predicted values for the differences on the next step, $\varphi_i^{(e)}(n+1)$, to be used for job 1 on the next step.
3. The calculation of predicted values for the dependent variables.

These are two situations when job 1 will have been done previous to the execution of this algorithm. If a step is being repeated, it is easiest to return the difference tables to the point they would be in just after step 1 using the formula

$$\varphi_i(n) = [\varphi_i^{(e)}(n+1) - \varphi_{i+1}^{(e)}(n+1)]/\beta_i(n+1), \quad i=0, 1, \ldots \tag{4.4}$$

which follows immediately from equation (2.6). If an interpolation to an off-step point is required at the end of a step, then it is best to convert the $\varphi_i^{(e)}$'s to φ_i's using equation (2.7) before doing the interpolation. We examine this point in more detail in the next section. Thus we introduce

$$\nu = \begin{cases} 0 \text{ if no update has occurred } (\varphi \text{ contains } \varphi^{(e)}\text{'s}) \\ 1 \text{ if there has been an update } (\varphi \text{ contains } \varphi\text{'s}) \end{cases} \tag{4.5}$$

The statement $T_{51} \leftarrow$ P64-I (for example) means there is a "go to" at P_{51}, and this "go to" is now to indicate a transfer to the line labeled with a P followed by the integer 64-I. The "go to" at P51 will then contain in braces the possible transfers and the conditions which determine the actual transfer to be used.

Additional notation used in the algorithm includes

$$\left.\begin{array}{ll} I & = d(\ell) - j(\ell) \text{ for the current value of } \ell. \\ \Sigma(i) & = \text{sum used in the formula for predicting } y^{(I+i)} \\ \tau_1 & = y_n^{(d-j)} - \varphi_0^{(e)}(n) \text{ for the current equation if } \nu = 0\text{; and } = 0 \text{ otherwise} \\ \tau_2 & = \text{used to contain } \varphi_k^*(n+1) \text{ (see equation 2.8) for the current equation} \\ \tau_3 & = \text{used to contain } \varphi_k^{(e)}(n+1) \text{ (see equation 2.6) for the current equation} \end{array}\right\} \tag{4.6}$$

For $q(\ell) > 1$, $\varphi_q^{(e)}$ is zero as far as the computation of the other differences is concerned (since τ_3 is initially set to 0), but at P23, $\varphi(q(\ell), \ell)$ is set equal to φ_q^* for use in the order selection algorithm later. (It is useful to have a difference with order one greater than is used in the corrector for purposes of order selection.) It is assumed that $\varphi(q(\ell), \ell)$ is set equal to 0 before coming back to this algorithm on the next step. This means that when beginning this algorithm (when $\nu = 0$)

$$\begin{aligned} &\varphi(q(\ell), \ell) = 0 \quad \text{if the order was not changed on the last step.} \\ &\varphi(q(\ell)-1, \ell) = 0 \quad \text{if the order was increased on the last step.} \end{aligned} \tag{4.7}$$

In addition, when the order is increased, $\varphi(\text{new } q(\ell), \ell)$ should be set equal to $-\varphi(\text{old } q(\ell), \ell)$ before setting $\varphi(\text{old } q(\ell), \ell)$ to zero in order that the correct value for $\varphi(q(\ell), \ell)$ be obtained by the algorithm below. Note that condition (4.7) can be used to detect if the order was increased prior to repeating a step. (Clearly, the order can not be allowed to increase on a step that is rejected. One must also replace $\xi(k)$ with $\xi(k+1)-h$, $k = n_h, n_h+1, \ldots,$ before executing the algorithm for obtaining new coefficients, when a step is being repeated.)

The case $q(\ell) = 1$ is treated the same as $q(\ell) = 2$ as far as computing other differences is concerned, but only $\varphi(0, \ell)$ is included in the computation of the sums. As indicated earlier, $q(\ell) = 1$ is treated as a special case in order to have an additional difference available to assist in order selection. Obvious simplifications can be made if this extra difference is not required. As given, the algorithm uses a first order predictor for $y_n^{(d-j)}$ when $q(\ell) = 1$, $j(\ell) > 0$, contrary to what is given in equation (3.6). (Also note that at P58 $\hat{y}^{(I)}(\ell) + \varphi(1, \ell) \equiv \varphi(0, \ell)$. The former is used because some implementations will want to carry y and $\hat{y}$ to more precision than φ.) The corrector formulas (3.7) and (3.11) should have $\varphi_0^{(e)}$ replaced by $\varphi_0^{(e)} - \varphi_1^{(e)}$ when $q(\ell) = 1$, if $q(\ell) = 1$ is being treated as a special case.

Finally, note that k_{min} should be set to 0 initially. (This can be done with the DATA initialization statement in FORTRAN.)

<u>Algorithm for Predicting and Updating Differences</u>

P1.	[Test if differences not updated yet.]	if $\nu=0$, go to P3.
P2.	[Set $\tau_1=0$, and transfers at P26 and P50.]	$\tau_1 \leftarrow 0$; $T_{50} \leftarrow T_{26} \leftarrow P_{35}$; go to P4.

P3.	[Set transfers at P26 and P50.]	$T_{50} \leftarrow T_{26} \leftarrow P37$.
P4.	[Initialize the equation counter.]	$\ell \leftarrow 1$.
P5.	[Set max. no. of repeated integrations.]	$I \leftarrow d(\ell) - j(\ell)$.
P6.	[Test if equation is stiff.]	if $j(\ell) = 0$, go to P15.
P7.	[Set τ_1 if $\nu = 0$.]	if $\nu = 0$, $\tau_1 \leftarrow e(\ell)$.
P8.	[Set transfers at P30, P33, P36, P39, P44, P51, and P59.]	$T_{39} \leftarrow T_{36} \leftarrow T_{33} \leftarrow T_{30} \leftarrow P44 - j(\ell)$; $T_{44} \leftarrow P49 - I$; $T51 \leftarrow P58 - j(\ell)$; $T_{59} \leftarrow P64 - I$.
P9.	[Go store 0 in diff. formula sums.]	go to $P14 - j(\ell)$.
P10.		$\Sigma(-4) \leftarrow 0$.
P11.		$\Sigma(-3) \leftarrow 0$.
P12.		$\Sigma(-2) \leftarrow 0$.
P13.		$\Sigma(-1) \leftarrow 0$.
P14.	[Go store 0 in integ. formula sums.]	go to P22-I.
P15.	[Set τ_1 if $\nu = 0$.]	if $\nu = 0$, $\tau_1 = f(\ell) - \varphi(0, \ell)$.
P16.	[Set transfers at P30, P33, P36, P39, and P51.]	$T_{39} \leftarrow T_{36} \leftarrow T_{33} \leftarrow T_{30} \leftarrow P49 - I$; $T51 \leftarrow P64 - I$.
P17.	[Go store 0 in integ. formula sums.]	go to P22-I.
P18.		$\Sigma(4) \leftarrow 0$.
P19.		$\Sigma(3) \leftarrow 0$.
P20.		$\Sigma(2) \leftarrow 0$.
P21.		$\Sigma(1) \leftarrow 0$.
P22.	[Set index for use in loop below.]	$k \leftarrow q(\ell) - 1$.
P23.	[Store φ^*_{k+1} (eq. (2.9)) into $\varphi(k+1, \ell)$.]	$\varphi(k+1, \ell) \leftarrow [\varphi(k+1, \ell) + \tau_1]\beta(k+1)$.
P24.	[Test if usual case of $q(\ell) > 1$.]	if $k > 0$, go to P26.
P25.	[Do special calculations when $q(\ell) = 1$.]	$\tau_3 \leftarrow \varphi(1, \ell)$; $\varphi(2, \ell) \leftarrow [\varphi(2, \ell) + \tau_1]$ $\beta(2)$; go to P38.
P26.	[Initialize τ_3, and test if only backward differences are required.]	$\tau_3 \leftarrow 0$; if $k \le n_h$, go to {P37 if $\nu = 0$; P35 if $\nu \ne 0$}.
P27.	[Set stopping index for mod. div. diffs. and test if τ_1 needed.]	$k_{min} \leftarrow n_h + 1$; if $\nu = 0$, go to P31.

P28.	[Set transfer at P50.]	$T_{50} \leftarrow P29$.
P29.	[Compute φ_k^*, $\nu \neq 0$, and $k > n_h$.]	$\tau_2 \leftarrow \varphi(k, \ell)\beta(k)$.
P30.	[Go form sums.]	go to {P44-j(ℓ) if j(ℓ) > 0; P49-I otherwise}.
P31.	[Set transfer at P50.]	$T_{50} \leftarrow P32$.
P32.	[Compute φ_k^*, $\nu = 0$, and $k > n_h$.]	$\tau_2 \leftarrow [\varphi(k, \ell) + \tau_1]\beta(k)$.
P33.	[Go form sums.]	go to {P44-j(ℓ) if j(ℓ) > 0; P49-I otherwise}.
P34.	[Set transfer at P50.]	$T_{50} \leftarrow P35$.
P35.	[Compute φ_k^*, $\nu \neq 0$, h constant.]	$\tau_2 \leftarrow \varphi(k, \ell)$.
P36.	[Go form sums.]	go to {P44-j(ℓ) if j(ℓ) > 0; P49-I otherwise}.
P37.	[Set transfer at P50.]	$T_{50} \leftarrow P38$.
P38.	[Compute φ_k^*, $\nu = 0$, h constant (usual case ?).]	$\tau_2 \leftarrow \varphi(k, \ell) + \tau_1$.
P39.	[Go form sums.]	go to {P44-j(ℓ) if j(ℓ) > 0; P49-I otherwise}.
P40.	[Form sums for differentiation	$\Sigma(-4) \leftarrow \Sigma(-4) + d(k, 4)\tau_2$.
P41.	formulas.]	$\Sigma(-3) \leftarrow \Sigma(-3) + d(k, 3)\tau_2$.
P42.		$\Sigma(-2) \leftarrow \Sigma(-2) + d(k, 2)\tau_2$.
P43.		$\Sigma(-1) \leftarrow \Sigma(-1) + d(k+1)\tau_2$.
P44.	[Go form sums for integrations (if any).]	go to {P49-I}.
P45.	[Form sums for integration	$\Sigma(4) \leftarrow \Sigma(4) + g(k, 4)\tau_2$.
P46.	formulas.]	$\Sigma(3) \leftarrow \Sigma(3) + g(k, 3)\tau_2$.
P47.		$\Sigma(2) \leftarrow \Sigma(2) + g(k, 2)\tau_2$.
P48.		$\Sigma(1) \leftarrow \Sigma(1) + g(k, 1)\tau_2$.
P49.	[Compute $\varphi_k^{(e)}$, see equation (2.6).]	$\tau_3 \leftarrow \tau_2 + \tau_3$; $\varphi(k, \ell) \leftarrow \tau_3$.
P50.	[Test for end of forming differences or end of forming mod. div. differences.]	$k \leftarrow k-1$; if $k \geq k_{min}$, go to {P29 if $\nu \neq 0$, $k_{min} \neq 0$; P32 if $\nu = 0$, $k_{min} \neq 0$; P35 if $\nu \neq 0$ $k_{min} = 0$; P38 if $\nu = 0$, $k_{min} = 0$}.

P51. [Test if done forming differences and sums.] if $k_{min}=0$, go to {P64-I if $j(\ell)=0$; P58-$j(\ell)$ otherwise}.

P52. [Set up to compute differences based $k_{min} \leftarrow 0$; if $\nu=0$, go to P37.

P53. on constant stepsize.] go to P34.

P54. [Compute y's using differentiation formulas.] $y^{(I+4)}(\ell) \leftarrow s(-4)\Sigma(-4)$.

P55. $y^{(I+3)}(\ell) \leftarrow s(-3)\Sigma(-3)$.

P56. $y^{(I+2)}(\ell) \leftarrow s(-2)\Sigma(-2)$.

P57. $y^{(I+1)}(\ell) \leftarrow s(-1)\Sigma(-1)$.

P58. $y^{(I)}(\ell) \leftarrow \hat{y}^{(I)}(\ell) + \varphi(1,\ell)$

go to {P64-I}.

P60. [Compute y's using integration formulas.] $y^{(I-4)}(\ell) \leftarrow \hat{y}^{(I-4)}(\ell) + H(1)\{\hat{y}^{(I-3)}(\ell) + H(2)[\hat{y}^{(I-2)}(\ell) + H(3)(\hat{y}^{(I-1)}(\ell) + h\Sigma(4))]\}$.

P61. $y^{(I-3)}(\ell) \leftarrow \hat{y}^{(I-3)}(\ell) + H(1)\{\hat{y}^{(I-2)}(\ell) + H(2)[\hat{y}^{(I-1)}(\ell) + h\Sigma(3)]\}$.

P62. $y^{(I-2)}(\ell) \leftarrow \hat{y}^{(I-2)}(\ell) + H(1)\{\hat{y}^{(I-1)}(\ell) + h\Sigma(2)\}$.

P63. $y^{(I-1)}(\ell) \leftarrow \hat{y}^{(I-1)}(\ell) + h\Sigma(1)$.

P64. [Test if more equations to be processed.] $\ell \leftarrow \ell+1$; if $\ell \leq n_e$, go to P5,

P65. [Set ν to indicate that φ is not updated.] $\nu \leftarrow 0$.

P66. [End of predicting and updating diff.] Exit.

5. Interpolation to Off-Step Points

A significant advantage of multistep methods over one-step methods is that a multistep method has sufficient information stored to enable one to get the solution at any point passed during the integration: with the same accuracy as is obtained at the end of the individual steps, without interfering with the integration process in any way, and without requiring any additional derivative evaluations.

Ordinarily one is interpolating to a point t which satisfies $t_{n-1} < t \le t_n$ where t_n is the value of t at the end of the current step. Sometimes it is necessary to extrapolate the solution ($t_n < t \le t_{n+1}$) because derivatives are impossible to compute at $t = t_{n+1}$. Finally, if the solution is being saved for later use, it is useful to know that the algorithm gives reasonable accuracy for $t_{n-q} \le t \le t_{n+1}$, where $q = \min\{q(\ell)\}$. It is assumed that values of $y^{(k)}$ from the current step have been stored in $\hat{y}^{(k)}$ (see equation 4.1) and that the interpolated values are to be stored in $y^{(k)}$, and in f.

In order that full accuracy be obtained, the $\varphi_i^{(e)}(n)$'s should be replaced by $\varphi_i(n)$'s before doing the interpolation. This can be done using equation (2.7) (with n+1 replaced by n); it should be done only if $\nu=0$, see equation (4.5); and if it is done, ν should be set equal 1.

Let

$$h_I = t - t_n$$
$$\rho = h_I / h_n, \qquad (5.1)$$

then h_I plays the same role as h_{n+1} in the predictor formulas (3.4)-(3.6), if the interpolation is looked at as just taking a new step. However the recursions (2.16) and (2.20) (with h_{n+1} replaced by h_I) can not be used since arbitrarily large values of α_1 occur as $h_I \to -h_n$. These recursions wouldn't give the desired coefficients anyway since from equations (2.8), (3.4)-(3.6) and the fact that we are now using $\varphi_i(n)$, it is clear that we should introduce

$$g_{i,k}^{(I)} = \beta_i^{(I)} g_{i,k}$$
$$d_{i,k}^{(I)} = \beta_i^{(I)} d_{i,k} \qquad (5.2)$$

for use in the interpolation, where $\beta_i^{(I)}$ is defined as in equation (2.4) with h_{n+1} replaced by h_I. Then equations (3.4)-(3.6) can be used for the interpolation if H and s are computed using h_I, φ_i^* is replaced by φ_i, g and d are replaced by $g^{(I)}$ and $d^{(I)}$, and it is understood that $d_{i,0}$ implicitly multiplied φ_i^* in equation (3.6). (We recommend using q=2 in the case that $q(\ell) = 1$.)

To avoid potential overflow in computing $s_{-k}=k!/h_I^k$, we suggest defining

$$\hat{d}_{i,k}^{(I)} = (k!/h_I^k)d_{i,k}^{(I)} \tag{5.3}$$

and using equation (3.5) with s_{-k} removed, and $d_{i,k}$ replaced by $\hat{d}_{i,k}^{(I)}$. (It is not a bad idea to simply compute $s_{-k}d_{i,k}$ and $s_k g_{i,k}$ instead of $d_{i,k}$ and $g_{i,k}$ when computing the coefficients for continuing an integration.)

With

$$\eta_i = h_I/\xi_i(n) \tag{5.4}$$

$$\gamma_i = \begin{cases} \rho & i = 0 \\ [h_I+\xi_{i-1}(n)]/\xi_i(n) & i > 0 \end{cases} \tag{5.5}$$

there follows immediately from equations (2.16), (2.20), and (5.2)-(5.5)

$$\hat{d}_{i,k}^{(I)} = \begin{cases} 0 & i < k \\ 1 & i=0,\ k=0 \\ \gamma_{i-1}\hat{d}_{i-1,0}^{(I)} & i=1,2,\dots,q-1;\ k=0 \\ \left(\dfrac{k}{\xi_{i-1}(n)}\right)\hat{d}_{i-1,k-1}^{(I)} + \gamma_{i-1}\hat{d}_{i-1,k}^{(I)} & i=k,k+1,\dots,q-1;\ k=1,2,\dots \end{cases} \tag{5.6}$$

$$g^{(I)}_{i,k} = \begin{cases} \gamma_{i-1} g^{(I)}_{i-1,k} - \eta_{i-1} g^{(I)}_{i-1,k+1} & i=q-1, q-2, \ldots, 1;\ k=m-1, \ldots, 1 \\ 1/k & k=m-1, \ldots, 1 \end{cases} \tag{5.7}$$

Of course, equation (5.7) is valid for any value of m, but we derive below a more efficient algorithm for computing $g^{(I)}_{i,m}$, where m is the maximum value of d-j (d and j defined as in equation (3.3)). If j=0, then $\hat{d}^{(I)}_{i,0}$ need not be computed, and f $(=y^{(d)})$ can be computed using the formula

$$f = \varphi_0 + \gamma_0[\varphi_1 + \gamma_1[\varphi_2 + \cdots \gamma_{q-2}\varphi_{q-1}]]\cdots] \tag{5.8}$$

The recursion to compute $g^{(I)}_{i,m}$ is obtained starting from the coefficients of the interpolating polynomial. Define

$$\varphi_{i,k}(n) = \begin{cases} \xi_0^k(n)\xi_0(n)\xi_1(n)\cdots\xi_{i-k-1}(n)\, w[t_n, \ldots, t_n, t_{n-1}, \ldots, t_{n-i+k}] & i \geq k+1 \\ \xi_0^k\, w[t_n, \ldots, t_n] & i = k \end{cases} \tag{5.9}$$

where the t_n's in the divided differences are repeated k+1 times. Clearly $\varphi_{i,0}(n) = \varphi_i(n)$ as defined in equation (2.4), and $\varphi_{i,i}$ gives scaled coefficients of the interpolating polynomial, so that

$$P_{q-1,n}(t) = \sum_{i=0}^{q-1} \varphi_{i,i}(n)[(t-t_n)/h_n]^i \tag{5.10}$$

From equations (2.2) and (5.9) it is easy to obtain

$$\varphi_{i,k}(n) = [h_n/\xi_{i-k}(n)]\varphi_{i,k-1}(n) + \varphi_{i+1,k}(n), \quad k=1, 2, \ldots, i;\ i=q-1, \ldots, 1 \tag{5.11}$$

where $\varphi_{q,k}(n) \equiv 0$.

(If one wants to integrate/interpolate/differentiate $P_{q-1,n}(t)$ for many different values of t, then it is most efficient to compute the $\varphi_{i,i}$ from the $\varphi_{i,0}$'s, and then to compute the desired result using equation (5.10). With $\varphi_{i,i}(n) = (h_n^i/i!)y_n^{(i)}$, i=0, 1, ..., q+d-2, one gets a method like ours in the Nordsieck formulation if one writes

$$\varphi_{i,i}(n+1) = \varphi_{i,i}^{(e)}(n+1) + (c_{q,n}^{(j-d+i)}(1))[(d-j)!/i!][\frac{h_{n+1}^{(d-j)}}{(d-j)!} y_{n+1}^{(d-j)} - \varphi_{d-j,d-j}^{(e)}(n+1)] \qquad (5.12)$$

where $c_{i,n}^{(k)}$ is defined as in section 2. Recursions for $\varphi_{i,i}^{(e)}(n+1)$ follow from considering $w[t_n, t_n, \ldots, t_n]$, $w[t_n, \ldots, t_n, t_{n+1}]$, $w[t_n, \ldots, t_n, t_{n+1}, t_{n+1}], \ldots$, $w[t_n, t_{n+1}, \ldots, t_{n+1}]$, $w[t_{n+1}, \ldots, t_{n+1}]$; the recursions obtained are identical to the computational shortcut due to Gear, [18]. Also see Thomas, [19]. In order for this Nordsieck type formulation to give a method equivalent to what we have described, some care is required when changing the order, see [14]. If many interpolations are to be performed during the integration, this Nordsieck formulation may be preferred to the use of modified divided differences if one is not interested in using different integration orders for different equations, and not too many equations are being integrated.)

The m-th integral of P is given by

$$\int_{t_n}^{t} \cdots \int_{t_n}^{t} P_{q-1,n}(t) = \sum_{i=0}^{q-1} a_{i,k}^{(m)}(n)\, \varphi_{i,k}(n) \qquad (5.13)$$

where it is clear from equation (5.10) that

$$a_{i,i}^{(m)} = h_n^m \rho^{i+m}/(i+m)\cdots(i+1) = h_I^m \rho^i/(i+m)\cdots(i+1) \qquad (5.14)$$

and from equations (5.11) and (5.13)

$$a_{i,k}^{(m)} = \begin{cases} \dfrac{h_n}{\xi_{i-k-1}(n)} a_{i,k+1}^{(m)} & k=0;\ i=1,2,\ldots,q-1 \\[2ex] \dfrac{h_n}{\xi_{i-k-1}(n)} a_{i,k+1}^{(m)} + a_{i-1,k}^{(m)} & k=i-j;\ i=q-1,\ldots,j+1;\ j=1,\ldots,q-2 \end{cases} \tag{5.15}$$

With the definition

$$\hat{a}_{i,k}^{(m)} = h_I^{-m}(m-1)!\,\rho^{-k} a_{i,k}^{(m)} \tag{5.16}$$

we have from equations (5.2), (5.4), (5.14)-(5.16)

$$\hat{a}_{i,k}^{(m)} = \begin{cases} \dfrac{(m-1)!}{(i+m)\cdots(i+1)} & k=i;\ i=0,1,\ldots,q-1 \\[2ex] \eta_{i-k-1}\hat{a}_{i,k+1}^{(m)} + \hat{a}_{i-1,k}^{(m)} & k=i-j;\ i=q-1,\ldots,j+1;\ j=1,\ldots,q-2 \\[2ex] \eta_{i-k-1}\hat{a}_{i,k+1}^{(m)} & k=0;\ i=1,2,\ldots,q-1 \end{cases} \tag{5.17}$$

and

$$g_{i,m}^{(I)} = \hat{a}_{i,0}^{(m)}, \qquad i=0,1,\ldots,q-1 \tag{5.18}$$

(Note that $a_{i,i}^{(m)}$ need be computed at most one time if m is not changed. Calculations can be arranged so that $d_{i,k}$, $g_{i,k}$, and $a_{i,k}$ ($k < i$) all occupy the same vector in storage.)

6. Concerning Some of the Messy Details

If one computes the difference table of a function with sufficient precision, for sufficiently small h, he will most likely find that for a given order q, differences of order q+k tend to decrease in magnitude as k takes the values 0, 1, ... Since in practice we almost always select the order q in such a way that the differences behave in the opposite way, it is reasonable to suspect that any theory based on results for $h \to 0$ and ignoring the effect of round-off error will be of limited value. We believe there is greater danger in attempting to apply rigorous mathematics to problems which do not satisfy the underlying assumptions, than there is in careful inductive reasoning from results on a selection of problems which individually are simple enough to understand, and collectively cover the types of difficulties found in real problems. Thus we have taken primarily an empirical approach. Most of our effort has been spent in poring over difference tables generated in the solution of a variety of problems while using a variety of algorithms for selecting the integration order. Most of what follows is either trivial or mere opinion, yet much of it is important in determining the effectiveness of a variable order algorithm for solving differential equations. The specific algorithms presented give an idea of what we have done, they are not intended as recommendations. For results from a variable order Adams method which makes use of some of the ideas below, see [16]; for a comparison with other methods, see [17].

6.1 General Design

The interface between algorithms and users, despite its importance in determining the effectiveness of an algorithm, has been given minimal attention in the numerical analysis literature. Because of the problems inherent in the addition of features not considered in the original design (probably by a person unfamiliar with the code) we believe it is a good idea to design for maximum flexibility if an algorithm is meant for general purpose use. Complex applications will require the flexibility, and it is a relatively simple matter to insert a flexible program into a package to be used by the unsophisticated user. Since code of wide generality is liable to contain much that is superfluous to some applications, we think it is also a good idea to make it easy for a user to remove that code he is not interested in using. A procedure for handling code which consists of many versions is described in [20].

We have found it difficult to anticipate user needs. Users can't know what they should have available when they don't know what the possibilities are, and thus they are not as much help in this area as one might think. The integrator in [15] is reasonably flexible and all of its features have been heavily used. But it proved to be insufficiently flexible for some users, leading to the design [21]. This in turn has had to be modified and extended to meet user needs that have surfaced since it was written. The results of a survey on the importance of various factors in a program for solving differential equations can be found in [22].

6.2 Some General Comments on Differences

In the case of the Adams method we have made it a practice to correct y, estimate errors, and select the integration order in the same block of code, thus making multiple use of the differences formed from predicted derivative values. Especially at low integration orders, we prefer such differences to those formed from corrected derivative values, since the former tend to converge less rapidly, giving a more conservative algorithm. For $q \le 2$ ($d=1$, $j=0$) it is possible for the difference table formed from corrected derivative values to converge nicely, while at the same time the numerical solution is diverging rapidly from an acceptable solution. Such problems do not occur when using predicted derivative values. Let

$$\varphi_k = \begin{cases} f(t_{n+1}, p_{n+1}, \ldots, p_{n+1}^{(d-1)}) - \varphi_0^{(e)}(n+1) & k = q \\ \varphi_k^{(e)}(n+1) + \varphi_q & k = q-1,\ q-2, \ldots \end{cases} \tag{6.2.1}$$

and let φ_{q+1} be computed as φ_q - [What is stored in the q-th difference location by the algorithm for predicting; see P23.], where q = max{2, integration order of the predictor}. Ordinarily, only the differences φ_{q-2}, φ_{q-1}, φ_q, and φ_{q+1} need be computed since they almost always (see below) provide sufficient information for selecting the order.

We have found the linear equation $y' = Gy$ (G a constant matrix) helpful in organizing some of our thoughts. For this case, if h is constant and G has distinct eigenvalues, $y_n = \Sigma c_i r_i^n$. Some of the r_i are extraneous to approximating the desired solution (See [23] for some background on this.), and for the cases d=1, j=0, q> 3; d=1, j=1, q> 1 it can be shown that the largest extraneous r_i has a negative real part. Since the k-th backward difference of r^n is given by $r^n[(r-1)/r]^k$, differences of an extraneous root tend to increase. By selecting the order at about that point where the differences start to increase, one is stopping at about that order where the influence of an extraneous root is starting to dominate the differences, which in turn guarantees a stable method.

A given difference may be small because the error is small, or because it happens to be passing through zero, or just as a fluke. Any decision made on the basis of one difference being small has a good chance of giving the wrong result. (In the sense for example that the order might be increased when it should not be; such wrong results do not necessarily do significant harm to the solution.) On the other hand, two small differences in succession will only rarely be misleading.

In judging the convergence of the differences, their signs are also important. For given magnitudes, alternating signs indicate the most rapid convergence.

6.3 Starting the Integration

Variable order methods do not require any special logic to start an integration, but still there are advantages to treating the start in a special way. Efficiency in the starting process can be critical for problems with frequent discontinuities. Because of their ability to find the proper stepsize quickly, we believe that variable order methods will frequently prove superior to good one-step methods on very short integrations. The Adams method starting procedure outlined below does not require much additional code, despite its apparent complexity. Initially, of course $q \equiv 1$.

Let p denote the predicted values at t_0+h, and c_1, c_2, c_3 successive corrected values. After $p^{(d)}$ is computed, compute c_1 and the estimated error on the first step. If the estimated error is too large (in the sense that such an error would ordinarily cause the step to be repeated), a new starting stepsize is selected on the assumption that the estimated error is proportional to the square of the stepsize. The resulting stepsize will usually give an acceptable error, and thus an initial value for the stepsize that is too large usually costs only one extra derivative evaluation. It is rare for more than two evaluations to be required. In estimating the error on the first step, we multiply our usual estimate by 1/4 since estimated errors would otherwise have a tendency to be much too large.

After obtaining a satisfactory c_1, compute $c_1^{(d)}$ and c_2. Estimate the error in $c_2^{(d-1)}$ by $4h(p^{(d)}-c_1^{(d)})$ and compute $s_1(\ell) = |h[p^{(d)}(\ell) - c_1^{(d)}(\ell)]/[p^{(d-1)}(\ell) - c_1^{(d-1)}(\ell)]|$. If the estimated error is too large (rarely happens) reduce the stepsize as in the preceding paragraph and start over. Otherwise check if $s_1 < 1/16$ for all equations, and if it is proceed to the next paragraph. If $s_1(\ell) > 1/16$ for some ℓ, then compute $c_2^{(d)}$ and c_3. Compute a new error and $s_2(\ell)$ using formulas like those above with p replaced by c_2. If the error is too large, start over; if not and $\min\{s_1(\ell), s_2(\ell)\} > 1/16$ for some ℓ then end the the starting phase. If none of these, then continue to the end of the first step immediately below.

Set y_1 = the final corrected value, and increase the order to 2. From this point until the end of the starting phase, no derivatives are computed after computing corrected values. (A PEC method is used.) This is justified since the second derivative evaluation is required primarily to improve the stability properties of the method. For low orders and small stepsizes, instability is usually not a problem. At the end of the second step increase the order to 3. At the end of the third step leave the order at 3, making φ_{q+1} available (and thus the usual order selection process possible) for the first time on the fourth step. If at any time the stepsize must be reduced, or the order must be reduced and the estimated error is too large to permit an increase in the stepsize, then end the starting phase. One can gain a little in the starting process by making it easier for the order to increase than would be prudent after getting started.

6.4 Determining if Too Much Precision Has Been Requested

We believe that some test for unreasonable accuracy requests should be considered an intrinsic part of any general purpose integration program. Many users have come to us because our integrator gave them a diagnostic to the effect that it could not get the accuracy they desired. These users appreciated being warned of a problem they would not have been aware of otherwise. Users of [15] rarely question us about this type of diagnostic now. We presume they have learned to select reasonable error tolerances, or to trust the diagnostic when they get it. Since a missed diagnostic of this type will usually only result in a less efficient integration, and since unjustified diagnostics make for bad relations with users, we recommend that one be conservative in the test which results in the diagnostic.

Clearly the absolute accuracy that can be obtained depends on the number of significant digits to which f is computed, and on the size of f. Thus in solving $y'' = -y$, $y(0) = 0$, $y'(0) = c$, an accuracy request that is impossible to meet for $c = 1$, may be easily met for $c = 10^{-20}$.

The test used in [15] assumes that round-off errors in f get magnified by 2^k in φ_k. (This is the magnification one gets if errors are of equal magnitude and alternate in sign from one step to the next. This assumption appears to work best, even though smaller magnification factors would appear more reasonable.) Thus, with

$$R_1 = [|\varphi_q| + |\varphi_{q+1}|]/[2^q|f|10^{-d_f}] \qquad (6.4.1)$$

where d_f is the number of significant digits in f, $R_1 < 1$ is an indication that round-off errors are limiting the accuracy. To a first approximation, the test in [15] gives a diagnostic if the estimated error is too large and $R_1 < 1$, where R_1 is computed on the assumption that d_f is approximately the number of significant digits in the computer's number system. We believe that it is too much of an imposition on the user to ask him to supply d_f; but even if it weren't, it is hard to justify trusting him for d_f when he cannot be trusted to supply a reasonable error tolerance. If round-off error is the primary component of the estimated error, then the estimated error for a larger stepsize tends to be too large. Thus in [15] the estimated error at twice the stepsize is reduced whenever $R_1 < 1$.

We have been experimenting recently with a test which is impossible to justify, and for which examples could be constructed which cause it to give unjustified diagnostics. It makes no assumptions about the precision to which f is computed, but rather makes some implicit assumptions about the regularity with which difference tables converge. An integrator using this test has been used on a few applications, and has resulted in three diagnostics. In two cases (once when the user thought everything was done in double precision) some calculations done in single precision, should have been done in double precision, and in the other case results were being interpolated from a table with insufficient accuracy. The integrator [15] would not detect these problems. This new test should not be trusted unless the order selection algorithm does a very good job of selecting the order, it should not be applied in the starting phase of an integration, and $q \leq 2$ should be treated as a special case.

Let

$$R_2 = [\,|\varphi_q| + |\varphi_{q+1}|\,]/r(q)[\,|\varphi_1^{(e)}| + |\varphi_2^{(e)}|\,] \qquad (6.4.2)$$

where

$$r(q) = \begin{cases} .1 & 3 \le q \le 4 \\ 2^{-[(q-1)^2-3]/3} & q > 4 \end{cases} \tag{6.4.3}$$

and let n_R be the number of steps for which the equation with the largest ratio of estimated error to requested error has had an $R_2 < 1$. If the estimated error is too large, and $n_R \ge 4$, then a diagnostic is indicated. The estimated error with the stepsize increased is reduced by the factor (See equations (3.28) and (4.1) for notation.)

$$\max\{R_2, \rho_i^{-k}\}, \text{ where } k = \max\{n_h, n_R\}-3\} \tag{6.4.4}$$

whenever $R_2 < 1$. When $R_2 < 1$, the integration order is increased if

$$|\varphi_{q+1}| < \min\{|\varphi_{q-1}|, |\varphi_q|\}. \tag{6.4.5}$$

6.5 Selection of Integration Order

The most important point to be made here is that a reasonably good job of selecting the order can be done with very little effort. For example, the algorithm used in [15] for $q > 2$ is given by the following.

Q1. $\tau_1 \leftarrow q+2$

Q2. if $|\varphi_{q-2}| < \tau_1 |\varphi_q|$, go to Q5

Q3. if $|\varphi_{q-1}| < \tau_1 |\varphi_{q+1}|$, go to Q8

Q4. increase q by one, and go to Q8

Q5. $\tau_1 \leftarrow .25\ \tau_1$

Q6. if $|\varphi_{q-2}| > \tau_1 |\varphi_q|$ or $|\varphi_{q-1}| > \tau_1 |\varphi_{q+1}|$, go to Q8

Q7. decrease q by 1

Q8. end of order selection

Note that the above algorithm uses 4 different differences and requires either 2 or 3 comparisons. If one does not want the order to oscillate unduly, then we believe that at least 4 differences must be examined.

In order to have a check for discontinuities it is sometimes necessary to check more than 4 differences. We now do this as follows. If

$$|\varphi_{q-2}| \le |\varphi_{q-1}| \le |\varphi_q| \le |\varphi_{q+1}|, \text{ and} \tag{6.5.1}$$

$$.875|\varphi_q - \varphi_{q-1}| < |\varphi_{q+1}| - |\varphi_q| \tag{6.5.2}$$

then q is reduced by 1; φ_{k+1} is replaced by φ_k for k = q-2, q-1; a new φ_{q-2} is computed; and the test above passed through again. The conditions for reducing q by more than 1 must be very stringent. For example, on a simple 2-body with eccentricity .6, using just the condition (6.5.1) for this test, we have seen the order reduced by about 6 over several steps, when a reduction of at most one was appropriate. If q starts out ≥ 5, is reduced to 3 and the differences still converge slowly, then a discontinuity is indicated and the integration restarted. Otherwise the final q is treated just as if it were the original as far as order selection is concerned.

The more factors that are taken into consideration, the better job one can do in selecting the order. For example, it helps to bias the test towards a low order when errors are decreasing, and vice versa when errors are increasing. (The former may increase the current error estimate, while decreasing the estimate of what the error would be with the stepsize increased.) Since the selection of integration order influences both the integration efficiency and tests such as discussed in 6.4 and 6.8, it is not easy to decide at what point additional effort does not justify the return. We think it is better to use a simple scheme such as the one in [15] on every step, rather than some more complicated scheme less frequently, and that at a minimum, some test for decreasing the order be made on every step.

The algorithm we are currently using works reasonably well, but is probably unduly complicated. It uses as a measure of the convergence of the difference table at φ_q (with appropriate safeguards to prevent overflow):

$$\theta_q = \begin{cases} |\varphi_{q+1}|/|\varphi_{q-1} - \varphi_q| & \text{if } \varphi_{q+1}\varphi_{q-1} \geq 0 \\ |\varphi_{q+1}|/|\varphi_{q-1}| & \text{if } \varphi_{q+1}\varphi_{q-1} < 0. \end{cases} \tag{6.5.3}$$

To a first approximation, the order is increased if

$$\theta_q + \theta_{q-1} < \begin{cases} .75 \text{ if h is closer to being decreased} \\ \rho_i^{-2} \text{ if h is closer to being increased,} \end{cases} \quad \text{and } \theta_q^2 < \theta_{q-1}, \text{ and } \theta_q < 2\theta_{q-1} \tag{6.5.3}$$

and it is decreased if

$$|\varphi_{q+1}| > |\varphi_q| > |\varphi_{q-1}| \tag{6.5.5}$$

Note that these tests differ from those used in [15] in that signs of the differences influence the tests, the order need not be increased when θ_q and θ_{q-1} are both very small, φ_{q-2} is not used in the test for decreasing the order, and the tests to not have an explicit dependence on q.

6.6 Error Estimation and the Selection of Stepsize

Many of our ideas in this area are given at the end of section 3. We currently estimate the error in $y^{(d-1)}$ for the Adams method by

$$|hg^*_{q,1}|[|\varphi_q| + |\varphi_{q+1}|]/(1 - \bar{\theta}) \tag{6.6.1}$$

where $\bar{\theta} = \min\{\theta_q, \theta_{q-1}, .75\}$, see equation (6.5.3). This reduces to equation (3.22) if φ_{q+1} and $\bar{\theta}$ are replaced by 0. The inclusion of $|\varphi_{q+1}|$ is for reasons discussed in 6.2, and the factor involving $\bar{\theta}$ is included because otherwise error estimates tend to be a little low when the differences decrease slowly. We currently use

$$|hg^*_{q,1}|\rho^q \max\{\sigma_q, 1\}[|\varphi_{q-1}| + |\varphi_q| + |\varphi_{q+1}|]/(1-\bar{\theta})^2 \tag{6.6.2}$$

for the estimated error with the stepsize increased by the factor ρ. The role of $\bar{\theta}$ in (6.6.2) is quite important since a slowly convergent difference table is liable to require a significantly lower order when the stepsize is increased, thus giving a larger error estimate.

As mentioned at the end of section 3, we think it is a good idea to vary the parameter one uses for making decisions on increasing the stepsize as a function of the increase or decrease of the error estimates. The way we do this is outlined below. Numbers in parentheses give the actual values of the parameters as currently implemented.

E   largest value in any equation of (estimated error)/(requested error).

ϵ_R(.3)   if $E > \epsilon_R$ the step is repeated with h reduced by the factor $\min\{1/2, \rho_d\}$.

E_A average value of E. Computed as $(E+E_A)/2$ at the end of every step, with an initial value of ϵ_R.

$\hat{E}_A$ value of E_A just before current E_A was computed.

E_x estimated value for E on the next step. $E_x = E\min\{2, E/\hat{E}_A\}$.

$\epsilon_x(.1)$ if $E_x > \epsilon_x$ the stepsize is reduced by the factor ρ_d.

$\overline{E}_I(.2)$ upper bound on E_I, see below.

$\underline{E}_I(.01)$ lower bound on E_I, see below.

E_I step is increased if in so doing the estimated E at the new stepsize is less than E_I. With an initial value of ϵ_x it is computed using

$$E_I = \begin{cases} \min\{\overline{E}_I, E_I\hat{E}_A/E_A\} & \text{if } E \le E_A \\ [E_I+\underline{E}_I]/2 & \text{if } E > E_A \end{cases}$$

Note that except for starting the integration, the stepsize is never reduced by more than $\min\{1/2, \rho_d\}$ no matter how large the error. If the step has to be repeated more than once we restart the integration. If the stepsize has been decreased, then increased, and is about to be decreased again, we reduce h by the factor $(1+\rho_d)/2$; this tends to reduce oscillations in the stepsize.

6.7 Treating $q(\ell) = 1$ As a Special Case When Using the Adams Method

Frequently when the order selection algorithm picks an order of 1, there is a stability problem. In this case the stability of the method can be improved by introducing a parameter ω into the corrector formula. Thus

$$y_{n+1}^{(d-k)} = p_{n+1}^{(d-k)} + \omega s_k g_{1,k} \varphi_1, \qquad 1/4 \le \omega \le 1. \qquad (6.7.1)$$

When solving $y' = \lambda y$ with $\omega = 1$, the above method is absolutely stable for $s = h\lambda \epsilon [-2, 0]$. The largest interval of absolute stability is obtained for $\omega = 1/4$, which gives a method absolutely stable for $s \epsilon [-8, 0]$. We have, with mixed success, attempted to approximate s, and for $s \le -2$ set $\omega = \max\{1/4, -2(1+s)/s^2\}$. This choice for ω zeros the root of the characteristic equation if $-4-2\sqrt{2} \le s \le -2$, assuming the equation is linear, and that s is approximated correctly.

Since $|g^*_{2,1}|$ is so much smaller than $|g^*_{1,1}|$ (1/12 vs. 1/2 for h constant), there is a tendency to overestimate the error when using (6.6.1) and (6.6.2). We have been substituting 1/8 for $\gamma^*_{1,1}$ in these expressions. We also increase the error estimates a little if $|s|$ is large or $\overline{\theta}$ is large. (A larger increase is used for (6.6.2) than for (6.6.1).) The increase in the estimate (6.6.1) is prudent; the increase in (6.6.2) quite important if one has a mildly stiff equation, the order has dropped to one, and one wants to avoid oscillations in the stepsize and minor irregularities in the solution.

6.8 Testing for Stiffness

It is frequently the case that only some of the differential equations in a large system are the cause of stiffness. Since the solution of a large stiff system requires considerable storage and computation for the iteration process and since users frequently do not know which, if any, of their equations are stiff, it is desirable to have a test for stiffness. With equations separated into those which are stiff and those which are not, fewer partial derivatives and smaller matrices are required. Since suggesting automatic classification of equations in [12], we have tried off and on without success to find a stiffness test which does not require computing partial derivatives. A test to tell when an equation being integrated as a stiff equation, could be integrated with the Adams method would also be useful, but should not be as difficult since in this case the Jacobian matrix is available.

Sometimes it is possible to say in advance which equations are stiff and which are not, in which case no tests are required. One problem at JPL was integrated with 9 equations treated as stiff and 5 as non-stiff. In this case, in addition to the savings in the iterations there was the bonus that partials connected with the stiff equations were easy to compute, while the other partials were extremely complicated.

It is worth noting that tests based on ratios such as $[f(y)-f(p)]/[y-p]$, where y and p are corrected and predicted values at the same time point, are not likely to be successful. For $y' = Gy$ as in 6.2,

$$y(t) \approx \sum_i c_{i,n} e^{\lambda_i (t-t_n)} \qquad (t \text{ near } t_n)$$

and with the onset of stiffness, the $c_{i,n}$ associated with λ_i's which have large negative real parts become very small. Then if the method is stable, which it will tend to be because of the way the integration order is selected, there is not enough of the solution in the direction of the eigenvectors associated with the large negative λ_i's to give useful ratios.

When integrating an equation which is stiff with an Adams method it is best if the stepsize is such that the method is absolutely stable, but not relatively stable. The order tends to come down gradually because of the relative instability, but there is no growth in the error since the method is absolutely stable. One problem with halving and doubling the stepsize with the Adams method is that as a result of doubling the stepsize it is possible to move from inside the region of relative stability to outside the region of absolute stability. When this occurs, the order is reduced, there is rapid error growth, and the stepsize must be halved. The algorithm is then relatively stable, the order is increased, the error estimates decrease and the cycle is repeated.

We think that monitoring R_2 of equation (6.4.2) (with suitable (?)

adjustments when q or h is changed) together with the flexibility in changing stepsize available through the use of modified divided differences, may provide what is needed to detect stiffness. There is of course also the possibility that we are attempting to extract information from difference tables that isn't there to begin with.

References

[1] Krogh, F. T.: Algorithms for Changing the Stepsize Used by a Multistep Method. JPL Internal Document, Section 314 TM No. 275, Jet Propulsion Laboratory, Pasadena, Calif. (Oct. 1970). (To appear in the SIAM Journal on Numerical Analysis.)

[2] Krogh, F. T.: A Variable Step Variable Order Multistep Method for the Numerical Solution of Ordinary Differential Equations. Information Processing 68 (Proceeding of the IFIP Congress 1968.) 194-199. North Holland Publishing Co., Amsterdam (1969).

[3] Forrington, C. V. D.: Extensions of the Predictor-Corrector Method for the Solution of Systems of Ordinary Differential Equations. Computer Journal 4 (1961-62), 80-84.

[4] Gabel, G. F.: A Predictor-Corrector Method Using Divided Differences. University of Toronto Department of Computer Science Technical Report No. 5 (October 1968).

[5] Piotrowski, P.: Stability, Consistency and Convergence of Variable k-Step Methods for Numerical Integration of Large Systems of Ordinary Differential Equations. Conference on the Numerical Solution of Differential Equations, Lecture Notes in Mathematics 109, 221-227. Springer-Verlag, Berlin (1969).

[6] Van Wyk, R.: Variable Mesh Multistep Methods for Ordinary Differential Equations. J. Comp. Physics 5 (1970), 244-264.

[7] Brayton, R. K., Gustavson, F. G. and Hachtel, G. D.: A New Algorithm for Solving Differential-Algebraic Systems Using Implicit Backward Differentiation Formulae. Proceedings of the IEEE 60 (1972), 98-108.

[8] Curtiss, C. F. and Hirschfelder, J. O.: Integration of Stiff Equations. Proc. of the Nat. Acad. of Sci. 38 (1952), 235-243.

[9] Gear, C. W.: The Automatic Integration of Stiff Ordinary Differential Equations. Information Processing 68 (Proceeding of the IFIP Congress 1968) 187-193. North Holland Publishing Co., Amsterdam (1969).

[10] Gear, C. W.: The Automatic Integration of Ordinary Differential Equations. Comm. of the ACM 14 (1971), 176-179.

[11] Rutishauser, H.: Bemerkungen zur Numerischen Integration Gewöhnlicher Differential Gleichunger n-ter Ordnung. Numerische Mathematik 2 (1960), 263-279.

[12] Krogh, F. T.: The Numerical Integration of Stiff Differential Equations. TRW Report No. 99900-6573-R000, TRW Systems, Redondo Beach, Calif. (March 1968).

[13] Klopfenstein, R. W.: Numerical Differentiation Formulas for Stiff Systems of Ordinary Differential Equations. RCA Review 32, (1971), 447-462.

[14] Shampine, L. F. and Gordon, M.K.: Local Error and Variable Order, Variable Step Adams Codes. (Submitted to the SIAM Journal on Numerical Analysis.)

[15] Krogh, F. T.: VODQ/SVDQ/DVDQ - Variable Order Integrators for the Numerical Solution of Ordinary Differential Equations. TU Doc. No. CP-2308, NPO-11643, Jet Propulsion Laboratory, Pasadena, Calif. (May 1969).

[16] Krogh, F. T.: On Testing a Subroutine for the Numerical Integration of Ordinary Differential Equations. JPL Internal Document, Section 314 TM No. 217 (revised), Jet Propulsion Laboratory, Pasadena, Calif. (Oct 1970). (To appear in the Journal of the ACM.)

[17] Hull, T. E., Enright, W. H., Fellen, B. M. and Sedgwick, A.E.: Comparing Numerical Methods for Ordinary Differential Equations. SIAM J. Numerical Analysis 9 (1972), 603-637.

[18] Gear, C. W.: The Numerical Integration of Ordinary Differential Equations. Math. of Comp. 21 (1967), 146-156.

[19] Thomas, L. H.: The Integration of Ordinary Differential Systems. The Ohio State University Engineering Experiment Station News 24 (1952), 8-9, 31-32.

[20] Krogh, F. T.: A Method for Simplifying the Maintenance of Software Which Consists of Many Versions. JPL Internal Document, Section 914 TM No. 314, Jet Propulsion Laboratory, Pasadena, Calif. (Sept. 1972).

[21] Krogh, F. T.: An Integrator Design. JPL TM No. 33-479, Jet Propulsion Laboratory, Pasadena, Calif. (May 1971).

[22] Krogh, F. T.: Opinions on Matters Connected with the Evaluation of Programs and Methods for Integrating Ordinary Differential Equations. SIGNUM Newsletter 7 No. 3 (Oct. 1972).

[23] Krogh, F. T.: A Test for Instability in the Numerical Solution of Ordinary Differential Equations. Journal of the ACM 14 (1967), 351-354.

THE ORDER OF DIFFERENTIAL EQUATION METHODS

J. C. Butcher
Department of Mathematics
University of Auckland
New Zealand

Introduction

During the last ten years, a number of methods for the initial value problem

$$y'(x) = f(y(x)),\ y(x_o) = y_o$$

have been suggested which fall outside the traditional classes of Runge-Kutta and Linear Multistep methods. For these methods it is usually clear how the basic convergence and error propagation theories (see for example Henrici [5]) can be modified to include them but there are reasons for seeking a completely new approach.

In this paper, we write the class of methods under consideration in the form used in Butcher [1]. That is, each method is represented by a pair of matrices A,B and the N approximations to a solution after n steps of length h are given by

$$y_i^{(n)} = \sum_{j=1}^{N} a_{ij} y_j^{(n-1)} + h \sum_{j=1}^{N} b_{ij}\ f(y_j^{(n)})$$

$(i = 1,2,\cdots,N)$

where a_{ij}, b_{ij} are the elements of A,B.

We say that a method is consistent if $As = s$ and $A\xi + Bs = \xi + s$ where ξ is some vector and s is the vector with every component equal to 1, and stable if $\|A^n\|$ is bounded for all positive integral n. It is shown in [1] that a method is convergent if and only if it is consistent and stable and we shall always assume these properties hold for methods considered in this paper.

Definition of Order

For the method characterised by A,B let M_h denote the function which acts upon a collection of N solution values to give the N results after one step of length h using the method. Also let S_h denote the function which acts upon an initial value and gives the collection of N results after one step of length h using a collection of N Runge-Kutta methods. We will think of these N Runge-Kutta methods as the starting methods for the multistep method we are considering.

Let $I^{(i)}$ $i = 1,2,\cdots,N$ denote the functions which act on a collection of N approximations and gives as a result the value of the ith of them.

Definition The method given by A,B is of order q relative to the stating method defined by S_h if for $i = 1,2,\cdots,N$ the Runge-Kutta method characterised by the function

(1) $$(I^{(i)} \circ S_h)^{-1} \circ I^{(i)} \circ M_h \circ S_h$$

is of order q.

We say that the method given by A,B is of order q if it is of order q relative to some starting method.

It is easy to see that a linear multistep or Runge-Kutta method of order q in the usual sense is of order q in the sense of this definition. It may happen, however, for example in the Runge-Kutta case, that the order may be higher in the new sense than in the traditional one (see Butcher [2]).

Analytic Interpretation.

The truncation error of a method is usually defined by substituting the exact solution into the formula relating the solution value at neighbouring steps. In our case we would compute

$$T_i = -\tilde{y}_i^{(n)} + \sum_{j=1}^{N} a_{ij}\tilde{y}_j^{(n-1)} + h\sum_{j=1}^{N} b_{ij}\, f(\tilde{y}_j^{(n)})$$

for $i = 1,2,\cdots,N$ where $\tilde{y}_1^{(n)}, \tilde{y}_2^{(n)}, \cdots, \tilde{y}_N^{(n)}$ correspond in some sense to the exact solution. However, the choice we make for $\tilde{y}_i^{(n)}$ is

$$(I^{(i)}\circ S_h)\ (y(x_n))$$

where $y(x_n)$ denotes the exact solution at $x_n = x_0 + nh$. Under the assumption that f satisfies certain smoothness conditions, the method is of order q if and only if $\|T_i\| = O(h^{q+1})$ for all $i = 1,2,\cdots,N$, as $h \to 0$.

Rate of Convergence.

To use a method characterised by A,B to compute a solution at a particular x we perform n steps each of length h where $nh = x - x_0$. To begin the computation we use

$$y_i^{(0)} = (I^{(i)}\circ S_h)\ (y_0)$$

for $i = 1,2,\cdots,N$ and for the final approximation to $y(x)$, we compute $(I^{(i)}\circ S_h)^{-1}\ (y_i^{(n)})$ for some particular choice of i.

It is shown (Butcher [3]), again under appropriate assumptions on f, that the error in the final approximation is $O(h^q)$ as $h \to 0$ where q is the order of the method relative to S_h. Thus, we have a simple generalization of the theory of global error as it applies in the standard special cases.

Algebraic Interpretation

It has been shown (Butcher [4]) that many of the properties of Runge-Kutta methods can be represented by algebraic properties in a certain group. Elements of this group are represented by real valued functions on the set of rooted-trees and to each Runge-Kutta method there corresponds a particular group element. The special group element p, corresponds to the "exact" method applied over a single unit step. The order of a Runge-Kutta method can be characterised by the property that if its corresponding group element is g, then $g(t) = p(t)$ for any rooted tree t with no more than q nodes.

Let $g_1, g_2, \cdots, g_N$ denote the group elements for the Runge-Kutta methods in the starting method used with the method defined by A,B. An analysis of the group elements for the methods given by (1) shows that the condition of the order definition is equivalent to

$$g_i(t) = \sum_{j=1}^{N} a_{ij}(p^{-1}g_j)(t) + \sum_{j=1}^{N} b_{ij}\, g_j'(t) \tag{2}$$

for all t with no more than q nodes. In (2), $g_j'(t)$ denotes the product $g_j(t_1)\, g_j(t_2) \cdots g_j(t_k)$ where $t_1, t_2, \cdots, t_k$ are the sub-trees that remain when the root of t is removed and $p^{-1}g_j$ is the group-theoretic product.

For example, the conditions for a method to be fourth order in the sense of this paper can be found by substituting for t each of the 8 trees of order 1,2,3 or 4 into (2) for each $i = 1,2,\cdots,N$ and eliminating the $g_i(t)$ from these equations. This result in four algebraic restrictions on the coefficients of the method in addition to those imposed by the consistency condition.

Examples

The third and fourth order Runge-Kutta methods

$$A = \begin{bmatrix} 0 & 0 & 0 & 1 \\ 0 & 0 & 0 & 1 \\ 0 & 0 & 0 & 1 \\ 0 & 0 & 0 & 1 \end{bmatrix}, \qquad B = \begin{bmatrix} 0 & 0 & 0 & 0 \\ \frac{1}{2} & 0 & 0 & 0 \\ -1 & 2 & 0 & 0 \\ \frac{1}{6} & \frac{2}{3} & \frac{1}{6} & 0 \end{bmatrix}$$

$$A = \begin{bmatrix} 0 & 0 & 0 & 0 & 1 \\ 0 & 0 & 0 & 0 & 1 \\ 0 & 0 & 0 & 0 & 1 \\ 0 & 0 & 0 & 0 & 1 \\ 0 & 0 & 0 & 0 & 1 \end{bmatrix}, \quad B = \begin{bmatrix} 0 & 0 & 0 & 0 & 0 \\ \frac{1}{2} & 0 & 0 & 0 & 0 \\ 0 & \frac{1}{2} & 0 & 0 & 0 \\ 0 & 0 & 1 & 0 & 0 \\ \frac{1}{6} & \frac{1}{3} & \frac{1}{3} & \frac{1}{6} & 0 \end{bmatrix}$$

may be altered slightly to give methods that require one less derivative calculation per step in each case.

$$A = \begin{bmatrix} 0 & 0 & 1 & 0 \\ 0 & 0 & 0 & 1 \\ 0 & 0 & 0 & 1 \\ 0 & 0 & 0 & 1 \end{bmatrix}, \quad B = \begin{bmatrix} 0 & 0 & 0 & 0 \\ \frac{1}{2} & 0 & 0 & 0 \\ -\frac{1}{2} & \frac{3}{2} & 0 & 0 \\ \frac{1}{6} & \frac{2}{3} & \frac{1}{6} & 0 \end{bmatrix}$$

$$A = \begin{bmatrix} 0 & 0 & 0 & 1 & 0 \\ 0 & 0 & 0 & 0 & 1 \\ 0 & 0 & 0 & 0 & 1 \\ 0 & 0 & 0 & 0 & 1 \\ 0 & 0 & 0 & 0 & 1 \end{bmatrix}, \quad B = \begin{bmatrix} 0 & 0 & 0 & 0 & 0 \\ \frac{1}{2} & 0 & 0 & 0 & 0 \\ 0 & \frac{1}{2} & 0 & 0 & 0 \\ \frac{1}{12} & \frac{1}{12} & \frac{5}{6} & 0 & 0 \\ \frac{1}{6} & \frac{5}{18} & \frac{7}{18} & \frac{1}{6} & 0 \end{bmatrix}$$

These methods fall into neither standard class but are of orders three and four respectively in the sense of this paper. Both require trivial starting computations to achieve this order of accuracy in practice but in each case the final stage of the method yields a result of the required accuracy without the need for a finishing formula.

References

[1] Butcher, J.C. *Math.Comp.* 20, 1-10 (1966).

[2] Butcher, J.C. *Conference on Numerical Solution of Differential Equations*, 133-139 (1969).

[3] Butcher, J.C. *Math. Comp.* to appear.

[4] Butcher, J.C. *Math. Comp.* 26, 79-106 (1972).

[5] Henrici, P., *Discrete Variable Methods in Ordinary Differential Equations*, Wiley, New York, (1962).

EQUATIONS OF CONDITION FOR HIGH ORDER RUNGE-KUTTA-NYSTRÖM FORMULAE

by

Dale G. Bettis

The University of Texas at Austin

Introduction

Runge-Kutta algorithms for approximating the solutions of second order differential equations were first introduced by E.J. Nyström [1925]. For the special case where the first derivative does not appear explicitly in the second order equations, E. Fehlberg [1972] has presented Runge-Kutta formula of orders four, five, six, seven, and eight which also include a stepsize control procedure. Fehlberg developed the equations of condition for formulas of order ten for this special case where the first derivatives are absent. In this paper the equations of condition of order eight will be given for the general system of equations of the form $\ddot{x} = f(t, x, \dot{x})$. For this general case, the number of equations of condition is considerably larger than for the special case when the first derivative is not present. Specifically, for orders two through eight, the number of equations for each order is 1, 1, 1, 2, 3, 5, 9 for the special case, and is 1, 1, 2, 5, 13, 34, 95 for the general case.

The Nyström Operator

The solution of the system of differential equations

$$(1) \qquad \ddot{x} = f(t, x, \dot{x}) \quad , \quad x(t_o) = x_o \quad , \quad \dot{x}(t_o) = \dot{x}_o$$

at $t = t_o + h$ is approximated by the basic Runge-Kutta-Nyström algorithm

(2)
$$x = x_o + h\dot{x}_o + h^2 \sum_{k=0}^{n} C_k f_k + O(h^P)$$

$$\dot{x} = \dot{x}_o + h \sum_{k=0}^{m} \dot{C}_k f_k + O(h^P)$$

where

(3)
$$f_o = f(t_o, x_o, \dot{x}_o)$$

$$f_k = f(t_o + \alpha_k h,\ x_o + \dot{x}_o \alpha_k h + h^2 \sum_{\lambda=0}^{k-1} \gamma_{k\lambda} f_\lambda,\ \dot{x}_o + h \sum_{\lambda=0}^{k-1} \beta_{k\lambda} f_\lambda)$$

where $k = 1,2,\ldots,\theta$, with θ the larger of n and m.

By expanding the solutions of (1) in a Taylor series and comparing term by term to the expansions of the right-hand-sides of (3), equations of conditions are produced that relate tha parameters α, β, γ, C, $\dot{C}$. In order to facilitate these expansions, the Nyström operator

$$D = \frac{\partial}{\partial t} + \dot{x}\frac{\partial}{\partial x} + f\frac{\partial}{\partial \dot{x}}$$

was utilized. For this operator

$$D(\phi + \psi) = D(\phi) + D(\psi) ,$$

$$D(\phi \cdot \psi) = \phi D(\psi) + \psi D(\phi) ,$$

$$D[D^n(\phi)] = D^{n+1}(\phi) + n[f\ D^{n-1}(\phi_x) + D(f)\ D^{n-1}(\phi_{\dot{x}})] .$$

Nyström [1925] used this operator to develop the equations of condition to order five. The extension of this development to higher orders is straight forward, although tedious. Many of these laborious computations were performed using the system SYMBAL [Engeli, 1969].

Runge-Kutta-Equations of Condition

The equations of condition are presented in the Table. The left side of the Table represents the equations of condition for x, and the right side gives the equations for $\dot{x}$. It is understood that C is replaced by $\dot{C}$ for the equations of $\dot{x}$. The terms h^P denote the order of the term of the corresponding Taylor series for the development of x and $\dot{x}$, on the left and the right sides, respectively, of the Table.

In order to facilitate the notation, the following abbreviations are introduced:

$$P_{\kappa\lambda} = \beta_{\kappa 1}\alpha_1^{\lambda} + \beta_{\kappa 2}\alpha_2^{\lambda} + \ldots + \beta_{\kappa\ \kappa-1}\ \alpha_{\kappa-1}^{\lambda}$$

$$Q_{\kappa\lambda} = \gamma_{\kappa 1}\alpha_1^{\lambda} + \gamma_{\kappa 2}\alpha_2^{\lambda} + \ldots + \gamma_{\kappa\ \kappa-1}\ \alpha_{\kappa-1}^{\lambda}$$

Acknowledgements

The partial support of the National Aeronautics and Space Administration, Marshall Space Flight Center, Contract NAS 8-27931, is gratefully acknowledged. The advice and insight provided by Dr. Erwin Fehlberg is also gratefully acknowledged.

TABLE

Equations of Conditions for Runge-Kutta-Nystrom Formulas

1.	h^2	$\frac{1}{2} = \sum_{\kappa=0}^{\theta} C_\kappa$	$= 1$	h
1.	h^3	$\frac{1}{6} = \sum_{\kappa=1}^{\theta} C_\kappa \alpha_\kappa$	$= \frac{1}{2}$	h^2
1.	h^4	$\frac{1}{12} = \sum_{\kappa=1}^{\theta} C_\kappa \alpha_\kappa^2$	$= \frac{1}{3}$	h^3
2.	h^4	$\frac{1}{24} = \sum_{\kappa=2}^{\theta} C_\kappa P_{\kappa 1}$	$= \frac{1}{6}$	h^3
1.	h^5	$\frac{1}{20} = \sum_{\kappa=1}^{\theta} C_\kappa \alpha_\kappa^3$	$= \frac{1}{6}$	h^4
2.	h^5	$\frac{1}{40} = \sum_{\kappa=2}^{\theta} C_\kappa \alpha_\kappa P_{\kappa 1}$	$= \frac{1}{8}$	h^4
3.	h^5	$\frac{1}{120} = \sum_{\kappa=2}^{\theta} C_\kappa Q_{\kappa 1}$	$= \frac{1}{24}$	h^4
4.	h^5	$\frac{1}{60} = \sum_{\kappa=2}^{\theta} C_\kappa P_{\kappa 2}$	$= \frac{1}{12}$	h^4
5.	h^5	$\frac{1}{120} = \sum_{\kappa=3}^{\theta} C_\kappa \sum_{\lambda=2}^{\kappa-1} \beta_{\kappa\lambda} P_{\lambda 1}$	$= \frac{1}{24}$	h^4
1.	h^6	$\frac{1}{30} = \sum_{\kappa=1}^{\theta} C_\kappa \alpha_\kappa^4$	$= \frac{1}{5}$	h^5
2.	h^6	$\frac{1}{180} = \sum_{\kappa=2}^{\theta} C_\kappa \alpha_\kappa Q_{\kappa 1}$	$= \frac{1}{30}$	h^5
3.	h^6	$\frac{1}{360} = \sum_{\kappa=2}^{\theta} C_\kappa Q_{\kappa 2}$	$= \frac{1}{60}$	h^5

TABLE continued

4. h^6	$\frac{1}{60} = \sum_{\kappa=2}^{\theta} C_\kappa \alpha_\kappa^2 P_{\kappa 1}$	$= \frac{1}{10}$	h^5
5. h^6	$\frac{1}{90} = \sum_{\kappa=2}^{\theta} C\kappa \alpha_\kappa P_{\kappa 2}$	$= \frac{1}{15}$	h^5
6. h^6	$\frac{1}{120} = \sum_{\kappa=2}^{\theta} C_\kappa \,{}_{\kappa 3}$	$= \frac{1}{20}$	h^5
7. h^6	$\frac{1}{120} = \sum_{\kappa=2}^{\theta} C_\kappa P_{\kappa 1}^2$	$= \frac{1}{20}$	h^5
8. h^6	$\frac{1}{720} = \sum_{\kappa=3}^{\theta} C_\kappa \sum_{\lambda=2}^{\kappa-1} \beta_{\kappa\lambda} Q_{\lambda 1}$	$= \frac{1}{120}$	h^5
9. h^6	$\frac{1}{240} = \sum_{\kappa=3}^{\theta} C_\kappa \sum_{\lambda=2}^{\kappa-1} \beta_{\kappa\lambda} \alpha_\lambda P_{\lambda 1}$	$= \frac{1}{40}$	h^5
10. h^6	$\frac{1}{720} = \sum_{\kappa=3}^{\theta} C_\kappa \sum_{\lambda=2}^{\kappa-1} \beta_{\kappa\lambda} P_{\lambda 2}$	$= \frac{1}{120}$	h^5
11. h^6	$\frac{1}{180} = \sum_{\kappa=3}^{\theta} C_\kappa \lambda_\kappa \sum_{\lambda=2}^{\kappa-1} \beta_{\kappa\lambda} P_{\lambda 1}$	$= \frac{1}{30}$	h^5
12. h^6	$\frac{1}{720} = \sum_{\kappa=3}^{\theta} C_\kappa \sum_{\lambda=2}^{\kappa-1} \gamma_{\kappa\lambda} P_{\lambda 1}$	$= \frac{1}{120}$	h^5
13. h^6	$\frac{1}{720} = \sum_{\kappa=4}^{\theta} C_\kappa \sum_{\lambda=3}^{\kappa-1} \beta_{\kappa\lambda} \sum_{\mu=2}^{\lambda-1} \beta_{\lambda\mu} P_{\mu 1}$	$= \frac{1}{120}$	h^5
1. h^7	$\frac{1}{42} = \sum_{\kappa=1}^{\theta} C_\kappa \alpha_\kappa^5$	$= \frac{1}{720}$	h^6
2. h^7	$\frac{1}{252} = \sum_{\kappa=2}^{\theta} C_\kappa \alpha_\kappa^2 Q_{\kappa 1}$	$= \frac{1}{36}$	
3. h^7	$\frac{1}{504} = \sum_{\kappa=2}^{\theta} C_\kappa \alpha_\kappa Q_{\kappa 2}$	$= \frac{1}{72}$	h^6

TABLE continued

4.	h^7	$\frac{1}{840} = \sum_{\kappa=2}^{\theta} C_\kappa Q_{\kappa 3}$	$= \frac{1}{720}$	h^6
5.	h^7	$\frac{1}{84} = \sum_{\kappa=2}^{\theta} C_\kappa \alpha_\kappa^3 P_{\kappa 1}$	$= \frac{1}{12}$	h^6
6.	h^7	$\frac{1}{210} = \sum_{\kappa=2}^{\theta} C_\kappa P_{\kappa 4}$	$= \frac{1}{30}$	h^6
7.	h^7	$\frac{1}{168} = \sum_{\kappa=2}^{\theta} C_\kappa \alpha_\kappa P_{\kappa 3}$	$= \frac{1}{24}$	h^6
8.	h^7	$\frac{1}{126} = \sum_{\kappa=2}^{\theta} C_\kappa \alpha_\kappa^2 P_{\kappa 2}$	$= \frac{1}{18}$	h^6
9.	h^7	$\frac{1}{252} = \sum_{\kappa=2}^{\theta} C_\kappa P_{\kappa 1} P_{\kappa 2}$	$= \frac{1}{36}$	h^6
10.	h^7	$\frac{1}{504} = \sum_{\kappa=2}^{\theta} C_\kappa P_{\kappa 1} Q_{\kappa 1}$	$= \frac{1}{72}$	h^6
11.	h^7	$\frac{1}{210} = \sum_{\kappa=2}^{\theta} C_\kappa \alpha_\kappa P_{\kappa 1}^2$	$= \frac{1}{24}$	h^6
12.	h^7	$\frac{1}{1260} = \sum_{\kappa=3}^{\theta} C_\kappa \sum_{\lambda=2}^{\kappa-1} \beta_{\kappa\lambda} \alpha_\lambda Q_{\lambda 1}$	$= \frac{1}{180}$	h^6
13.	h^7	$\frac{1}{2520} = \sum_{\kappa=3}^{\theta} C_\kappa \sum_{\lambda=2}^{\kappa-1} \beta_{\kappa\lambda} Q_{\lambda 2}$	$= \frac{1}{360}$	h^6
14.	h^7	$\frac{1}{420} = \sum_{\kappa=3}^{\theta} C_\kappa \sum_{\lambda=2}^{\kappa-1} \beta_{\kappa\lambda} \alpha_\lambda^2 P_{\lambda 1}$	$= \frac{1}{60}$	h^6
15.	h^7	$\frac{1}{630} = \sum_{\kappa=3}^{\theta} C_\kappa \sum_{\lambda=2}^{\kappa-1} \beta_{\kappa\lambda} \alpha_\lambda P_{\lambda 2}$	$= \frac{1}{90}$	h^6
16.	h^7	$\frac{1}{840} = \sum_{\kappa=3}^{\theta} C_\kappa \sum_{\lambda=2}^{\kappa-1} \beta_{\kappa\lambda} P_{\lambda 3}$	$= \frac{1}{120}$	h^6
17.	h^7	$\frac{1}{840} = \sum_{\kappa=3}^{\theta} C_\kappa \sum_{\lambda=2}^{\kappa-1} \beta_{\kappa\lambda} P_{\lambda 1}^2$	$= \frac{1}{120}$	h^6

TABLE continued

18.	h^7	$\frac{1}{1008} = \sum_{\kappa=3}^{\theta} C_\kappa \alpha_\kappa \sum_{\lambda=2}^{\kappa-1} \beta_{\kappa\lambda} Q_{\lambda 1}$	$= \frac{1}{144}$	h^6
19.	h^7	$\frac{1}{336} = \sum_{\kappa=3}^{\theta} C_\kappa \alpha_\kappa \sum_{\lambda=2}^{\kappa-1} \beta_{\kappa\lambda} \alpha_\lambda P_{\lambda 1}$	$= \frac{1}{48}$	h^6
20.	h^7	$\frac{1}{504} = \sum_{\kappa=3}^{\theta} C_\kappa \alpha_\kappa \sum_{\lambda=2}^{\kappa-1} \beta_{\kappa\lambda} P_{\lambda 2}$	$= \frac{1}{72}$	h^6
21.	h^7	$\frac{1}{252} = \sum_{\kappa=3}^{\theta} C_\kappa \alpha_\kappa^2 \sum_{\lambda=2}^{\kappa-1} \beta_{\kappa\lambda} P_{\lambda 1}$	$= \frac{1}{36}$	h^6
22.	h^7	$\frac{1}{5040} = \sum_{\kappa=3}^{\theta} C_\kappa \sum_{\lambda=2}^{\kappa-1} \gamma_{\kappa\lambda} Q_{\lambda 1}$	$= \frac{1}{720}$	h^6
23.	h^7	$\frac{1}{1680} = \sum_{\kappa=3}^{\theta} C_\kappa \sum_{\lambda=2}^{\kappa-1} \gamma_{\kappa\lambda} \alpha_\lambda P_{\lambda 1}$	$= \frac{1}{240}$	h^6
24.	h^7	$\frac{1}{2520} = \sum_{\kappa=3}^{\theta} C_\kappa \sum_{\lambda=2}^{\kappa-1} \gamma_{\kappa\lambda} P_{\lambda 2}$	$= \frac{1}{360}$	h^6
25.	h^7	$\frac{1}{1008} = \sum_{\kappa=3}^{\theta} C_\kappa \alpha_\kappa \sum_{\lambda=2}^{\kappa-1} \gamma_{\kappa\lambda} P_{\lambda 1}$	$= \frac{1}{144}$	h^6
26.	h^7	$\frac{1}{504} = \sum_{\kappa=3}^{\theta} C_\kappa P_{\kappa 1} \sum_{\lambda=2}^{\kappa-1} \beta_{\kappa\lambda} P_{\lambda 1}$	$= \frac{1}{72}$	h^6
27.	h^7	$\frac{1}{5040} = \sum_{\kappa=4}^{\theta} C_\kappa \sum_{\lambda=3}^{\kappa-1} \beta_{\kappa\lambda} \sum_{\mu=2}^{\lambda-1} \beta_{\lambda\mu} Q_{\mu 1}$	$= \frac{1}{720}$	h^6
28.	h^7	$\frac{1}{1680} = \sum_{\kappa=4}^{\theta} C_\kappa \sum_{\lambda=3}^{\kappa-1} \beta_{\kappa\lambda} \sum_{\mu=2}^{\lambda-1} \beta_{\lambda\mu} \alpha_\mu P_{\mu 1}$	$= \frac{1}{240}$	h^6
29.	h^7	$\frac{1}{2520} = \sum_{\kappa=4}^{\theta} C_\kappa \sum_{\lambda=3}^{\kappa-1} \beta_{\kappa\lambda} \sum_{\mu=2}^{\lambda-1} \beta_{\lambda\mu} P_{\mu 2}$	$= \frac{1}{360}$	h^6
30.	h^7	$\frac{1}{1260} = \sum_{\kappa=4}^{\theta} C_\kappa \sum_{\lambda=3}^{\kappa-1} \beta_{\kappa\lambda} \alpha_\lambda \sum_{\mu=2}^{\lambda-1} \beta_{\lambda\mu} P_{\mu 1}$	$= \frac{1}{180}$	h^6

TABLE continued

31.	h^7	$\frac{1}{5040} = \sum_{\kappa=4}^{\theta} C_\kappa \sum_{\lambda=3}^{\kappa-1} \beta_{\kappa\lambda} \sum_{\mu=2}^{\lambda-1} \gamma_{\lambda\mu} P_{\mu 2}$	$= \frac{1}{720}$	h^6
32.	h^7	$\frac{1}{1008} = \sum_{\kappa=4}^{\theta} C_\kappa \alpha_\kappa \sum_{\lambda=3}^{\kappa-1} \beta_{\kappa\lambda} \sum_{\mu=2}^{\lambda-1} \beta_{\lambda\mu} P_{\mu 1}$	$= \frac{1}{144}$	h^6
33.	h^7	$\frac{1}{5040} = \sum_{\kappa=4}^{\theta} C_\kappa \sum_{\lambda=3}^{\kappa-1} \gamma_{\kappa\lambda} \sum_{\mu=2}^{\lambda-1} \beta_{\lambda\mu} P_{\mu 1}$	$= \frac{1}{720}$	h^6
34.	h^7	$\frac{1}{5040} = \sum_{\kappa=5}^{\theta} C_\kappa \sum_{\lambda=4}^{\kappa-1} \beta_{\kappa\lambda} \sum_{\mu=3}^{\lambda-1} \beta_{\lambda\mu} \sum_{\nu=2}^{\mu-1} \beta_{\mu\nu} P_{\nu 1}$	$= \frac{1}{720}$	h^6
1.	h^8	$\frac{1}{56} = \sum_{\kappa=1}^{\theta} C_\kappa \alpha_\kappa^6$	$= \frac{1}{7}$	h^7
2.	h^8	$\frac{1}{336} = \sum_{\kappa=2}^{\theta} C_\kappa \alpha_\kappa^3 Q_{\kappa 1}$	$= \frac{1}{42}$	h^7
3.	h^8	$\frac{1}{1680} = \sum_{\kappa=2}^{\theta} C_\kappa Q_{\kappa 4}$	$= \frac{1}{210}$	h^7
4.	h^8	$\frac{1}{1120} = \sum_{\kappa=2}^{\theta} C_\kappa \alpha_\kappa Q_{\kappa 3}$	$= \frac{1}{140}$	h^7
5.	h^8	$\frac{1}{672} = \sum_{\kappa=2}^{\theta} C_\kappa \alpha_\kappa^2 Q_{\kappa 2}$	$= \frac{1}{84}$	h^7
6.	h^8	$\frac{1}{2016} = \sum_{\kappa=2}^{\theta} C_\kappa Q_{\kappa 1}^2$	$= \frac{1}{252}$	h^7
7.	h^8	$\frac{1}{112} = \sum_{\kappa=2}^{\theta} C_\kappa \alpha_\kappa^4 P_{\kappa 1}$	$= \frac{1}{14}$	h^7
8.	h^8	$\frac{1}{336} = \sum_{\kappa=2}^{\theta} C_\kappa P_{\kappa 5}$	$= \frac{1}{42}$	h^7
9.	h^8	$\frac{1}{280} = \sum_{\kappa=2}^{\theta} C_\kappa \alpha_\kappa P_{\kappa 4}$	$= \frac{1}{35}$	h^7
10.	h^8	$\frac{1}{224} = \sum_{\kappa=2}^{\theta} C_\kappa \alpha_\kappa^2 P_{\kappa 3}$	$= \frac{1}{28}$	h^7

TABLE continued

11.	h^8	$\frac{1}{168} = \sum_{\kappa=2}^{\theta} C_\kappa \alpha_\kappa^3 P_{\kappa 2}$	$= \frac{1}{21}$	h^7
12.	h^8	$\frac{1}{448} = \sum_{\kappa=2}^{\theta} C_\kappa P_{\kappa 1} P_{\kappa 3}$	$= \frac{1}{56}$	h^7
13.	h^8	$\frac{1}{504} = \sum_{\kappa=2}^{\theta} C_\kappa P_{\kappa 2}^2$	$= \frac{1}{63}$	h^7
14.	h^8	$\frac{1}{1344} = \sum_{\kappa=2}^{\theta} C_\kappa P_{\kappa 1} Q_{\kappa 2}$	$= \frac{1}{168}$	h^7
15.	h^8	$\frac{1}{1008} = \sum_{\kappa=2}^{\theta} C_\kappa P_{\kappa 2} Q_{\kappa 1}$	$= \frac{1}{126}$	h^7
16.	h^8	$\frac{1}{336} = \sum_{\kappa=2}^{\theta} C_\kappa \alpha_\kappa P_{\kappa 1} P_{\kappa 2}$	$= \frac{1}{42}$	h^7
17.	h^8	$\frac{1}{672} = \sum_{\kappa=2}^{\theta} C_\kappa \alpha_\kappa P_{\kappa 1} Q_{\kappa 1}$	$= \frac{1}{84}$	h^7
18.	h^8	$\frac{1}{224} = \sum_{\kappa=2}^{\theta} C_\kappa \alpha_\kappa^2 P_{\kappa 1}^2$	$= \frac{1}{28}$	h^7
19.	h^8	$\frac{1}{448} = \sum_{\kappa=2}^{\theta} C_\kappa P_{\kappa 1}^3$	$= \frac{1}{56}$	h^7
20.	h^8	$\frac{1}{2016} = \sum_{\kappa=3}^{\theta} C_\kappa \sum_{\lambda=2}^{\kappa-1} \beta_{\kappa\lambda} \alpha_\lambda^2 Q_{\lambda 1}$	$= \frac{1}{252}$	h^7
21.	h^8	$\frac{1}{4032} = \sum_{\kappa=3}^{\theta} C_\kappa \sum_{\lambda=2}^{\kappa-1} \beta_{\kappa\lambda} \alpha_\kappa Q_{\lambda 2}$	$= \frac{1}{504}$	h^7
22.	h^8	$\frac{1}{6720} = \sum_{\kappa=3}^{\theta} C_\kappa \sum_{\lambda=2}^{\kappa-1} \beta_{\kappa\lambda} Q_{\lambda 3}$	$= \frac{1}{840}$	h^7
23.	h^8	$\frac{1}{672} = \sum_{\kappa=3}^{\theta} C_\kappa \sum_{\lambda=2}^{\kappa-1} \beta_{\kappa\lambda} \alpha_\lambda^3 P_{\lambda 1}$	$= \frac{1}{84}$	h^7
24.	h^8	$\frac{1}{1680} = \sum_{\kappa=3}^{\theta} C_\kappa \sum_{\lambda=2}^{\kappa-1} \beta_{\kappa\lambda} P_{\lambda 4}$	$= \frac{1}{210}$	h^7

TABLE continued

25.	h^8	$\frac{1}{1344} = \sum_{\kappa=3}^{\theta} C_\kappa \sum_{\lambda=2}^{\kappa-1} \beta_{\kappa\lambda}\alpha_\lambda P_{\lambda 3}$	$= \frac{1}{168}$	h^7
26.	h^8	$\frac{1}{1008} = \sum_{\kappa=3}^{\theta} C_\kappa \sum_{\lambda=2}^{\kappa-1} \beta_{\kappa\lambda}\alpha_\lambda^2 P_{\lambda 2}$	$= \frac{1}{126}$	h^7
27.	h^8	$\frac{1}{2016} = \sum_{\kappa=3}^{\theta} C_\kappa \sum_{\lambda=2}^{\kappa-1} \beta_{\kappa\lambda} P_{\lambda 1} P_{\lambda 2}$	$= \frac{1}{252}$	h^7
28.	h^8	$\frac{1}{4032} = \sum_{\kappa=3}^{\theta} C_\kappa \sum_{\lambda=2}^{\kappa-1} \beta_{\kappa\lambda} P_{\lambda 1} Q_{\lambda 1}$	$= \frac{1}{504}$	h^7
29.	h^8	$\frac{1}{1344} = \sum_{\kappa=3}^{\theta} C_\kappa \sum_{\lambda=2}^{\kappa-1} \beta_{\kappa\lambda}\alpha_\lambda P_{\lambda 1}^2$	$= \frac{1}{168}$	h^7
30.	h^8	$\frac{1}{1680} = \sum_{\kappa=3}^{\theta} C_\kappa \alpha_\kappa \sum_{\lambda=2}^{\kappa-1} \beta_{\kappa\lambda}\alpha_\lambda Q_{\lambda 1}$	$= \frac{1}{210}$	h^7
31.	h^8	$\frac{1}{3360} = \sum_{\kappa=3}^{\theta} C_\kappa \alpha_\kappa \sum_{\lambda=2}^{\kappa-1} \beta_{\kappa\lambda} Q_{\lambda 2}$	$= \frac{1}{420}$	h^7
32.	h^8	$\frac{1}{560} = \sum_{\kappa=3}^{\theta} C_\kappa \alpha_\kappa \sum_{\lambda=2}^{\kappa-1} \beta_{\kappa\lambda}\alpha_\lambda^2 P_{\lambda 1}$	$= \frac{1}{70}$	h^7
33.	h^8	$\frac{1}{840} = \sum_{\kappa=3}^{\theta} C_\kappa \alpha_\kappa \sum_{\lambda=2}^{\kappa-1} \beta_{\kappa\lambda}\alpha_\lambda P_{\lambda 2}$	$= \frac{1}{105}$	h^7
34.	h^8	$\frac{1}{1120} = \sum_{\kappa=3}^{\theta} C_\kappa \alpha_\kappa \sum_{\lambda=2}^{\kappa-1} \beta_{\kappa\lambda} P_{\lambda 3}$	$= \frac{1}{140}$	h^7
35.	h^8	$\frac{1}{1120} = \sum_{\kappa=3}^{\theta} C_\kappa \alpha_\kappa \sum_{\lambda=2}^{\kappa-1} \beta_{\kappa\lambda} P_{\lambda 1}^2$	$= \frac{1}{140}$	h^7
36.	h^8	$\frac{1}{1344} = \sum_{\kappa=3}^{\theta} C_\kappa \alpha_\kappa^2 \sum_{\lambda=2}^{\kappa-1} \beta_{\kappa\lambda} Q_{\lambda 1}$	$= \frac{1}{168}$	h^7
37.	h^8	$\frac{1}{448} = \sum_{\kappa=3}^{\theta} C_\kappa \alpha_\kappa^2 \sum_{\lambda=2}^{\kappa-1} \beta_{\kappa\lambda}\alpha_\lambda P_{\lambda 1}$	$= \frac{1}{56}$	h^7
38.	h^8	$\frac{1}{672} = \sum_{\kappa=3}^{\theta} C_\kappa \alpha_\kappa^2 \sum_{\lambda=2}^{\kappa-1} \beta_{\kappa\lambda} P_{\lambda 2}$	$= \frac{1}{84}$	h^7

TABLE continued

39.	h^8	$\frac{1}{336} = \sum_{\kappa=3}^{\theta} C_\kappa \alpha_\kappa^3 \sum_{\lambda=2}^{\kappa-1} \beta_{\kappa\lambda} P_{\lambda 1}$	$= \frac{1}{42}$	h^7
40.	h^8	$\frac{1}{10080} = \sum_{\kappa=3}^{\theta} C_\kappa \sum_{\lambda=2}^{\kappa-1} \gamma_{\kappa\lambda} \alpha_\lambda Q_{\lambda 1}$	$= \frac{1}{1260}$	h^7
41.	h^8	$\frac{1}{20160} = \sum_{\kappa=3}^{\theta} C_\kappa \sum_{\lambda=2}^{\kappa-1} \gamma_{\kappa\lambda} Q_{\lambda 2}$	$= \frac{1}{2520}$	h^7
42.	h^8	$\frac{1}{3360} = \sum_{\kappa=3}^{\theta} C_\kappa \sum_{\lambda=2}^{\kappa-1} \gamma_{\kappa\lambda} \alpha_\lambda^2 P_{\lambda 1}$	$= \frac{1}{420}$	h^7
43.	h^8	$\frac{1}{5040} = \sum_{\kappa=3}^{\theta} C_\kappa \sum_{\lambda=2}^{\kappa-1} \gamma_{\kappa\lambda} \alpha_\lambda P_{\lambda 2}$	$= \frac{1}{630}$	h^7
44.	h^8	$\frac{1}{6720} = \sum_{\kappa=3}^{\theta} C_\kappa \sum_{\lambda=2}^{\kappa-1} \gamma_{\kappa\lambda} P_{\lambda 3}$	$= \frac{1}{840}$	h^7
45.	h^8	$\frac{1}{6720} = \sum_{\kappa=3}^{\theta} C_\kappa \sum_{\lambda=2}^{\kappa-1} \gamma_{\kappa\lambda} P_{\lambda 1}^2$	$= \frac{1}{840}$	h^7
46.	h^8	$\frac{1}{6720} = \sum_{\kappa=3}^{\theta} C_\kappa \alpha_\kappa \sum_{\lambda=2}^{\kappa-1} \gamma_{\kappa\lambda} Q_{\lambda 1}$	$= \frac{1}{840}$	h^7
47.	h^8	$\frac{1}{2240} = \sum_{\kappa=3}^{\theta} C_\kappa \alpha_\kappa \sum_{\lambda=2}^{\kappa-1} \gamma_{\kappa\lambda} \alpha_\lambda P_{\lambda 1}$	$= \frac{1}{280}$	h^7
48.	h^8	$\frac{1}{3360} = \sum_{\kappa=3}^{\theta} C_\kappa \alpha_\kappa \sum_{\lambda=2}^{\kappa-1} \gamma_{\kappa\lambda} P_{\lambda 2}$	$= \frac{1}{420}$	h^7
49.	h^8	$\frac{1}{1344} = \sum_{\kappa=3}^{\theta} C_\kappa \alpha_\kappa^2 \sum_{\lambda=2}^{\kappa-1} \gamma_{\kappa\lambda} P_{\lambda 1}$	$= \frac{1}{168}$	h^7
50.	h^8	$\frac{1}{2016} = \sum_{\kappa=3}^{\theta} C_\kappa \sum_{\lambda=2}^{\kappa-1} [\beta_{\kappa\lambda} P_{\lambda 1}]^2$	$= \frac{1}{252}$	h^7
51.	h^8	$\frac{1}{2688} = \sum_{\kappa=3}^{\theta} C_\kappa P_\kappa \sum_{\lambda=2}^{\kappa-1} \beta_{\kappa\lambda} Q_{\lambda 1}$	$= \frac{1}{336}$	h^7
52.	h^8	$\frac{1}{896} = \sum_{\kappa=3}^{\theta} C_\kappa P_{\kappa 1} \sum_{\lambda=2}^{\kappa-1} \beta_{\kappa\lambda} \alpha_\lambda P_{\lambda 1}$	$= \frac{1}{112}$	h^7

TABLE continued

53.	h^8	$\frac{1}{1344} = \sum_{\kappa=3}^{\theta} C_\kappa P_{\kappa 1} \sum_{\lambda=2}^{\kappa-1} \beta_{\kappa\lambda} P_{\lambda 2}$	$= \frac{1}{168}$	h^7
54.	h^8	$\frac{1}{1008} = \sum_{\kappa=3}^{\theta} C_\kappa P_{\kappa 2} \sum_{\lambda=2}^{\kappa-1} \beta_{\kappa\lambda} P_{\lambda 1}$	$= \frac{1}{126}$	h^7
55.	h^8	$\frac{1}{672} = \sum_{\kappa=3}^{\theta} C_\kappa \alpha_\kappa P_{\kappa 1} \sum_{\lambda=2}^{\kappa-1} \beta_{\kappa\lambda} P_{\lambda 1}$	$= \frac{1}{84}$	h^7
56.	h^8	$\frac{1}{2688} = \sum_{\kappa=3}^{\theta} C_\kappa P_{\kappa 1} \sum_{\lambda=2}^{\kappa-1} \gamma_{\kappa\lambda} P_{\lambda 1}$	$= \frac{1}{336}$	h^7
57.	h^8	$\frac{1}{2016} = \sum_{\kappa=3}^{\theta} C_\kappa Q_{\kappa 1} \sum_{\lambda=2}^{\kappa-1} \beta_{\kappa\lambda} P_{\lambda 1}$	$= \frac{1}{252}$	h^7
58.	h^8	$\frac{1}{10080} = \sum_{\kappa=4}^{\theta} C_\kappa \sum_{\lambda=3}^{\kappa-1} \beta_{\kappa\lambda} \sum_{\mu=2}^{\lambda-1} \beta_{\lambda\mu} \alpha_\mu Q_{\mu 1}$	$= \frac{1}{1260}$	h^7
59.	h^8	$\frac{1}{20160} = \sum_{\kappa=4}^{\theta} C_\kappa \sum_{\lambda=3}^{\kappa-1} \beta_{\kappa\lambda} \sum_{\mu=2}^{\lambda-1} \beta_{\lambda\mu} Q_{\mu 2}$	$= \frac{1}{2520}$	h^7
60.	h^8	$\frac{1}{3360} = \sum_{\kappa=4}^{\theta} C_\kappa \sum_{\lambda=3}^{\kappa-1} \beta_{\kappa\lambda} \sum_{\mu=2}^{\lambda-1} \beta_{\lambda\mu} \alpha_\mu^2 P_{\mu 1}$	$= \frac{1}{420}$	h^7
61.	h^8	$\frac{1}{5040} = \sum_{\kappa=4}^{\theta} C_\kappa \sum_{\lambda=3}^{\kappa-1} \beta_{\kappa\lambda} \sum_{\mu=2}^{\lambda-1} \beta_{\lambda\mu} \alpha_\mu P_{\mu 2}$	$= \frac{1}{630}$	h^7
62.	h^8	$\frac{1}{6720} = \sum_{\kappa=4}^{\theta} C_\kappa \sum_{\lambda=3}^{\kappa-1} \beta_{\kappa\lambda} \sum_{\mu=2}^{\lambda-1} \beta_{\lambda\mu} P_{\mu 3}$	$= \frac{1}{840}$	h^7
63.	h^8	$\frac{1}{6720} = \sum_{\kappa=4}^{\theta} C_\kappa \sum_{\lambda=3}^{\kappa-1} \beta_{\kappa\lambda} \sum_{\mu=2}^{\lambda-1} \beta_{\lambda\mu} P_{\mu 1}^2$	$= \frac{1}{840}$	h^7
64.	h^8	$\frac{1}{8064} = \sum_{\kappa=4}^{\theta} C_\kappa \sum_{\lambda=3}^{\kappa-1} \beta_{\kappa\lambda} \alpha_\lambda \sum_{\mu=2}^{\lambda-1} \beta_{\lambda\mu} Q_{\mu 1}$	$= \frac{1}{1008}$	h^7
65.	h^8	$\frac{1}{2688} = \sum_{\kappa=4}^{\theta} C_\kappa \sum_{\lambda=3}^{\kappa-1} \beta_{\kappa\lambda} \alpha_\lambda \sum_{\mu=2}^{\lambda-1} \beta_{\lambda\mu} \alpha_\mu P_{\mu 1}$	$= \frac{1}{336}$	h^7

TABLE   continued

66. h^8	$\dfrac{1}{4032} = \displaystyle\sum_{\kappa=4}^{\theta} C_\kappa \sum_{\lambda=3}^{\kappa-1} \beta_{\kappa\lambda}\alpha_\lambda \sum_{\mu=2}^{\lambda-1} \beta_{\lambda\mu}P_{\mu 2}$	$= \dfrac{1}{504}$	h^7
67. h^8	$\dfrac{1}{2016} = \displaystyle\sum_{\kappa=4}^{\theta} C_\kappa \sum_{\lambda=3}^{\kappa-1} \beta_{\kappa\lambda}\alpha_\lambda^2 \sum_{\mu=2}^{\lambda-1} \beta_{\lambda\mu}P_{\mu 1}$	$= \dfrac{1}{252}$	h^7
68. h^8	$\dfrac{1}{40320} = \displaystyle\sum_{\kappa=4}^{\theta} C_\kappa \sum_{\lambda=3}^{\kappa-1} \beta_{\kappa\lambda} \sum_{\mu=2}^{\lambda-1} \gamma_{\lambda\mu}Q_{\mu 1}$	$= \dfrac{1}{5040}$	h^7
69. h^8	$\dfrac{1}{13440} = \displaystyle\sum_{\kappa=4}^{\theta} C_\kappa \sum_{\lambda=3}^{\kappa-1} \beta_{\mu\lambda} \sum_{\mu=2}^{\lambda-1} \gamma_{\lambda\mu}\alpha_\mu P_{\mu 1}$	$= \dfrac{1}{1680}$	h^7
70. h^8	$\dfrac{1}{20160} = \displaystyle\sum_{\kappa=4}^{\theta} C_\kappa \sum_{\lambda=3}^{\kappa-1} \beta_{\kappa\lambda} \sum_{\mu=2}^{\lambda-1} \gamma_{\lambda\mu}P_{\mu 2}$	$= \dfrac{1}{2520}$	h^7
71. h^8	$\dfrac{1}{8064} = \displaystyle\sum_{\kappa=4}^{\theta} C_\kappa \sum_{\lambda=3}^{\kappa-1} \beta_{\kappa\lambda}\alpha_\lambda \sum_{\mu=2}^{\lambda-1} \gamma_{\lambda\mu}P_{\mu 1}$	$= \dfrac{1}{1008}$	h^7
72. h^8	$\dfrac{1}{4032} = \displaystyle\sum_{\kappa=4}^{\theta} C_\kappa \sum_{\lambda=3}^{\kappa-1} \beta_{\kappa\lambda}P_{\lambda 1} \sum_{\mu=2}^{\lambda-1} \beta_{\lambda\mu}P_{\mu 1}$	$= \dfrac{1}{504}$	h^7
73. h^8	$\dfrac{1}{6720} = \displaystyle\sum_{\kappa=4}^{\theta} C_\kappa\alpha_\kappa \sum_{\lambda=3}^{\kappa-1} \beta_{\kappa\lambda} \sum_{\mu=2}^{\lambda-1} \beta_{\lambda\mu}Q_{\mu 1}$	$= \dfrac{1}{840}$	h^7
74. h^8	$\dfrac{1}{2240} = \displaystyle\sum_{\kappa=4}^{\theta} C_\kappa\alpha_\kappa \sum_{\lambda=3}^{\kappa-1} \beta_{\kappa\lambda} \sum_{\mu=2}^{\lambda-1} \beta_{\lambda\mu}\alpha_\mu P_{\mu 1}$	$= \dfrac{1}{280}$	h^7
75. h^8	$\dfrac{1}{3360} = \displaystyle\sum_{\kappa=4}^{\theta} C_\kappa\alpha_\kappa \sum_{\lambda=3}^{\kappa-1} \beta_{\kappa\lambda} \sum_{\mu=2}^{\lambda-1} \beta_{\lambda\mu}P_{\mu 2}$	$= \dfrac{1}{420}$	h^7
76. h^8	$\dfrac{1}{1680} = \displaystyle\sum_{\kappa=4}^{\theta} C_\kappa\alpha_\kappa \sum_{\lambda=3}^{\kappa-1} \beta_{\kappa\lambda}\alpha_\lambda \sum_{\mu=2}^{\lambda-1} \beta_{\lambda\mu}P_{\mu 1}$	$= \dfrac{1}{210}$	h^7
77. h^8	$\dfrac{1}{6720} = \displaystyle\sum_{\kappa=4}^{\theta} C_\kappa\alpha_\kappa \sum_{\lambda=3}^{\kappa-1} \beta_{\kappa\lambda} \sum_{\mu=2}^{\lambda-1} \gamma_{\lambda\mu}\mathbf{P}_{\mu 1}$	$= \dfrac{1}{840}$	h^7
78. h^8	$\dfrac{1}{1344} = \displaystyle\sum_{\kappa=4}^{\theta} C_\kappa\alpha_\kappa^2 \sum_{\lambda=3}^{\kappa-1} \beta_{\kappa\lambda} \sum_{\mu=2}^{\lambda-1} \beta_{\lambda\mu}P_{\mu 1}$	$= \dfrac{1}{168}$	h^7
79. h^8	$\dfrac{1}{40320} = \displaystyle\sum_{\kappa=4}^{\theta} C_\kappa \sum_{\lambda=3}^{\kappa-1} \gamma_{\kappa\lambda} \sum_{\mu=2}^{\lambda-1} \beta_{\lambda\mu}Q_{\mu 1}$	$= \dfrac{1}{5040}$	h^7

TABLE continued

80. h^8	$\frac{1}{13440} = \sum_{\kappa=4}^{\theta} C_\kappa \sum_{\lambda=3}^{\kappa-1} \gamma_{\kappa\lambda} \sum_{\mu=2}^{\lambda-1} \beta_{\lambda\mu}\alpha_\mu P_{\mu 1}$	$= \frac{1}{1680}$	h^7
81. h^8	$\frac{1}{20160} = \sum_{\kappa=4}^{\theta} C_\kappa \sum_{\lambda=3}^{\kappa-1} \gamma_{\kappa\lambda} \sum_{\mu=2}^{\lambda-1} \beta_{\lambda\mu} P_{\mu 2}$	$= \frac{1}{2520}$	h^7
82. h^8	$\frac{1}{10080} = \sum_{\kappa=4}^{\theta} C_\kappa \sum_{\lambda=3}^{\kappa-1} \gamma_{\kappa\lambda}\alpha_\lambda \sum_{\mu=2}^{\lambda-1} \beta_{\lambda\mu} P_{\mu 1}$	$= \frac{1}{1260}$	h^7
83. h^8	$\frac{1}{40320} = \sum_{\kappa=4}^{\theta} C_\kappa \sum_{\lambda=3}^{\kappa-1} \gamma_{\kappa\lambda} \sum_{\mu=2}^{\lambda-1} \gamma_{\lambda\mu} P_{\mu 1}$	$= \frac{1}{5040}$	h^7
84. h^8	$\frac{1}{6720} = \sum_{\kappa=4}^{\theta} C_\kappa\alpha_\kappa \sum_{\lambda=3}^{\kappa-1} \gamma_{\kappa\lambda} \sum_{\mu=2}^{\lambda-1} \beta_{\lambda\mu} P_{\mu 1}$	$= \frac{1}{840}$	h^7
85. h^8	$\frac{1}{2688} = \sum_{\kappa=4}^{\theta} C_\kappa P_{\kappa 1} \sum_{\lambda=3}^{\lambda-1} \beta_{\kappa\lambda} \sum_{\mu=2}^{\lambda-1} \beta_{\lambda\mu} P_{\mu 1}$	$= \frac{1}{346}$	h^7
86. h^8	$\frac{1}{40320} = \sum_{\kappa=5}^{\theta} C_\kappa \sum_{\lambda=4}^{\kappa-1} \beta_{\kappa\lambda} \sum_{\mu=3}^{\lambda-1} \beta_{\lambda\mu} \sum_{\nu=2}^{\mu-1} \beta_{\mu\nu} Q_{\nu 1}$	$= \frac{1}{5040}$	h^7
87. h^8	$\frac{1}{13440} = \sum_{\kappa=5}^{\theta} C_\kappa \sum_{\lambda=4}^{\kappa-1} \beta_{\kappa\lambda} \sum_{\mu=3}^{\lambda-1} \beta_{\lambda\mu} \sum_{\nu=2}^{\mu-1} \beta_{\mu\nu}\alpha_\nu P_{\nu 1}$	$= \frac{1}{1680}$	h^7
88. h^8	$\frac{1}{20160} = \sum_{\kappa=5}^{\theta} C_\kappa \sum_{\lambda=4}^{\kappa-1} \beta_{\kappa\lambda} \sum_{\mu=3}^{\lambda-1} \beta_{\lambda\mu} \sum_{\nu=2}^{\mu-1} \beta_{\mu\nu} P_{\nu 2}$	$= \frac{1}{2520}$	h^7
89. h^8	$\frac{1}{10080} = \sum_{\kappa=5}^{\theta} C_\kappa \sum_{\lambda=4}^{\kappa-1} \beta_{\kappa\lambda} \sum_{\mu=3}^{\lambda-1} \beta_{\lambda\mu}\alpha_\mu \sum_{\nu=2}^{\mu-1} \beta_{\mu\nu} P_{\nu 1}$	$= \frac{1}{1260}$	h^7
90. h^8	$\frac{1}{40320} = \sum_{\kappa=5}^{\theta} C_\kappa \sum_{\lambda=4}^{\kappa-1} \beta_{\kappa\lambda} \sum_{\mu=3}^{\lambda-1} \beta_{\lambda\mu} \sum_{\nu=2}^{\mu-1} \gamma_{\mu\nu} P_{\nu 1}$	$= \frac{1}{5040}$	h^7
91. h^8	$\frac{1}{8064} = \sum_{\kappa=5}^{\theta} C_\kappa \sum_{\lambda=4}^{\kappa-1} \beta_{\kappa\lambda}\alpha_\lambda \sum_{\mu=3}^{\lambda-1} \beta_{\lambda\mu} \sum_{\nu=2}^{\mu-1} \beta_{\mu\nu} P_{\nu 1}$	$= \frac{1}{1008}$	h^7
92. h^8	$\frac{1}{40320} = \sum_{\kappa=5}^{\theta} C_\kappa \sum_{\lambda=4}^{\kappa-1} \beta_{\kappa\lambda} \sum_{\mu=3}^{\lambda-1} \gamma_{\lambda\mu} \sum_{\nu=2}^{\mu-1} \beta_{\mu\nu} P_{\nu 1}$	$= \frac{1}{5040}$	h^7
93. h^8	$\frac{1}{6720} = \sum_{\kappa=5}^{\theta} C_\kappa\alpha_\kappa \sum_{\lambda=4}^{\kappa-1} \beta_{\kappa\lambda} \sum_{\mu=3}^{\lambda-1} \beta_{\lambda\mu} \sum_{\nu=2}^{\mu-1} \beta_{\mu\nu} P_{\nu 1}$	$= \frac{1}{840}$	h^7

TABLE continued

94. h^8	$\frac{1}{40320} = \sum_{\kappa=5}^{\theta} C_\kappa \sum_{\lambda=4}^{\kappa-1} \gamma_{\kappa\lambda} \sum_{\mu=3}^{\lambda-1} \beta_{\lambda\mu} \sum_{\nu=2}^{\mu-1} \beta_{\mu\nu} P_{\nu 1} = \frac{1}{5040}$	h^7
95. h^8	$\frac{1}{40320} = \sum_{\kappa=6}^{\theta} C_\kappa \sum_{\lambda=5}^{\kappa-1} \beta_{\kappa\lambda} \sum_{\mu=4}^{\lambda-1} \beta_{\lambda\mu} \sum_{\nu=3}^{\mu-1} \beta_{\mu\nu} \sum_{\eta=2}^{\nu-1} \beta_{\nu\eta} P_{\eta 1} = \frac{1}{5040}$	h^7

References

1. Engeli, Max E., "Users Manual for the Formula Manipulation Language SYMBAL," The University of Texas at Austin Computation Center No. TRM8.01, July 1969.

2. Fehlberg, Erwin, "Classical Eighth- and Lower- Order Runge-Kutta-Nyström Formula with Stepsize Control for Special Second-Order Differential Equations," NASA Technical Report R-381, March 1972.

3. Nyström, I.J., "Über die Numerische Integration von Differtialgleichungen," Acta Soc. Sci. Fenn. 50 No. 13, 1925.

ON THE NON-EQUIVALENCE OF MAXIMUM POLYNOMIAL DEGREE NORDSIECK-GEAR AND CLASSICAL METHODS

by

Roy Danchick

Aerojet ElectroSystems Corp.

Azusa, California

Introduction

The intent of this paper is to show that maximum polynomial degree (p.d.) Nordsieck-Gear methods (References 2 and 3) are not equivalent, in the sense of Descloux and Gear (References 1 and 2), to corresponding Adams-Moulton methods. The notion of derivative interpolating polynomials is introduced and used to derive both the correct initial step polynomial degree maximizing coefficients and explicit, initial step truncation-error formulas. The associated error constants are shown to be vanishingly smaller than their classical analogues in the sense that the ratio of the maximal polynomial degree Nordsieck-Gear initial step error constant to the equal polynomial degree Adams-Moulton error constant tends to zero as the polynomial degree tends to infinity. Most importantly, it is shown that even with correct initial step polynomial degree maximizing coefficient, non-trivial stabilizing correction of retained information induces a reduction of one in method polynomial degree after the initial step, if the degree maximizing coefficient is held constant. Then, a generalized "error mop-up" theorem follows. The theorem's conclusion provides the foundation for a new method based on the recursive computation of the p.d. maximizing coefficients. The use of these varying coefficients preserves maximal degree in the solution throughout the numerical integration from starting through all intermediate stepsize changes. This preservation of maximal p.d. by coefficient variation distinguishes the new method from the standard Nordsieck-Gear methods. Lastly, though exact scaled derivative starting values are postulated in the proofs, all conclusions remain valid if, instead, starting errors of order $O(h^{k+2})$ in each scaled derivative are assumed. A starting procedure which guarantees this order of accuracy has been developed and will be described in a forthcoming paper.

DISCUSSION

Preliminaries

The k-predictor/(k+1)-corrector version of the Nordsieck (k+1)-value method for the numerical integration of the pth order differential equation,

$$y^{(p)}(x) = f\,(x,y,y^{(1)}, \ldots, y^{(p-1)})$$

can be written in Gear's compact notational form as the vector difference equation:

$$(1) \qquad Y_{n+1} = AY_n + \ell\left\{(h^p/p!)f((n+1)h,\ AY_n) - e_p^{\,T}AY_n\right\}$$

where
h = step size

$Y_n = \left\{y(nh),\ hy^{(1)}(nh), \ldots,\ h^k y^{(k)}(nh)/k!\right\}^T,$

A = the (k+1) x (k+1) Pascal Triangle Matrix,

e_p = the vector whose p+1 component is one with all other components zero.

The vector $\ell = (\ell_0, \ell_1, \ldots, \ell_k)^T$ is chosen so that

(a) The matrix $(I - \ell e_p^{\,T})A$ has characteristic equation $p(x) = (x-1)^p x^{k+1-p}$

and

(b) The relation defined by Equation (1) is exact in the first p+1 components for all polynomials of degree $\leq$ k+1, given the exact values for the components of Y_0 and the exact value of $y^{(p)}(h)$.

In this formulation, abbreviated here as (k, p), Gear has shown that p(x) is independent of $\ell_0, \ell_1, \ldots \ell_{p-1}$; thus, these parameters can be chosen to maximize polynomial degree in the sense of (b) above.

The method is optimally stable in the sense of (a) above, and the initial step truncation error is $O(h^{k+2})$ in each of the first p+1 components of $Y_0(h)$ if $y \in C^{k+2}[0,h]$ and exact starting values are given. Gear's insight into the stabilizing effect of properly correcting all retained information, and his constructive proof of the existence of a stabilizing ℓ for each k, has established a basis for hurdling the Dahlquist stability problem. Unfortunately, however, a logical oversight by the late Doctor Nordsieck in his trailblazing paper, has led to some logically incorrect conclusions about the Nordsieck-Gear Methods. The main purpose here is to lay to rest these mistaken, yet widely held, notions and to reveal a unique family of truly one-step, maximum-order, optimally stable methods.

A Counterexample

Gear's choice of ℓ_0 (Y in Nordsieck's notation) does not actually maximize the initial step polynomial degree of the method if the k+1 dimensional vector of retained information is $(y(x), hy^{(1)}(x), h^2y^{(2)}(x)/2!\ldots, h^ky^{(k)}(x)/k!)$. This can be shown by the simple counterexample, $y(x) = x^3$, in the evaluation of Nordsieck's corrector form for $k = 2$:

(2) $$y(x+h) = y(x) + h\left\{f(x) + a(x) + \frac{5}{12}\left[f(x+h) - f_p(x+h)\right]\right\}$$

where
$$f(x) = y^{(1)}(x)$$
$$a(x) = hy^{(2)}(x)/2$$
$$f_p(x) = f(x) + 2a(x).$$

For $y(x) = x^3$, when $y(x)$, $y^{(1)}(x)$, $y^{(2)}(x)$, and $y^{(1)}(x+h)$ are given exactly, the right-hand side (R.H.S.) of Equation 2 above can be written:

(3) $$x^3 + h\left\{3x^2 + 3hx + \frac{5}{12}\left[3(x^2 + 2hx + h^2) - (3x^2 + 6hx)\right]\right\}$$
$$= x^3 + h\left\{3x^2 + 3hx + \frac{5}{4}h^2\right\}$$
$$= (x + h)^3 + h^3/4 \neq (x + h)^3$$

The logical inconsistency in Equation (3) occurs because the coefficient Y = 5/12 is inappropriate for the initial step maximization of polynomial degree. The transformation of predictor-basis matrices, such as the Pascal triangle, though preserving the roots of characteristic stability polynomial at h = 0, does not, in general, preserve maximal initial step polynomial degree. The correct initial step polynomial degree maximizing coefficients for Gear's p = 1, and p = 2 (k+1)-value (k scaled derivatives) methods are, for example:

(4)

p = 1	p = 2
$\ell_0 = 1/(k+1)$	$\ell_0 = 2/[k(k+1)]$
	$\ell_1 = 2/k$

To vividly contrast the effects of using the correct and the incorrect initial step coefficient, consider the initial value problem

(5)
$$y^{(1)}(x) = 3y/x$$
$$y(x_0) = x_0^3$$
$$\left(|h|<|x_0|,\ x_0 \neq 0\right)$$

whose solution is simply $y(x) = x^3$. If, starting with the exact value of $6x_0$ for $y^{(2)}(x_0)$, the Nordsieck-Gear 3-value method is iterated, then

(6)
$$y_0(x_0+h) = x_0^3+h\left\{3x_0^2+3hx_0\right\} = (x_0+h)^3-h^3$$
$$y_1(x_0+h) = (x_0+h)^3 + h^3/4$$

(7)
$$y_2(x_0+h) = x_0^3 +h\left\{3x_0^2+3hx_0+\frac{5}{12}\left[\frac{3\left[(x_0+h)^3+h^3/4\right]}{(x_0+h)} - 3x_0^2-6hx_0\right]\right\}$$
$$= x_0^3+h\left\{3x_0^2+3hx_0+\frac{5}{4}\left[h^2+h^3/4(x_0+h))\right]\right\}$$
$$= (x_0+h)^3 + h^3/4 + 5h^4/\left[16(x_0+h)\right]$$

It readily follows, by induction on n, that

(8)
$$y_n(x_0+h) = y(x_0+h) + (h^3/4)\sum_{i=0}^{n} (\bar{h})^i$$

where
$$\bar{h} = 5h/\left[4(x_0+h)\right]$$

By contrast, using the correct coefficient (1/3),

$$y_0^*(x_0+h) = (x_0+h)^3 - h^3$$

(9)
$$y_1^*(x_0+h) = x_0^3+h\left[3x_0^2+3hx_0+\frac{1}{3}\left\{\frac{3\left[(x_0+h)^3-h^3\right]}{(x_0+h)} - 3x_0^2-6hx_0\right\}\right]$$
$$= (x_0+h)^3 - h^4/(x_0+h)$$
$$y_2^*(x_0+h) = x_0^3+h\left[3x_0^2+3hx_0+\frac{1}{3}\left\{\frac{3\left[(x_0+h)^3-h^4/(x_0+h)\right]}{(x_0+h)} - 3x_0^2-6hx_0\right\}\right]$$
$$= (x_0+h)^3 - h^5/(x_0+h)^2$$

And, thus, again by induction

(10) $$y_n^*(x_0+h) = (x_0+h)^3 - h^3\left\{h/(x_0+h)\right\}^n$$

Thus, for all $\left|h\right|$ sufficiently small

$$y_n^*(x_0+h) \longrightarrow y(x_0+h),$$

but for no $\left|h\right|$ does

$$y_n(x_0+h) \longrightarrow y(x_0+h).$$

The Derivative-Interpolating Polynomial

The correct initial-step coefficients can be obtained by inspection from the fundamental, unique, derivative-interpolating polynomials:

(11) $$p_h(t) = T_k(t)+h\left[y^{(1)}(x+h)-T_k^{(1)}(x+h)\right](t-x)^{k+1}\Big/\left[(k+1)h^{k+1}\right]$$

$$(p = 1)$$

(12) $$q_h(t) = T_k(t)+(h^2/2)\left[y^{(2)}(x+h)-T_k^{(2)}(x+h)\right]2(t-x)^{k+1}\Big/\left[(k+1)kh^{k+1}\right]$$

$$(p = 2)$$

where

$T_k(t) = y(x)+y^{(1)}(x)(t-x)+ \dots . +y^{(k)}(x)\ (t-x)^k/k!$ and

$y^{(i)}(\ )$ is the ith derivative of

$y \in C^{k+2}\left[x,x+h\right]$ $(i = 0, 1, 2\dots, k)$.

By virtue of their construction, p_h and q_h have the following derivative-interpolating properties:

$$p_h^{(i)}(x) = y^{(i)}(x) \tag{13}$$

$$p_h^{(1)}(x+h) = y^{(1)}(x+h)$$

$$q_h^{(i)}(x) = y^{(i)}(x)$$

$$q_h^{(2)}(x+h) = y^{(2)}(x+h)$$

$$(i = 0, 1, 2, \ldots, k)$$

More generally, the corresponding fundamental interpolating polynomial, $r_h(t)$, for a (k+1)-value direct m^{th} order method, is given by

$$r_h(t) = T_k(t)+(h^m/m!)\left[y^{(m)}(x+h)-T_k^{(m)}(x+h)\right](t-x)^{k+1}\Big/\left[\binom{k+1}{m} h^{k+1}\right] \tag{14}$$

The derivative-interpolating polynomials $p_h(t)$, $q_h(t)$, and $r_h(t)$ are fundamental in the sense that they not only generate the corresponding polynomial degree maximizing initial-step coefficients but their introduction readily leads to an explicit characterization of the initial-step truncation error in terms of $y^{k+2}(\xi)$ $(x < \xi < x+h)$. As will be shown in the development to follow, the corresponding error constants are $C_1^{k+2} = -1/[(k+1)(k+2)!]$ (p=1) and $C_2^{k+2} = -2/[k(k+2)!]$ (p=2). It should be noted that these error constants are much smaller than those of the corresponding Adams-Moulton and Cowell methods, respectively. For example, for the 5th order Adams-Moulton method the ratio is $\frac{2}{81}$, while for the 5th order Cowell, the ratio is $\frac{1}{6}$.

In fact, it is easy to show that

$$\lim_{k\to\infty} C_1^{k+2}\big/C_{am}^{k+2} = \lim_{k\to\infty} C_2^{k+2}\big/C_c^{k+2} = 0$$

where C_{am}^{k+2} and C_c^{k+2} are the equivalent polynomial degree Adams-Moulton and Cowell error constants, respectively.

Derivation of the Initial Single-Step Truncation Error

Suppose $y \in C^{k+2}[x,x+h]$. Let $p_n(t)$ be the derivative-interpolating polynomial introduced in the previous Subsection, "agreeing" with y and its higher order derivatives at t = x and agreeing with $y^{(1)}$ at t = x+h. Let $H_1(t) = [t - x - (k+2)h/(k+1)]\,(t-x)^{k+1}$. $H_1(t)$ is monic, vanishes together with its first k derivatives at t = x, vanishes nowhere else in [x, x+h]. In addition, $H_1^{(1)}(t)$ vanishes at t = x+h.

Let $\bar{t} \in [x,x+h]$. Choose $C(\bar{t}) \ni$: $F(t)=y(t)-p_h(t)-C(\bar{t})H_1(t)$ vanishes at $t = \bar{t}$. Such a choice is always possible as $H_1(t) \neq 0$ in $(x,x+h]$ and $F(x) = 0$ for any choice of $C(x)$. Thus, $F(x) = F(\bar{t}) = 0$.
$\therefore \exists \zeta_1 \in (x,\bar{t}) \ni$: $F^{(1)}(\zeta_1) = 0$. Since $F^{(1)}(x+h) = 0, \exists \zeta_2 \in (\zeta_1,x+h) \ni$: $F^{(2)}(\zeta_2) = 0$. Since also $F^{(1)}(x) = 0, \exists \zeta_2^1 \in (x,\zeta_1) \ni$: $F^{(2)}(\zeta_2^1) = 0$.
Proceeding recursively, we show that $F^{(i)}(t)$ vanishes at least twice in $(x,x+h)$ for $i = 2, 3,\ldots, k-1$. Thus, $F^{(k)}(t)$ vanishes at least three times in $[x,x+h]$. Therefore, $F^{(k+1)}(t)$ vanishes at least twice and hence, finally, $\exists \zeta \in (x,x+h) \ni$:

$$F^{(k+2)}(\zeta) = y^{(k+2)}(\zeta) - (k+2)!\, C(\bar{t}) = 0$$

$$\therefore C(\bar{t}) = y^{(k+2)}(\zeta)/(k+2)!$$

In particular, for $\bar{t} = x+h, \exists \zeta \in (x,x+h) \ni$:

$$y(x+h) - p_h(x+h) = y^{(k+2)}(\zeta)H_1(x+h)/(k+2)! \qquad (15)$$

$$= [-1/(k+1)]\, y^{(k+2)}(\zeta)\, h^{k+2}/(k+2)!$$

The above proof goes through directly for the $p = 2$ direct second-order method by showing that $G^{(3)}(t)$ vanishes at least three times in $[x,x+h]$, where $G(t) = y(t) - q_h(t) - C(\bar{t})H_2(t)$, $y(t) \in C^{k+2}[x,x+h]$, etc. In this case, $H_2(t) = \dfrac{(t-x-(k+2)h)(t-x)^{k+1}}{k}$ and the $p = 2$ initial single-step truncation error formula is obtained according to

$$y(x+h)-q_h(x+h) = [-2/k]y^{(k+2)}(\eta)\, h^{k+2}/(k+2)! \qquad (16)$$

$$(\eta \in (x,x+h))$$

<u>Reduction and Preservation of Order (Polynomial Degree)</u>

The $(k,1)$ case where $k > 1$ will be covered. The generalization to the case $k > p > 1$ is immediate. Let Y_n, $n=0, 1, 2,\ldots$ be the sequence of approximating vectors for the solution of the differential equation $y^{(1)}(x) = f(x,y)$ whose solution $\bar{y}(x) \in C^{k+2}[0, X]$, and such that $f_y(x,y)$ is continuous in a neighborhood of the graph $(x,\bar{y}(x))$, $x \in [0,X]$. It is further assumed that $Y_0 = \bar{Y}_0$, i.e., the starting values are exact. Thus,

$$(17)\qquad AY_0 = \bar{Y}_1 - \left[h^{k+1}/(k+1)!\right]\left[\bar{y}^{(k+1)}\left(\zeta_1^0\right), \binom{k+1}{1}\bar{y}^{(k+1)}\left(\zeta_1^1\right), \ldots \binom{k+1}{k}\bar{y}^{(k+1)}\left(\zeta_1^k\right)\right]^T$$

where

$$\zeta_1^j \in (0,h),\ j = 0,1,2,\ldots,\ k$$

$$(18)\qquad y(h) = \bar{y}(h) - h^{k+1}\,\bar{y}^{(k+1)}\left(\zeta_1^0\right)/(k+1)!$$

$$+ \frac{1}{k+1}\left\{ hf\left(h, \bar{y}(h) - h^{k+1}\,\bar{y}^{(k+1)}\left(\zeta_1^0\right)\ (k+1)!\right) - \left[hf(h,\bar{y}(h)) - (k+1)h^{k+1}\,\bar{y}^{(k+1)}\left(\zeta_1^1\right)/(k+1)!\right]\right\}$$

$$= \bar{y}(h) + \left[h^{k+1}/(k+1)!\right]\left[\bar{y}^{(k+1)}\left(\zeta_1^1\right) - \bar{y}^{k+1}\left(\zeta_1^0\right)\right]$$

$$- \frac{1}{k+1}h^{k+2}\,f_y\,(h,\theta)\bar{y}^{(k+1)}\left(\zeta_1^0\right)/(k+1)!$$

$$= \bar{y}(h) + \rho h^{k+2}\bar{y}^{(k+2)}(\psi)/(k+1)! - \frac{1}{k+1}h^{k+2}f_y(h,\theta_1) \cdot \bar{y}^{(k+1)}\left(\zeta_1^0\right)/(k+1)!$$

where it follows from the derivation of Equation (15) that

$$(19)\qquad \rho\bar{y}^{k+2}(\psi_1) = \bar{y}^{(k+2)}(\psi_1')/\left[(k+1)(k+2)\right]$$

$$\psi_1,\ \psi_1' \in (0,h) \text{ and } \theta_1 = \bar{y}(h) - \lambda_1 h^{k+1}\,\bar{y}^{(k+1)}\left(\zeta_1^0\right)/(k+1)!,$$

with $0 < \lambda_1 < 1$

$$(20)\qquad \therefore\ e^{(0)}(h) = \bar{y}(h) - y(h) = O(h^{k+2})$$

It should be noted that the third term in the R.H.S. of Equation (18) above can be made as small in absolute value as desired for sufficiently small h by corrector iteration. Similarly, it follows that

$$(21)\qquad e^{(1)}(h) = h\bar{y}^{(1)}(h) - hy^{(1)}(h) = O(h^{k+2}),$$

However, since $\ell_j < \binom{k+1}{j}\Big/(k+1)$ for $j \geq 2$, it follows, by the same reasoning, that

(22)
$$e^{(j)}(h) = \frac{-h^{k+1}}{(k+1)!}\binom{k+1}{j}\bar{y}^{(k+1)}\left(\zeta_1^0\right)$$
$$+\ \ell_j \frac{h^{k+1}}{(k+1)!}\left(\binom{k+1}{1}\bar{y}^{(k+1)}\left(\zeta_1^1\right)\right)$$
$$+\ O(h^{k+2})$$
$$= O(h^{k+1})$$
$$(j = 2, 3, \ldots, k)$$

Proceeding from h to 2h, it follows that the second-step equation takes the form

(23)
$$y(2h) = \bar{y}(2h)$$
$$+\left(-h^{k+1}\Big/(k+1)!\right)\sum_{j=2}^{k}\left[1 - j/(k+1)\right]\binom{k+1}{j}\bar{y}^{(k+1)}\left(\zeta_1^j\right)$$
$$+\left(h^{k+1}\Big/(k+1)!\right)\bar{y}^{(k+1)}\left(\zeta_1^1\right)\ell_j\left[k+1\right]\sum_{j=2}^{k}\left[1 - j/(k+1)\right]$$
$$+\ O(h^{k+2}) = \bar{y}(2h) + O(h^{k+1})$$

Hence, as was to be shown,

(24)
$$e^{(0)}(2h) = O(h^{k+1})$$

and it is easily seen that the order of $e^{(0)}(2h)$ cannot be raised to $O(h^{k+2})$ by corrector iteration. This result immediately generalizes to any multivalue method characterized by nonzero stabilizing coefficients. That is, every such initial-step maximal-polynomial-degree method must suffer a decrease of one in the order of the error in proceeding from h to 2h. This phenomenon will, of course, not be observable in the behavior of non-maximal-degree initial-step methods, where the initial single-step error is already $O(h^{k+1})$.

Next is exhibited a method for preserving order in the fixed step-size context by appropriately changing the degree maximizing coefficient from step-to-step for not more than k+1 steps. This order-preservation method will be a consequence of the following "error mop-up lemma" and a fixed-point theorem.

Lemma(*)

For fixed h and N let $\alpha_0, \alpha_1, \ldots, \alpha_N$ be N real numbers, not all zero, such that $\sum_{i=0}^{N} \alpha_i = 0$. Let $y(x) \in C^{k+2}[0, Nh]$ with $x_i \in [0, Nh]$, $i = 0, 1, 2 \ldots, N$.

Then

$$\sum_{i=0}^{N} \alpha_i y^{(k+1)}(x_i) = O(h) \tag{25}$$

Proof

For each $i = 0, 1, 2, \ldots N$

$$y^{(k+1)}(x_i) = y^{(k+1)}(0) + x_i y^{(k+2)}(\xi_i) \tag{26}$$

$$\therefore \sum_{i=0}^{N} \alpha_i y^{(k+1)}(x_i) = y^{(k+1)}(0) \sum_{i=0}^{N} \alpha_i + \sum_{i=0}^{N} \alpha_i x_i y^{(k+2)}(\xi_i) \tag{27}$$

$$= \sum_{i=0}^{N} \alpha_i x_i y^{(k+2)}(\xi_i) = O(h)$$

(*) This simple proof kindly provided by Dr. C.W. Gear

The following observation is now made by way of Equations (17) through (23). Suppose S_n is defined as the vector of $O(h^{k+1})$ constant error coefficient sums at step n. Thus, for example, from Equation (22)

$$(28) \qquad S_1 = \left[h^{k+1}/(k+1)!\right] \begin{bmatrix} 0 \\ 0 \\ (k+1)\,\ell_2 - \binom{k+1}{2} \\ \vdots \\ (k+1)\,\ell_k - \binom{k+1}{k} \end{bmatrix},$$

and from Equation (23)

$$(29) \qquad S_2 = \left[h^{k+1}/(k+1)!\right] \begin{bmatrix} \sum_{j=2}^{k} 1 - \left[j/(k+1)\right]\left[(k+1)\,\ell_j - \binom{k+1}{j}\right] \\ \vdots \end{bmatrix}$$

Then, it follows by induction on n that, up to the proportionality constant, $h^{k+1}/(k+1)!$,

$$(30) \qquad S_{n+1} = \left[I - \ell e_1^T\right]\left[AS_n - \left\{\binom{k+1}{i}\right\}\right],$$

where

$$\left\{\binom{k+1}{i}\right\} = \begin{bmatrix} 1 \\ \binom{k+1}{1} \\ \binom{k+1}{2} \\ \vdots \\ \binom{k+1}{k} \end{bmatrix}$$

Thus, if at step n+l it is possible to choose an $\ell_0^{(n+1)}$ so that

$$e_0^T AS_n = e_0^T AS_n - 1 + \left(k+1-e_1^T AS_n\right)\ell_0^{(n+1)} = 0,$$

the $e^{(0)}((n+1)h)$ $O(h^{k+1})$ constant error coefficients sum to zero, the lemma applies and $e^{(0)}((n+1)h) = O(h^{k+2})$. The following lemma shows that if such a choice of the $\ell_0^{(n+1)}$ is possible, then the sequence terminates after not more than k+l steps; i.e., $\ell_0^{(k)} = \ell_0^{(k+1)} = \dots$.

Lemma

If $k+1 - e_1^T AS_n \neq 0$ for n = 0, 1, 2,.... k, then S_k is the unique fixed point of the mapping:

$$\varphi(S) = \left(\Pi - \ell^* e_1^T\right)\left(AS - \left\{\binom{k+1}{i}\right\}\right), \tag{31}$$

where

$$\Pi = \begin{bmatrix} 0..0...0 \\ . & \\ . & \\ 0 & I \\ . & \\ . & \\ 0 & \end{bmatrix},$$

and

$$\ell^* = \begin{bmatrix} 0 \\ 1 \\ \ell_2 \\ . \\ . \\ \ell_k \end{bmatrix}.$$

This system of linear equations can be expanded, component by component, to yield the system of k-l equations

$$\sum_{i=1}^{k-j} \binom{j+i}{j} s_{j+i} - \ell_j \sum_{r=2}^{k} r s_r = \binom{k+1}{j} - (k+1)\ell_j \tag{32}$$

$$(j = 2, 3, \dots, k-1)$$

and

$$- \ell_k \sum_{r=2}^{k} r s_r = (k+1)(1-\ell_k)$$

Which, since $\ell_k \neq 0$, has the unique solution

$$s_0 = 0 \tag{33}$$

$$s_1 = 0$$

$$\begin{bmatrix} s_2 \\ s_3 \\ \cdot \\ \cdot \\ \cdot \\ s_k \end{bmatrix} = \binom{j+i}{j}^{-1} \left[\binom{k+1}{j} - (k+1)\ell_j/\ell_k \right]$$

Proof

Suppose $k+1 - e_1{}^T A S_n \neq 0$ for $n = 0, 1, 2, \ldots, k$. Then, for each n, S_n has zero first and second components and

$$\begin{aligned} S_{k+1} - S_k &= \Pi(S_{k+1} - S_k) \\ &= \Pi(I - \ell^{k+1} e_1{}^T)A\Pi(S_k - S_{k-1}) - \Pi(\ell^{k+1} - \ell^k)e_1{}^T\binom{k+1}{i} \\ &= (\Pi - \ell^* e_1{}^T)A\Pi(S_k - S_{k-1}) = (\Pi - \ell^* e_1{}^T)A(S_k - S_{k-1}) \end{aligned} \tag{34}$$

Hence, $S_{k+1} - S_k = \left[\Pi(I - \ell^* e_1{}^T)A\right]^{k+1} S_1 = 0$ since the characteristic polynomial of $\pi(I - \ell^* e_1{}^T)A$ is x^{k+1}.

Hence,

$$S_k = S_{k+1} = \ldots. = S. \tag{35}$$

Thus, if $k+1 - e_1{}^T A S_n \neq 0$ for all $n \leq k$

$$\ell_0^{(n+1)} = (1 - e_0{}^T A S_n)/(k+1 - e_1{}^T A S_n) \tag{36}$$

provides the required ℓ_0 sequence.

That the above lemma is not vacuous in content can be shown for the cases k = 2, 3, 4, 5, by direct computation, yielding the following (k, l) table:

$k/\ell_0^{(n)}$	$\ell_0^{(1)}$	$\ell_0^{(2)}$	$\ell_0^{(3)}$	$\ell_0^{(4)}$	$\ell_0^{(5)}$
2	1/3	5/12			
3	1/4	11/30	3/8		
4	1/5	469/1440	959/2760	251/720	
5	1/6	5659/19440	10871/33480	28249/85680	95/288

It should be noted that the terminating coefficients 5/12, 3/8, etc. are identical to Gear's. Thus, naturally, the following conjecture: For each k and for each p a terminating order-optimizing coefficient sequence can be constructed. The terminal coefficient is equal to Gear's which is, in turn, identical to the classical Adams-Moulton for p=1.

The above conclusions and the conjecture apply in the fixed-step context. In the general case, if the ratio of new step size to old is ρ_n in going from step n to n+1, then it is easy to show that the order-preserving coefficient formula generalizes to:

$$\ell_0^{(n+1)} = \left[1 - e_0^T A D_n S_n / \rho_n^{k+1}\right] \Big/ \left[k+1 - e_1^T A D_n S_n / \rho_n^{k+1}\right] \tag{37}$$

where

$$D_n = \begin{bmatrix} 1 & & & & 0 \\ & \rho_n & & & \\ 0 & & \rho_n^2 & & \\ & & & \ddots & \\ & & & & \rho_n^k \end{bmatrix},$$

provided the denominator does not vanish. Since the denominator vanishes for at most k+1 distinct ρ_n it is thus always possible to find a suitable p_n for step-size expansion or contraction which preserve both stability and maximum order.

Conclusions

It follows, from the results, that the notion of equivalence among numerical methods in the sense of Gear and Descloux is not applicable when starting values are sufficiently accurate or in the presence of stepsize change. However, the important, fundamental core of Gear's work, the matrix theoretical construction of maximally stable k+1 value methods, remains intact. This foundation has been built upon by removing some logical inconsistencies. The removal of these inconsistencies has revealed the intrinsic properties of a family of truly optimally stable, one-step, maximum p.d. methods believed to have great unexplored potential for accuracy, stability, speed, error control, ease of interpolation, and implementability. Results on the development of starting, error estimation, and control procedures and experimental results with variable-step versions will be described in a later paper.

Acknowledgements

I would like to express my appreciation to Dr. J. Dyer, and my colleagues, Drs. I. Tarnove, D. Pope and D. Carta, for their invaluable advice and encouragement in the preparation of this paper. Also, a special thanks to Ralph Kahn for his programming support.

REFERENCES

1. Descloux, J., "Note on a Paper by Nordsieck, "Department of Computer Science Report No. 131, University of Illinois, Urbana, Illinois.

2. Gear, C. W., "The Numerical Integration of Ordinary Differential Equations", Math. Comp., 21, pp 146-156.

3. Nordsieck, A. (1962), "On the Numerical Integration of Ordinary Differential Equations," Math. Comp., 16, pp 22-49.

PHASE SPACE ANALYSIS IN NUMERICAL INTEGRATION OF ORDINARY DIFFERENTIAL EQUATIONS

by

Bernard E. Howard

University of Miami

Coral Gables, Florida

Introduction

The trouble with solving differential equations on a digital computer is the distressing frequency of strange results. The purpose of this paper is to describe how I attack a problem to find out what is going on. The procedure is summarized in this introduction, and each step is detailed in subsequent sections. The key is phase space topology, where phase space is defined to be the space of dependent variables.

To solve $x' = f(x)$, where $x, x' \in E_n$, $f: E_n \to E_n$, and $' = d/dt$, proceed as follows:

1) Determine the Jacobian J of f.
2) Find the critical points c, where $f(c) = 0$.
3) Solve the algebraic eigenproblem for J(c) for each c.
4) Integrate along each real eigenvector corresponding to a positive real eigenvalue starting $\pm\varepsilon$ distance from its critical point. Repeat for negative real eigenvalues using reversed time; i.e., by solving $x' = -f(x)$
5) Determine plane of real oscillation corresponding to each pair of conjugate complex eigenvalues, and analyze oscillation in that plane.
6) Combine above results to determine tentative topology of phase space of system.
7) Conduct a posteriori error analysis to determine optimum integration step-size.

8) Determine in like manner topology of phase space of approximating finite difference equation, which may well be different from topology of phase space of differential system whose solution is sought.

These steps will be illustrated by a simple numerical example due to Lyn Lane. The illustrative example was chosen for pedagogical reasons, but the method is of great power and generality. E.g., it has been applied to a 24 dimensional problem (in phase space), which is the magnitude of a 4 body problem in astronomy.

1. The Jacobian

We consider a system of n first order (in general nonlinear) ordinary differential equations. Many higher order equations, including the equations of the k-body problem in astronomy, can be changed to such a system by simple substitution of new letters for derivatives, and the form $x' = f(x)$ yields considerable insight into the qualitative behavior of the system through phase space analysis.

In component form the vector equation $x' = f(x)$ is

$$x_i' = f_i(x_1, x_2, \ldots x_n) \quad , \; i = 1,\ldots,n$$

The Jacobian of the system is the $n \times n$ matrix

$$J(f) = \frac{\partial(f_1,\ldots,f_n)}{\partial(x_1,\ldots,x_n)} = \frac{\partial f_i}{\partial x_j} \quad ; \; i = 1,\ldots,n \;\; ; \;\; j = 1,\ldots,n$$

The Jacobian is useful in obtaining the critical points and in determining local behavior in the neighborhood of these points.

Lyn Lane Example:

$$x_1' = x_1^2 - 2x_2^2 + 9 \qquad\qquad J = \begin{pmatrix} 2x_1 & -4x_2 \\ 2x_1 & -2x_2 \end{pmatrix}$$

$$x_2' = x_1^2 - x_2^2$$

2. Critical Points

A critical point is by definition a point c where $f(c) = 0$. In simple cases it may be possible to obtain all such points algebraically, as in the illustrative example where the $f_i = 0$ are the four intersections of an ellipse and a degenerate hyperbola, viz $(\pm 3, \pm 3)$.

In more complicated cases where it is not feasible to obtain the critical points by solving $f(c) = 0$ in closed form, a generalized Newton-Raphson iteration is useful. Start with an initial guess $x = \alpha_1$ and solve the following system of simultaneous linear equations (say by Gauss elimination) for $\delta\alpha$:

$$J\delta\alpha = -f$$

where J and f are each evaluated at α_1. Form a second approximation $\alpha_2 = \alpha_1 + \delta\alpha$. Repeat the process; under appropriate conditions (e.g., if the initial guess α_1 is close enough to a simple root) the sequence of α_i will converge to the desired critical point c. The derivation of the formula is based on a simple application of Taylor series which will not be reproduced here. (See Conte and deBoor, 1972)

Example

Suppose $\alpha_1 = (2,4)$ in the Lyn Lane example. Then $J\delta\alpha = -f$ becomes:

$$\begin{pmatrix} 4 & -16 \\ 4 & -8 \end{pmatrix} \begin{pmatrix} \delta\alpha_1 \\ \delta\alpha_2 \end{pmatrix} = \begin{pmatrix} 19 \\ 12 \end{pmatrix}$$

whose solution is $\delta\alpha = (5/4, -7/8)$, whence $\alpha_2 = (13/4, 25/8)$ which is closer to $c = (3,3)$ than α_1.

3. Algebraic Eigenproblem

Local behavior in the neighborhood of a critical point is determined by solving the algebraic eigenproblem for the Jacobian of f evaluated at the critical point. The algebraic eigenproblem may be stated as follows: Given a matrix J, find a scalar λ and a vector v such that

$$Jv = \lambda v$$

There are many ways of attacking this problem (see Wilkinson 1965, e.g.). Two of the more satisfactory are the QR method, and Householder reduction to

Hessenberg form followed by bisection or determinental evaluation. Let us just assume that there is a standard package in the computer center program library which produces satisfactory results, and proceed from there.

Example

Quadrant 1: $c = (3,3)$

$$J = \begin{pmatrix} 6 & -12 \\ 6 & -6 \end{pmatrix}$$

The eigenproblem $Jv = \lambda v$ may be written $(J - \lambda I)v = 0$ which is a system of linear homogenous equations in the v_i. A necessary condition for a solution other than $v = 0$ is that $\det (J - \lambda I) = \lambda^2 + 36 = 0$ or $\lambda = \pm 6i$, whence

$$(J - \lambda I)v = \begin{pmatrix} 6 \mp 6i & -12 \\ 6 & -6 \mp 6i \end{pmatrix} \begin{pmatrix} v_1 \\ v_2 \end{pmatrix} = \begin{pmatrix} 0 \\ 0 \end{pmatrix}$$

giving $v = (1 \pm i, 1) = (1,1) \pm i(1,0)$

Q2: $c = (-3,3)$

$$J = \begin{pmatrix} -6 & -12 \\ -6 & -6 \end{pmatrix}$$

Characteristic equation: $\det (J - \lambda I) = \lambda^2 + 12\lambda - 36 = 0$

Eigenvalues: $\lambda = -6 \pm 6\sqrt{2}$

For $\lambda = -6 + 6\sqrt{2}$:

$$(J - \lambda I)v = \begin{pmatrix} -6\sqrt{2} & -12 \\ -6 & -6\sqrt{2} \end{pmatrix} \begin{pmatrix} v_1 \\ v_2 \end{pmatrix} = \begin{pmatrix} 0 \\ 0 \end{pmatrix}$$

giving $v = (-\sqrt{2}, 1)$

For $\lambda = -6 - 6\sqrt{2}$, $v = (\sqrt{2}, 1)$

Q3: $c = (-3,-3)$

$$J = \begin{pmatrix} -6 & 12 \\ -6 & 6 \end{pmatrix}$$

Characteristic equation: $\lambda^2 + 36 = 0$

Eigenvalues: $\lambda = \pm 6i$

Eigenvectors: $v = (1 \mp i,1) = (1,1) \mp i(1,0)$

Q4: $c = (3,-3)$

$$J = \begin{pmatrix} 6 & 12 \\ 6 & 6 \end{pmatrix}$$

Eigenvalues: $\lambda = 6 \pm 6\sqrt{2}$

Eigenvectors: $v = (\pm\sqrt{2},1)$

4. Real Eigenvalues

The first step in the location of boundaries which separate regions of different behavior is to identify all real eigenvalues. Pick a positive real eigenvalue, and identify its corresponding eigenvector and critical point. Select a starting point a very small distance from the critical point along the eigenvector. Integrate along this eigenvector to see where it goes. The resulting special solution trajectory is called a separatrix. Repeat, starting on the other side of the critical point (since an eigenvector is really an undirected line) and obtain another separatrix. Repeat for all positive real eigenvalues.

Now for the negative real eigenvalues and their corresponding eigenvectors. The negative sign on a real eigenvalue means that the solution trajectory approaches exponentially and terminates on the critical point; we want to see where it comes from. To do that, reverse time, and repeat the procedure of the last paragraph, only this time integrating the equation $x' = -f(x)$, which of course is a trivial modification of a computer program. We now have an entire set of one type of separatrix.

Example

Figure 1 is the phase space of the Lyn Lane example at this stage of the game. Since the vector x has two components, the phase space is a plane. The four critical points (±3,±3), one in each quadrant are shown. Eight separatrices are shown, obtained in the following order. First, the two emanating from the critical point in Q2 corresponding to the positive real eigenvalue $6(\sqrt{2}-1)$; these two separatrices leave the critical point (-3,3) in the directions $(-\sqrt{2},1)$ and $(\sqrt{2},-1)$ respectively (these are the two opposing directions along the same eigenvector). The starting points fed into the computer were (-3.00028284,3.0002) and (-2.99971716, 2.9998). Next, the two outgoing separatrices (as indicated by arrowheads) emanating from the critical point in Q4. Finally, the remaining four separatrices, two from each of the above critical points, obtained by integrating $x' = -f(x)$; i.e., $x_1' = -(x_1^2 - 2x_2^2 + 9)$ and $x_2' = -(x_1^2 - x_2^2)$.

5. Complex Eigenvalues

The eight separatrices determined in the previous section and shown in Figure 1 appear at first glance to nail down the topology of the phase space (plane) of this system. However, we note that the eigenvalues of the critical points in Q1 and Q3 are pure imaginary, and this bears further investigation. With a pair of complex conjugate eigenvalues, the presence of the imaginary part indicates an oscillation, in this case of angular velocity 6, or period $2\pi/6 = 1.0472$. A positive real part would indicate exponential growth, a negative real part exponential decay toward the critical point, the same as for a pure real eigenvalue. But a zero real part indicates a simple closed curve (harmonic motion) as far as the linear terms are concerned, and we must look to the higher order terms to determine local behavior (Kamke, 1956).

In the present example there are two possibilities. Either Q1 is stable and Q3 is unstable as appears to be the case from superficial examination of Figure 1, or the reverse is true, in which case we have a stable limit cycle encircling the critical point in Q1, and an unstable limit cycle encircling the critical point in Q3. (Robert Pearson has produced an example of the latter type.)

To study local behavior in the neighborhood of a critical point with complex eigenvalues, we need to determine the plane of real oscillation. Corresponding to a pair of complex conjugate eigenvalues of a real matrix, there is a pair of complex conjugate eigenvectors. The real part and the imaginary part of the eigenvectors, taken separately as two real vectors, span the plane of real oscillation of the corresponding complex eigenvalues. In the example, the entire phase space consists of only one plane, but in higher order examples this is a highly useful result. The proof involves equations similar to those on p. 630 of Wilkinson.

Example

The nontrivial procedure is illustrated with the Lyn Lane example. Let us start with Q1. We first translate the origin to the critical point by substituting $x_1 = 3 + y_1$, $x_2 = 3 + y_2$ in the original differential equations and obtain

$$y_1' = 6y_1 - 12y_2 + y_1^2 - 2y_2^2$$

$$y_2' = 6y_1 - 6y_2 + y_1^2 - y_2^2$$

Note that the linear terms are $J_c y$ as they should be.

Next we introduce the linear transformation $y = Tz$, where

$$T = \begin{pmatrix} 1 & 1 \\ 0 & 1 \end{pmatrix}, \quad T^{-1} = \begin{pmatrix} 1 & -1 \\ 0 & 1 \end{pmatrix}$$

In component form, $y_1 = z_1 + z_2$, $y_2 = z_2$. The columns of T are the imaginary and real parts of the complex eigenvectors, and are the axes of the new coordinate system. Substituting in the differential equations for y and solving for z_1', z_2' produces

$$z_1' = -6z_2 - z_2^2$$

$$z_2' = 6z_1 + z_1^2 + 2z_1z_2$$

In this form we see that the linear terms are the equations of simple harmonic motion of frequency 6. To determine the effect of the nonlinear terms upon otherwise closed circles in the z plane, we transform to polar coordinates. Let

$$r^2 = z_1^2 + z_2^2 \qquad rr' = z_1 z_1' + z_2 z_2'$$

$$\tan\theta = z_2/z_1 \qquad r^2\theta' = z_1 z_2' - z_2 z_1'$$

Upon making this transformation and applying some trigonometric identities the z equations become:

$$r' = r^2 \sin 2\theta \cos(\theta - \pi/4)/\sqrt{2}$$

$$\theta' = 6 + \sqrt{2}r\,[\sin 2\theta \cos(\theta + \pi/4)/2 + \cos(\theta - \pi/4)]$$

The above is a system of differential equations of precisely the type we have been discussing. Applying the general theory, we first compute the critical points and discover that they are the other three in the original equation, which is comforting. In particular, the Jacobian evaluated at the other focal point has eigenvalues $\pm 6i$, as it should.

Knowing that the present origin is a regular point, we can compute sample trajectories in its neighborhood with confidence. If $r(0) = 0$, it remains there and θ increases at a constant rate of 6, which is a valid but uninformative result. Taking any non-zero value of r, say $r(0) = 0.3$, $\theta(0) = 0$, and integrating around one cycle (to $t = 1.0472$) we obtain the information we seek, namely, that there is a 7% net decrease in r in one cycle, and that the critical point in quadrant 1 is thus a stable focal point, and there are no limit cycles.

Another approach is merely to tabulate and plot the two functions of θ appearing in the expressions for r' and θ'. This has been done and is shown in Figure 2. We see there are two regions, one where $\theta' > 6$ and the other where $\theta' < 6$. In region 1 the angular velocity is faster than the average 6; in this region the mean value of r' is positive, and contrariwise in region 2. Since over one cycle the trajectory spends a longer time in the region of negative r' than in the region of positive r', the net change in one cycle is a decrease in r.

No matter how great the dimension of the phase space of a system, we analyze each oscillation in its own plane in turn, so the procedure illustrated here is quite general.

6. Phase Space Topology

The qualitative behavior of a physical system corresponding to $x' = f(x)$ is compactly describable by the topology of the phase space of its differential equation. This topology is determined by the location and nature of (local behavior near) the critical points, and the way they are connected together. Through every regular point passes one and only one solution trajectory; such a trajectory over the domain $(-\infty < t < +\infty)$ is called an orbit. Then orbits which can be transformed continuously and smoothly into one another represent qualitatively similar behavior patterns of the physical system. The connection of the phase space is defined to be the number of topologically distinct orbits. The connection of the phase space thus identifies the number of qualitatively distinct behavior patterns the corresponding physical system possesses.

Example. The connection of the phase plane of the Lyn Lane example is 5. The five distinct families of orbits are:

1) Starting at ∞ in Q4, ending at ∞ in Q2, circling below all separatrices in Q3.
2) Starting at ∞ in Q4, ending at the stable focal point in Q1.
3) Starting at the unstable focal point in Q3, ending at the stable focal point in Q1.
4) Starting at the unstable focal point in Q3, ending at ∞ in Q2.
5) Starting at ∞ in Q4, ending at ∞ in Q2, circling above all separatrices in Q1.

The slope of the orbits going to ∞ is about $-.83$; i.e., the orbits are asymptotic to the negative root of $2m^3-m^2-m+1$, obtained by setting $x_2 = mx_1$ in dx_2/dx_1, and letting $x_1 \to \pm\infty$.

The phase plane can be mapped onto a sphere to picture the role of the point at infinity. A sketch on a ping-pong ball with a ball-point pen is enlightening.

7. Error Analysis

In the preceding discussion, the phrase "integrate along a particular path" was blithely used. Now if one could do this with a wave of the wand, as the Walrus said, it would be grand! Unfortunately on the one side of the looking glass is a differential equation, on the other side an approximating finite difference equation, and the distortion of the glass is such that the solution to the one may well bear little resemblance to the solution of the other. That, after all, is what the fuss is all about. So let us proceed with a diagnosis and treatment of the ills of discretization.

The first step is an a priori estimate of a reasonable stepsize to use; this parameter depends upon the computer and the integration method. Then after a tentative determination of the phase space topology, the next step is an a posteriori error analysis of optimum stepsize and error bound. Some programs have variable stepsize built in, but I have chosen a fixed stepsize for illustrative purposes.

First, select the highest frequency (in this case 6, associated with the foci), and an integration method. For the example, the popular Runge-Kutta fourth order was selected, although Hull et al 1972 have since concluded that "in general, Runge-Kutta methods are not competitive." The equations $x' = -6y$, $y' = 6x$ of simple harmonic motion were integrated around one cycle for subdivisions of 8, 16, 32,..., 2048. The difference between the radius $\sqrt{x^2 + y^2}$ at the end of one cycle, and the correct value of 1 was used as a measure of error. It was found that n = 128 or stepsize h = .008 produced minimum error of $1.5 * 10^{-8}$, down in the noise level of the 36 bit word Univac 1106.

Next, select a starting point for integration along the unstable real eigenvectors. E.g., one starting point will be $(3 + \varepsilon\sqrt{2}, -3 + \varepsilon)$, associated with the critical point (3,-3) and the eigenvector $(\sqrt{2},1)$. The idea is to select ε small enough so the nonlinear terms will not affect the starting point being on the linear eigenvector, yet not any closer to the critical point than necessary, so the orbit won't take too long getting started. In the Lyn Lane example, $(3 + \varepsilon\sqrt{2})^2$ is the dominant term.. A binary representation of this number in scientific notation is $(.1001 + 2^{-3}\ 3\varepsilon\sqrt{2} + 2^{-3}\ \varepsilon^2) * 2^4$. When x^2 is computed we want the nonlinear term not to affect the lowest order bit in the 27 bit fractional part of the Univac 1106 single precision word. That is, we want $2^{-3}\varepsilon^2 < 2^{-27}$ or $\varepsilon < 2^{-12} = 2.436 * 10^{-4}$. An ε of $2 * 10^{-4}$ was chosen, giving a starting point of (3.00028284, -2.99980000).

The separatrix produced by this starting point is clearly (from Figure 1) the key one to use in an error analysis. The output of the Runge-Kutta program produced this separatrix as values of x and y for each t; in addition, the quantities $r = \sqrt{(x-3)^2 + (y-3)^2}$ (the distance from the focal point in the first quadrant), the velocity $v = \sqrt{x'^2 + y'^2}$, and the angular velocity v/r were produced. The integration was performed for a sequence of stepsizes surrounding the tentative value of h = .008, as shown in Table 1. The somewhat strange looking higher values were selected because the maximum value of r appeared to be around t = .78, so stepsizes commensurate with this were selected for the *a posteriori* error analysis.

The idea behind the *a posteriori* error analysis is that maximum accuracy is achieved when discretization and roundoff errors are about equal. The cumulative error is of the form $\alpha h^m + \beta/h$. The first term is the discretization error (often confusingly called truncation error), which is usually exhibited in a textbook presentation of each specific method. For example, in Runge-Kutta fourth order, the single step discretization error is $O(h^5)$; the cumulative discretization error over a fixed time interval is $O(h^4)$, since the number of steps is inversely proportional to h. The second term is the roundoff error, which is proportional to the number of steps, or inversely proportional to the stepsize. The minimum error occurs when $h = (\beta/m\alpha)^{1/(m+1)}$ for sgn (α) = sgn (β) (obtained by setting dE/dh = 0,) and when $h = (\beta/\alpha)^{1/(m+1)}$ for sgn (α) = -sgn (β). The difficulty is that α and β aren't known. The latter depends upon computer wordsize, the former upon the equations being integrated. For a problem with known answer as standard of reference a plot of log |Error| vs log stepsize (for fixed t) consists of a straight-line portion of slope m where discretization error is dominant (large h), a straight-line portion of slope -1 where roundoff error is dominant (small h) and a curvilinear portion (typically less than an octave wide) joining the other two sections, where the two types of errors are about equal. If the error changes sign, the central portion is a deep sharp v, with the apex at $-\infty$ on log-log paper. Clearly the optimum h in this case is the one producing zero error. If the error does not change sign, the central portion is a u-shaped curve, with lowest point indicating optimum stepsize and minimum error.

For a problem whose correct answer is not known (which of course is what we are faced with), the standard of reference needed for a log E vs log h plot must be obtained by extrapolation to zero error, using either of the linear portions of the graph. When discretization error is dominant, set $q_t = q_1 + \alpha h_1^m = q_2 + \alpha h_2^m$, eliminate α and obtain:

$$q_t = \frac{q_2 - q_1(h_2/h_1)^m}{1 - (h_2/h_1)^m}$$

where q_1 and q_2 are the computed values of some quantity for stepsizes h_1 and h_2 respectively, q_t is the extrapolated true value, and m is the assumed known order of the method. Where roundoff error is dominant, set $q_t = q_1 + \beta/h_1 = q_2 + \beta/h_2$, eliminate β and obtain:

$$q_t = \frac{q_2 - q_1(h_1/h_2)}{1 - (h_1/h_2)}$$

The integration is performed for a geometric progression of values of h, and an extrapolation made for each successive pair of computed values. When two successive extrapolations agree, the three computed points lie on a straight line on the log-log graph. An agreement in extrapolation from the two opposing sides gives credence to that value for use in computing error for the log E vs log h plot. A final check is that the measured slopes of the plot agree with those assumed in deriving the extrapolation formulae.

The results for the Lyn Lane example are shown in Table 1. Since the r is decreasing monotonically with stepsize, there will be a change of sign in the error. It would be hard to know how to interpret the tabulated figures without the above analysis, but with it we interpolate to $h = \infty$ from the small values of h (roundoff error dominant) and to $h = 0$ from the large values of h (discretization error dominant); the results are shown in Table 1. The small random variations about a mean of 15.116974 of the values extrapolated from small h is to be expected, since roundoff error is a random variable.

From these results an approximate true value of 15.117 was selected, the final column of estimated error computed in Table 1, and graphed in Figure 3. From Figure 3 we observe that (a) the measured slopes of -1 between .0025 and .01625 and -4 between .039 and .078 agree with the hypotheses of the extrapolation formulae; (b) there is a characteristic sharp v around the change of sign; (c) for stepsizes greater than about .08 (just 10 times the *a priori* estimate) the slope of the log E vs. log h curve differs significantly from -4, and hence we have an upper bound on stepsizes for this method for this problem (d) the optimum stepsize is $.026 < h < .02$, producing an answer with relative error less than 1 part in 15,000. The curves actually shown in Figure 3 were produced using a stepsize of .02, or 5/2 times the *a priori* estimate.

The error analysis has been described in terms of a simple numerical example in the interests of effective communication, but of course is quite general. It has been my experience that often an elegant formula from textbook or paper doesn't work out in practice the way it is supposed to. The procedure described herein is one I use constantly, and have found to be quite effective. It provides insight into a given problem, and is a revealing basis for comparison of different integration methods.

8. Further Adventures in Phase Space

We are interested in the phase space of the original vector differential equation, and that is what we have been pretending we have in Figure 1. But of course in reality we never do solve the differential equation; rather, we solve a finite difference equation which purports to approximate the given differential equation in some reasonable way. The question now is, how well does the solution to the difference equation approximate the solution to the differential equation. In particular, what happens to the phase space topology? If there were only a mild quantitative deformation, but the qualitative structure remained unchanged, we should be most happy. But unfortunately the phase space of the finite difference equation can differ from the phase space of the differential equation in the following non-trivial ways:

(1) The number of the critical points can change

(2) The character of one or more critical points can change (e.g., from a stable node to a saddle)

(3) The general structure can change; i.e., the number of separatrices and their termini, or the way the critical points "connect up" can change.

These phenomena will be illustrated briefly for the Lyn Lane example; using a modified Euler method which is less complicated to describe than the Runge-Kutta. The modified Euler is the simplest of the predictor-corrector methods, consisting of a simple Euler predictor and trapezoid corrector. The formulae associated with $x' = f(x)$ are:

$$x_p = x + hf(x)$$

$$x_c = x + h(f(x) + f(x_p))/2$$

Applied to the Lyn Lane example, this method produces

$$x_c = x + h\dot{x} + h^2\,(x\dot{x} - 2y\dot{y}) - 0.5h^3(\dot{x}^2 - 2\dot{y}^2)$$

$$y_c = y + h\dot{y} + h^2\,(x\dot{x} - y\dot{y}) - 0.5h^3(\dot{x}^2 - \dot{y}^2)$$

where $\dot{x} = x^2 - 2y^2 + 9$, $\dot{y} = x^2 - y^2$

A critical point is a point such that $x_c = x$, $y_c = y$, as in the differential system. We see that for the simple Euler (the first two terms in the two equations for x_c and y_c), the critical points are the same as for the differential system, but for the modified Euler the critical points are the intersections of two quartics in x and y -- in general as many as 16 critical points (the solutions are functions of h)!

To determine the topology of the phase space of the difference equations, one proceeds as in the previous sections of this paper.

First, the Jacobian of the difference equations is determined. In general this will not be the same as the Jacobian of the differential equations.

Second, the critical points of the difference equations are determined, using the Jacobian and Newton-Raphson iteration if need be.

Third, local behavior in the neighborhood of each critical point is determined by solving the algebraic eigenproblem for the Jacobian at each critical point. Assuming local behavior is dominated by the linear term, it is given by $x_c = Jx$. The condition for unconditional stability of a critical point is that all eigenvalues are interior to the unit circle, as opposed to the condition in the differential system that they all lie on the left half plane.

Fourth, fifth, sixth and seventh, the real and complex eigenvalues are studied as in the differential system, the phase plane topology of the difference system is determined, and an a posteriori error analysis is conducted. To present the details even for a simple illustrative example would be another paper (Barbary and Mira, 1970). E.g., for methods based on truncated Taylor series, the eigenvalues of J(3,3) for methods of order 1,2,3,4, are $\lambda = 1 \pm i6h$, $1-18h^2 \pm i6h$, $1-18h^2 \pm i6h(1-6h^2)$, and $1-18h^2 + 54h^4 \pm i6h(1-6h^2)$ respectively, Thus the stable focus in Q1 is unstable for all h for simple and modified Euler, but is stable for 3rd order Runge-Kutta for $h < \sqrt{3}/6$, and for 4th order Runge-Kutta for $h < \sqrt{2}/3$.

A final useful idea is that of stabilization by global feedback (Hochfeld et al 1957, Baumgarte and Stiefel 1972). Suppose we wish to solve the equation x $x' = f(x)$; $x, x' \varepsilon E_n$; f: $E_n \rightarrow E_n$ and it is known that the solution should lie on some hypersurface $g(x) = 0$, g: $E_n \rightarrow E_1$. The function g might represent one of the physical conservation laws, such as conservation of mass, momentum, or energy. Then one mechanizes the equation

$$x' = f(x) - \mu g \nabla g$$

where μ is some positive constant roughly equal to $\sqrt{n}$ divided by the average norm of Δg over the region of interest. The latter equation has as stable limit cycle the desired solution to the former; that is, if the solution drifts away (and all computer integration schemes produce errors which increase in time) from the hypersurface $g = 0$, the solution is driven back by the feedback term.

Conclusion

To study a differential equation of the form $x' = f(x)$; $x, x' \varepsilon E_n$; $f: E_n \to E_n$ (and most higher order equations can be reduced to this form), it is not as fruitful to select initial conditions at random and display time dependent coordinates, as it is to determine the phase space topology. The procedure described in this paper reveals the different qualitative behavior patterns possible for the system, and the regions of sensitivity to computational errors and to qualitative changes in the behavior of the system, Moreover, this is a useful way of comparing different methods of numerical integration.

Table 1

$x_o = 3.00028284$ $\qquad$ $y_o = -2.9998$ $\qquad$ $t = .78$

h	$r = \sqrt{(x-3)^2 + (y-3)^2}$	extrapolated	r - 15.117
		to h = ∞	
.0025	15.124307		.00731
		15.116829	
.005	15.120568		.00351
		15.117253	
.01	15.118911		.00191
		15.116688	
.012	15.118540		.00154
		15.116874	
.015	15.118207		.00121
		15.117224	
.01625	15.118131		.00113
		ave = 15.116974	
.02	15.117459		.000459
.026	15.114921		-.00208
		to h = 0	
.03	15.111560		-.00544
		15.120	
.039	15.094931		-.0221
		15.1161	
.078	14.777181		-.340
		15.8279	
.0975	14.704119		-.413

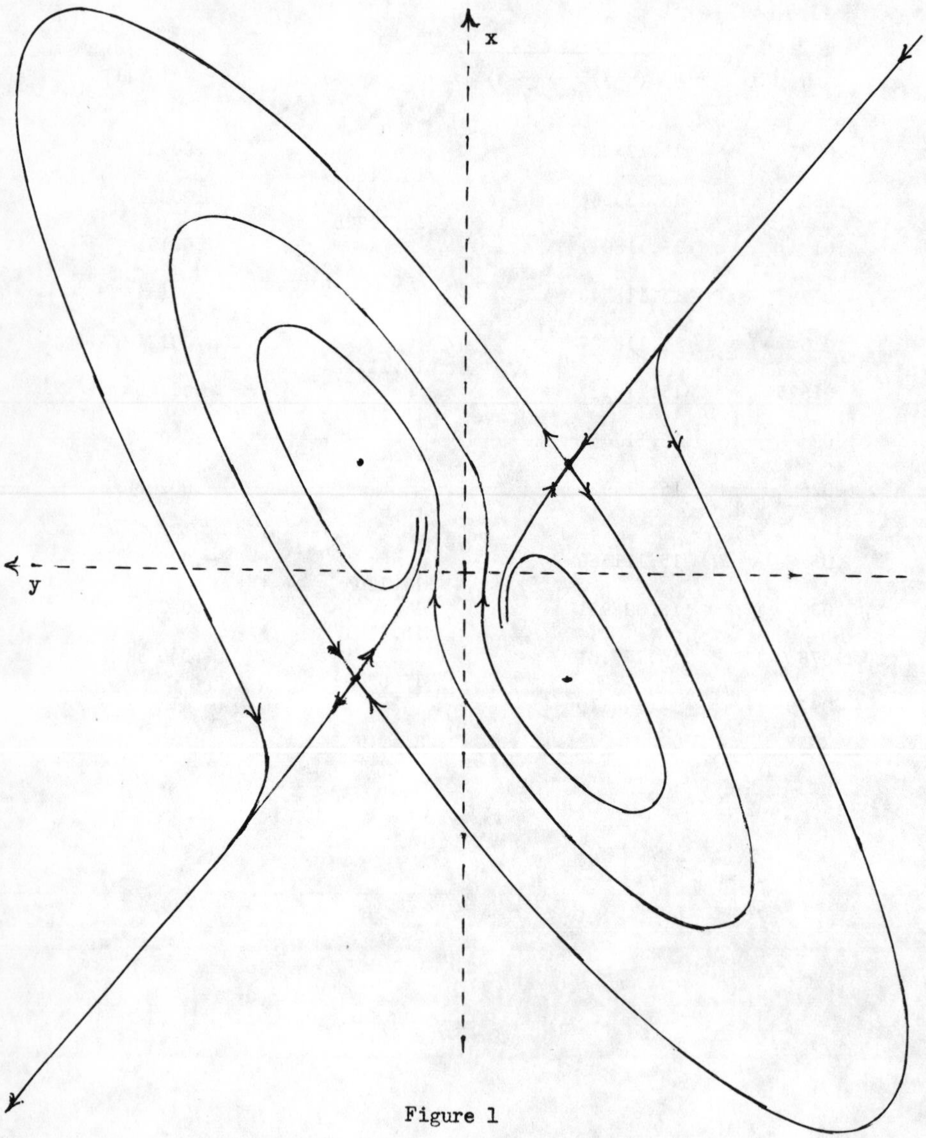

Figure 1

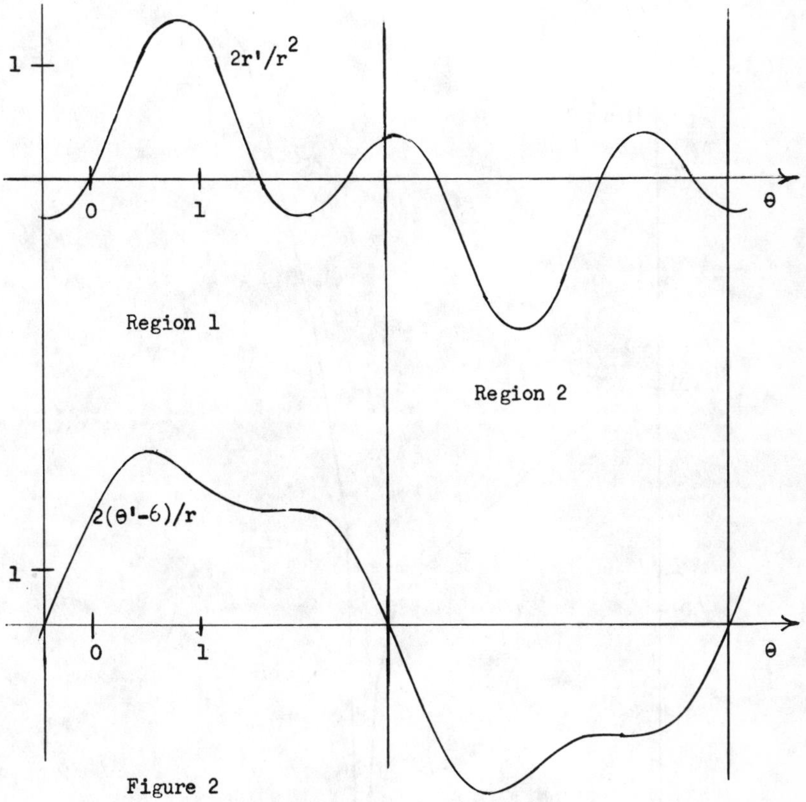

Figure 2

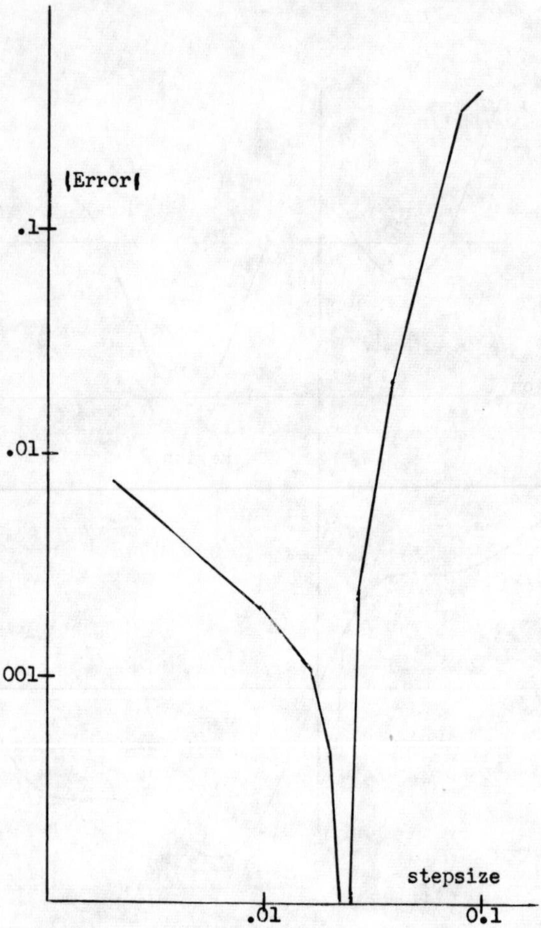

Figure 3

References

1. Jean-Pierre Barbary and Christian Mira, Problèmes de Discrétisation d'une Équation Différentialle Antonome du Deuxième Ordre, Proc. First Formator Symposium on Mathematical Methods for Large Scale Systems, (Ustav Teorie Informace a Automatizace, Libice 1970) Czechoslovak Acad. Sci., Prague, 1970.

2. J. Baumgarte and E. Stiefel, "Examples of Transformations Improving the Numerical Accuracy of the Integration of Differential Equations," these Proceedings,

3. S.D. Conte and Carl deBoor, Elementary Numerical Analysis, McGraw-Hill, 2nd ed., 1972

4. T.V. Davies and Eleanor M. James, Nonlinear Differential Equations, Addison-Wesley, 1966

5. E. Hochfeld (ed.), Stabilization of Computer Circuits, U. of Chicago, 1957. Available from U.S. Government Printing Office as WADC TR 57-425 (AD 155740).

6. B.E. Howard, Nonlinear System Simulation, Simulation, Oct. 1966.

7. E. Kamke, Differentialgleichungen Losungsmethoden und Losungen, Akademishe Verlagsgesellschaft, Leipzig, 1956.

8. J.H. Wilkinson, The Algebraic Eigenvalue Problem, Clarendon Press, Oxford, 1965

MULTI-OFF-GRID METHODS IN MULTI-STEP INTEGRATION OF ORDINARY DIFFERENTIAL EQUATIONS†

by

Dr. Paul R. Beaudet

COMPUTER SCIENCES CORPORATION
6565 Arlington Boulevard
Falls Church, Virginia

ABSTRACT

New off-grid methods, that are not limited by the Dahlquist stability theorem, are introduced for the numerical integration of first- and second-order systems of differential equations. The methods are characterized by having all derivative evaluations performed at locations off the grid of final solution values. All on-grid solution values and off-grid derivative evaluations are used over m back steps to calculate the solution value at the next on-grid location. There are n off-grid derivative evaluations associated with each of these intervals located at fractional positions $\gamma_1 h, \ldots, \gamma_n h$ relative to the on-grid locations. Off-grid geometric parameters are found which give highly stable integrators of maximum possible order, $O\left(h^{m+mn+n-1}\right)$. These integrators also have the property that their coefficients decrease geometrically with remoteness from the current interval of integration; and although the second mean value theorem is not applicable in obtaining an exact error expression for these methods, the error coefficients are generally several orders of magnitude less than comparable on-grid Adams or Cowell error coefficients. Comparisons are made with other integration techniques for satellite orbital calculations and other systems or ordinary differential equations.

---

†Work supported by NASA Goddard Space Flight Center Under Contract No. NAS 5-11790

I. INTRODUCTION

Multi-off-grid methods of numerical integration are multi-step generalizations of Gaussian Quadratures. These methods are devised for solving first and second order systems of differential equations. Figure 1 shows the geometry of these integrators. In the multi-off-grid methods, all derivatives are evaluated at off-grid locations, designated by an "X". The "Os" of Figure 1 designate on-grid points. The basic philosophy which led to the concept of multi-off-grid methods of integration stems from the realization that derivative (force) evaluations for orbital calculations can be extremely costly; therefore, derivative evaluations should not necessarily be performed at the on-grid locations just because very accurate state values happen to be available there. In the multi-off-grid philosophy, all derivatives are evaluated at off-grid points.

REFINERS (MULTI-OFF-GRID)

PROBLEM: $Y' = F(Y, t);\; Y(t_o) = Y_o$

$Y'' = F(Y, t);\; Y(t_o) = Y_o;\; Y'(t_o) = Y'_o$

GEOMETRY:

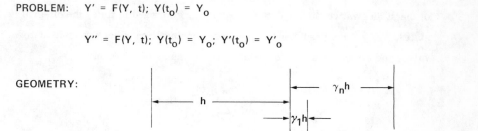

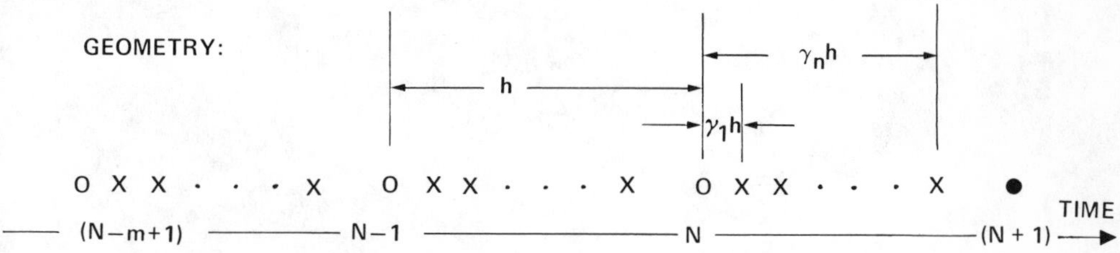

O GRID OF FINAL SOLUTION POINTS

X OFF–GRID DERIVATIVE EVALUATIONS

NOTATION: (m, n) METHOD USES
m BACK INTERVALS AND
n OFF–GRID DERIVATIVES PER INTERVAL

CALCULATIONAL FORMULA:

$$Y_{N+1} = \sum_{j=0}^{m-1} \alpha_j Y_{N-j} + h^{(1,2)} \sum_{j=0}^{m-1} \sum_{\ell=1}^{n} \beta_{\ell,j} F_{N-j+\gamma_\ell}$$

NUMBER OF INDEPENDENT VARIABLES: $(m + mn + n)$

Figure 1. Geometry of the Multi-Off-Grid Method of Numerical Integration

In the multi-off-grid methods the formula for calculating the on-grid state is called a refiner. The refiner uses m on-grid back states and mn off-grid derivatives located at distances $\gamma_1 h$ to $\gamma_n h$ from the on-grid points. There are m+mn+n undetermined coefficients in the formula; n off-grid location parameters, (γ_1 to γ_n), m on-grid state coefficients, ($\alpha_0 \ldots \alpha_{m-1}$), and mn off-grid derivative coefficients, ($\beta_{\ell,j}$). Stable refiner formulas exist to maximum possible order, $O\left(h^{m+mn+n-1}\right)$.

An important theorem for the Classical on-grid methods which use information over k back steps was given by Dahlquist (1) in 1956. This theorem shows that the order of the method cannot exceed k+2 for even k, and k+1 for odd k. This is surprising because coefficients can be found which will exactly correct polynomials of order 2k. These methods have order (2k-1), but are unstable. This limitation of on-grid methods is overcome by the multi-off-grid methods presented in this paper, thereby achieving a more efficient and economical numerical integration procedure for a given level of accuracy.

Multi-off-grid methods require off-grid state predictors for the evaluation of the n derivatives at each step. Progressing forward in time, the off-grid states, $\vec{Y}_{N+\gamma_\ell}$ are predicted using a linear combination of back on-grid state values and off-grid derivative evaluations; formulas of the form

$$\vec{Y}_{N+\gamma_{\ell'}} = \vec{Y}_N + h \sum_{j=0}^{k} \sum_{\ell=1}^{n} b_{\ell,j}{}^{(\ell')} \vec{F}_{N-j+\gamma_\ell}$$

for class I problems and

$$\vec{Y}_{N+\gamma_{\ell'}} = (1+\gamma_{\ell'}) \vec{Y}_N - \gamma_{\ell'} \vec{Y}_{N-1} + h^2 \sum_{j=0}^{k} \sum_{\ell=1}^{n} b_{\ell,j}{}^{(\ell')} \vec{F}_{N-j+\gamma_\ell}$$

for class II problems are used. The derivatives at these γ_ℓ off-grid points are evaluated, and the on-grid state value, $\vec{Y}_{N+1}$, is calculated using a linear combination of m back $\vec{Y}$'s and mn back derivative evaluations. For m=1, these methods have the Gaussian quadrature geometries.

II. THE SYSTEM OF ORDER EQUATIONS

To derive the system of order equations which are satisfied by the αs , βs , and γs , the calculational formula of Figure 1 is expanded in a Taylor series about any convenient point. Typically the point N is chosen. The result is

$$\sum_{i=0}^{\infty} \vec{Y}_N^{(i)} \frac{h^i}{i!} = \sum_{j=0}^{m-1} \alpha_j \sum_{i=0}^{\infty} \vec{Y}_N^{(i)} \frac{(-jh)^i}{i!}$$

$$+ h^c \sum_{j=0}^{m-1} \sum_{\ell=1}^{n} \beta_{\ell,j} \sum_{i=0}^{\infty} \frac{\vec{F}_N^{(i)}}{i!} \left[(\gamma_\ell - j)h\right]^i$$

where c is 1 for class I (first order) problems, and 2 for class II (second order) problems; and a superscript enclosed by parentheses, indicates the corresponding derivative with respect to time (the independent variable), e.g., (i) indicates the ith derivative. Next, for consistency with the differential equation, replace $\vec{F}_N^{(i)}$ by $\vec{Y}_N^{(i+c)}$ and require the coefficients of $\frac{\vec{Y}_N^{(i)} h^i}{i!}$ to be zero for i=0, 1, 2, ... to the maximum possible value of i consistent with either stability or the number of degrees of freedom in the method. The resulting system of order equations is illustrated below.

For class I problems

$$0 = -1 + \sum_{j=0}^{m-1} \alpha_j (-j)^i + i \sum_{j=0}^{m-1} \sum_{\ell=1}^{n} \beta_{\ell,j} (\gamma_\ell - j)^{i-1}$$

$$i=0,\ 1,\ 2,\ \ldots,\ m+mn+n-1$$

For class II problems

$$0 = -1 + \sum_{j=0}^{m-1} \alpha_j (-j)^i + i(i-1) \sum_{j=0}^{m-1} \sum_{\ell=1}^{n} \beta_{\ell,j} (\gamma_\ell - j)^{i-2}$$

$$i=0,\ 1,\ 2,\ \ldots,\ m+mn+n-1$$

Solution of these equations defines the α , β , γ coefficients to be used in the refiner formula. The approach which has been used to solve this system of order equations for the refiner α , β , and γ coefficients is as follows: The α coefficients are eliminated analytically by multiplying the order equations by Stirling numbers of the first kind and summing. The determinants formed by the resulting coefficients of the β's are used to form n equations in the n unknown γ's . There exist extraneous roots of the form $(\gamma_\ell = \gamma_{\ell'} \pm \text{integer})$ which are numerically extracted. Then Newton-Raphson is used to obtain the remaining roots. The β's are obtained by triangularization, back-substitution, and iterative correction, of the resulting linear system of equations. The α's are obtained by back substitution into the Stirling formulas. These values are then used as the initial guess in a Newton-Raphson iteration on the system of order equations.

III. ALGORITHMS AND COEFFICIENTS

The predictor-corrector methods of classical Adams/Cowell processes permit many algorithms depending on the number, q , of corrector cycles used and whether or not a final derivative evaluation, E , is performed. These algorithms are designated by $PE(CE)^q$ and $P(EC)^q$. The multi-off-grid methods permit a similar variety of algorithms by correcting and re-evaluating derivatives at the off-grid points. The multi-off-grid algorithms are designated by $\left[PE(CE)^q\right]^n R$ where the square brackets $\left[\ \right]^n$ designate the n off-grid locations and R designates the refiner process. Experience with these methods have shown that for class I problems the $\left[PECE\right]^n R$ algorithm and for class II problems the $\left[PE\right]^n R$ algorithm are the most efficient.

The following class I and class II set of coefficients are recommended for general use.

Class I $\left[PECE\right]^2 R$ algorithm.

Geometry and Refiner Coefficients

```
GAMMA( 1) =   0.7954421046336664 D 00
GAMMA( 2) =   0.1271258027986261 D 01
ALPHA ( 0) = -0.1297664704422448 D-02  PLUS  1.0
ALPHA ( 1) =  0.1297664704422448 D-02
BETA( 1, 1) =   0.4475454414672187 D 00
```

```
BETA( 2, 1) =     0.6981174877253335 D-02
BETA( 1, 2) =     0.2907895784612232 D-01
BETA( 2, 2) =     0.5176920905138280 D 00
```

Note that the α_0 coefficient is expressed as a perturbation from the nominal value 1.0.

Predictor Coefficients for the γ_1 Location

```
A( 1) =    0.1000000000000000 D 01
B( 1, 1) = -0.3803476922261824 D 01
B( 2, 1) =  0.2000967550605683 D 01
B( 1, 2) = -0.6171650115402375 D 01
B( 2, 2) =  0.6289502282856881 D 01
B( 1, 3) = -0.1241029862503715 D 01
B( 2, 3) =  0.3562947918775003 D 01
B( 1, 4) =  0.0
B( 2, 4) =  0.1581812525640118 D 00
```

Corrector Coefficients for the γ_1 Location

```
A( 1) =    0.1000000000000000 D 01
B0( 1) =    0.1581812525640056 D 00
B( 1, 1) = -0.2887152834602417 D 00
B( 2, 1) =  0.7673443677220461 D 00
B( 1, 2) = -0.1601244085106864 D 00
B( 2, 2) =  0.2779765759652276 D 00
B( 1, 3) = -0.7406679620039631 D-02
B( 2, 3) =  0.4818627997335474 D-01
```

Predictor Coefficients for the γ_2 Location

```
A( 1) =    0.1000000000000000 D 01
B0( 1) =    0.1608130823371966 D 01
B( 1, 1) =  0.4928938751469921 D 01
B( 2, 1) = -0.2612053902995243 D 01
B( 1, 2) =  0.2930871758028034 D 01
B( 2, 2) = -0.4809023013186763 D 01
B( 1, 3) =  0.1456194798433876 D 00
B( 2, 3) = -0.9212258685450413 D 00
```

Corrector Coefficients for the γ_2 Location

```
A( 1) =    0.1000000000000000 D 01
B0( 1) =    0.6910987603905592 D 00
B0( 2) =    0.1456194798433880 D 00
B( 1, 1) =  0.2235597922274106 D 00
B( 2, 1) =  0.2847138318844550 D 00
B( 1, 2) =  0.3410402314834434 D-01
B( 2, 2) = -0.1036440539442582 D 00
B( 1, 3) =  0.0
B( 2, 3) = -0.4193805563638096 D-02
```

Class II $[PE]^2R$ algorithm.

Geometry and Refiner Coefficients

```
GAMMA( 1) =   0.1246689505465554 D 01
GAMMA( 2) =   0.1729980056318249 D 01
ALPHA( 0)  =  2.0
ALPHA( 1)  = -1.0
BETA( 1, 1) =    0.5078959139263864 D-03
BETA( 2, 1) =   -0.2189045058284332 D-04
BETA( 1, 2) =    0.4015686098810537 D 00
BETA( 2, 2) =    0.1086431032080018 D 00
BETA( 1, 3) =    0.9692720927557860 D-01
BETA( 2, 3) =    0.3923750721720225 D 00
```

Predictor Coefficients for the γ_1 Location

```
A( 1) =   0.2246689505465554 D 01
A( 2) =  -0.1246689505465554 D 01
B( 1, 1) =  -0.5409444123317008 D-01
B( 2, 1) =   0.4314771806556548 D 00
B( 1, 2) =  -0.1222967538968725 D 01
B( 2, 2) =   0.1603903743214676 D 01
B( 1, 3) =  -0.5305934504235123 D 00
B( 2, 3) =   0.1047820552758178 D 01
B( 1, 4) =  -0.1938002013244933 D-01
B( 2, 4) =   0.1442960883810990 D 00
```

Predictor Coefficients for the γ_2 Location

```
A( 1) =   0.2729980056318249 D 01
A( 2) =  -0.1729980056318249 D 01
B0( 1) =   0.3661288385784120 D 00
B( 1, 1) =   0.1523083491877451 D 01
B( 2, 1) =   0.8626120629436270 D-01
B( 1, 2) =   0.9280839854793630 D 00
B( 2, 2) =  -0.2696134693305231 D 00
B( 1, 3) =   0.1096133059761196 D 00
B( 2, 3) =  -0.3695254054045009 D 00
B( 1, 4) =   0.0
B( 2, 4) =  -0.1262642768211445 D-01
```

In the index, (ℓ , j) , ℓ designates the relative off-grid location (1 or 2) and j designates the associated back interval, $j = 0, 1 \ldots m - 1$.

IV. PROPERTIES OF THE REFINER

The resulting refiner has a number of interesting properties.

- Each off-grid parameter can be used to increase the order of the refiner formula.

- Stable formulas exist to maximum order possible.
- For each topology, (m , n) , there are many such stable solutions.
- The local error coefficient of the refiner is extremely small.
- The α , β cooefficients decrease geometrically with remoteness from the current interval of integration; this implies little round-off error if the sums are approximately ordered.
- Extraneous roots of the characteristic polynomial are near zero.
- The refiner formula is so accurate that it is limited only by the accuracy of the off-grid predictions.

Figure 2 shows the comparison of the refiner error coefficients with the classical on-grid corrector error coefficients. In some cases multi-off-grid

ERROR COEFFICIENTS

	MULTI-OFF-GRID (m, n)		ON-GRID ADAMS/COWELL	ORDER
CLASS I	(1, 2)	.0001	.02	4
	(1, 3)	.00003	.01	6
	(2, 2)	.00004	.01	7
	(4, 1)	.0003	.009	8
	(2, 3)	4×10^{-10}	.007	10
	(6, 1)	6×10^{-6}	.005	12
	(4, 2)	3×10^{-11}	.005	13
CLASS II	(2, 1)	.08	.004	4
	(3, 1)	.009	.004	6
	(2, 2)	.00006	.003	7
	(4, 1)	.001	.003	8
	(3, 2)	1×10^{-8}	.002	10
	(4, 2)	1×10^{-10}	.001	13

Figure 2. Comparison of On-Grid and Multi-Off-Grid Error Coefficients for Class I and Class II Methods of the Same Order

methods have error coefficients which are eight orders of magnitude smaller than those of corresponding on-grid methods.

Figures 3 and 4 show some of the refiner geometries of class I and class II methods for solving first order systems of differential equations. The shaded area follows the magnitude of the coefficients and is intended to show that these coefficients decrease geometrically with remoteness from the current interval of integration. The "O's" are on-grid points; the "Δ's" designate the off-grid locations.

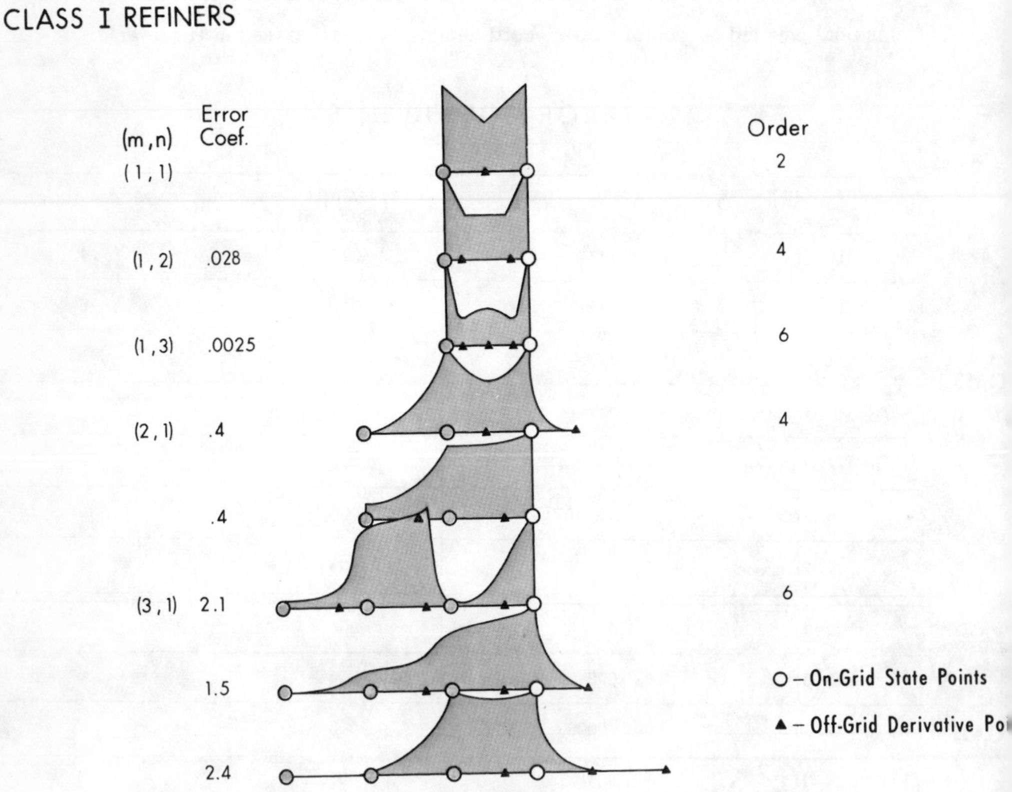

Figure 3. Class I Refiners. (Shaded area contours the magnitude of the coefficients) (1 of 5)

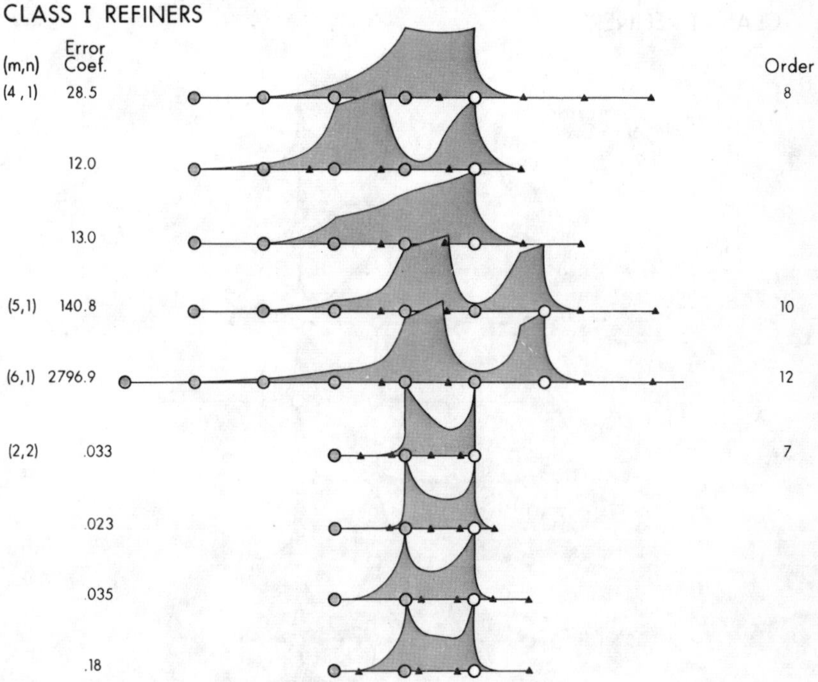

Figure 3. Class I Refiners. (Shaded area contours the magnitude of the coefficients) (2 of 5)

CLASS I REFINERS

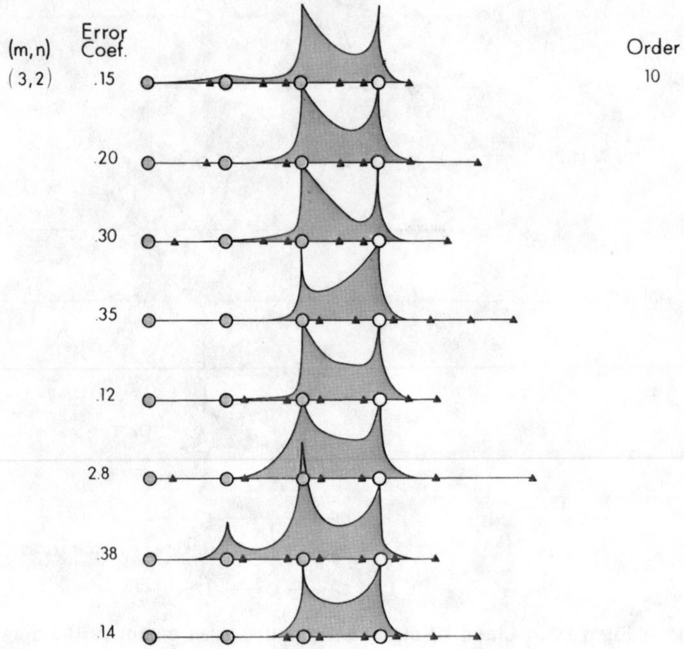

Figure 3. Class I Refiners. (Shaded area contours the magnitude of the coefficients) (3 of 5)

CLASS I REFINERS

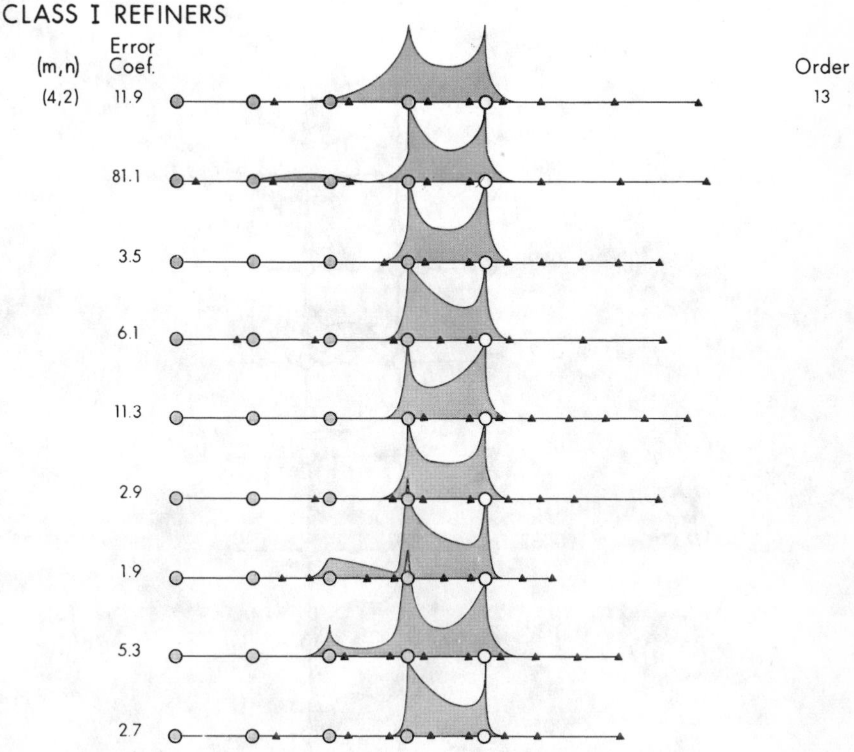

Figure 3. Class I Refiners. (Shaded area contours the magnitude of the coefficients) (4 of 5)

CLASS I REFINERS

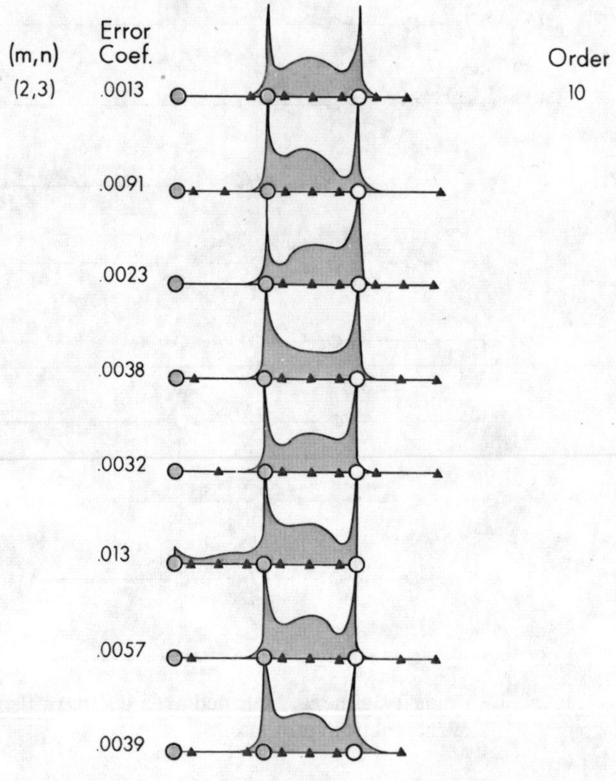

Figure 3. Class I Refiners. (Shaded area contours the magnitude of the coefficients) (5 of 5)

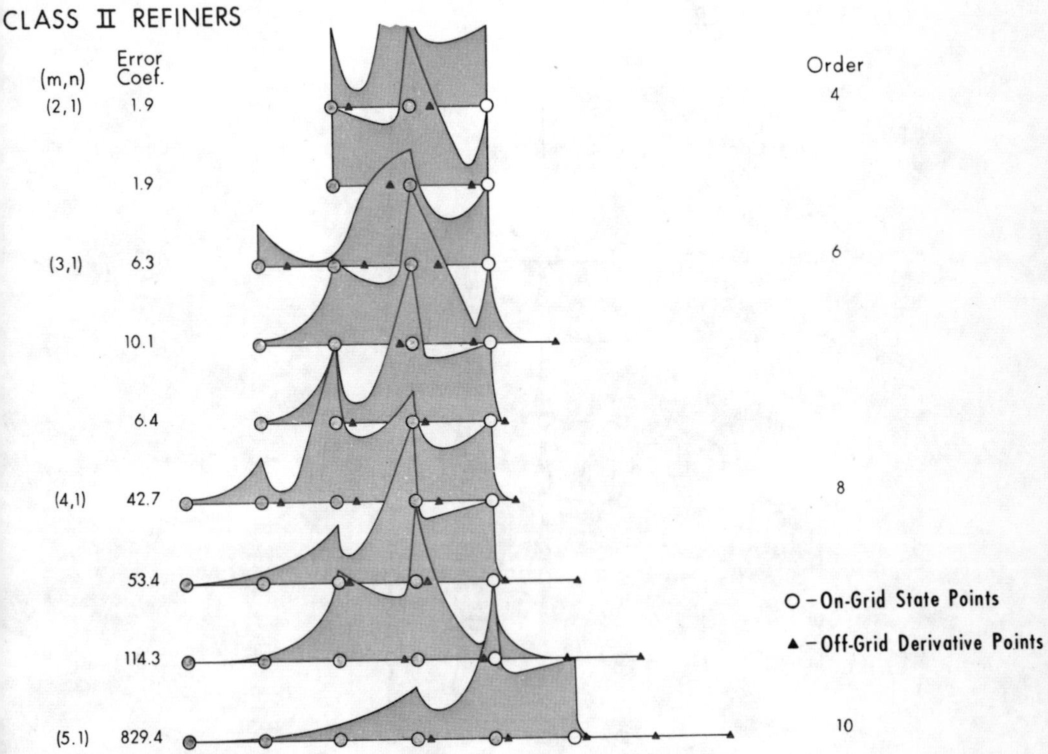

Figure 4. Class II Refiners. (Shaded area contours the magnitude of the coefficients) (1 of 3)

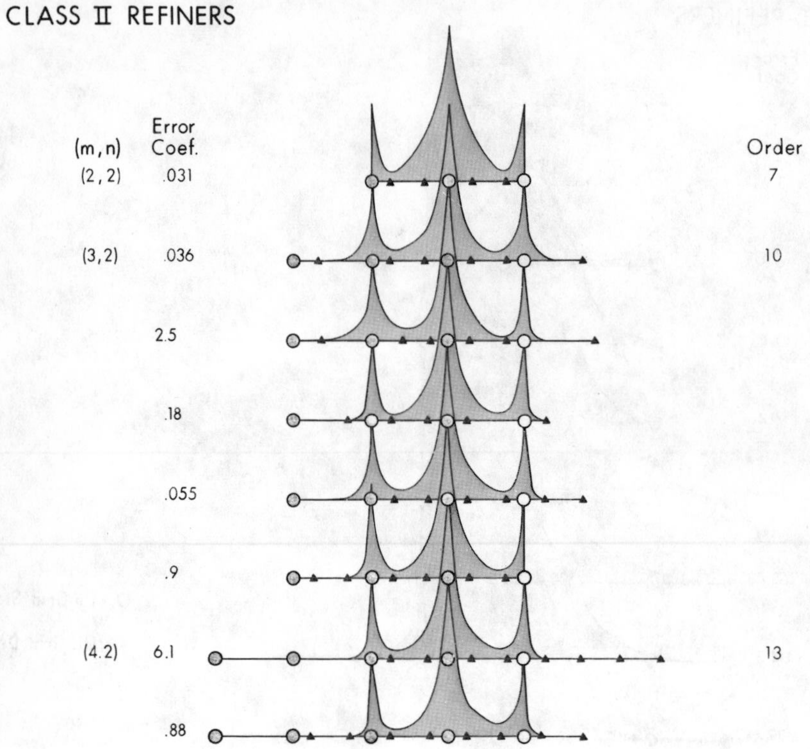

Figure 4. Class II Refiners. (Shaded area contours the magnitude of the coefficients) (2 of 3)

CLASS II REFINERS

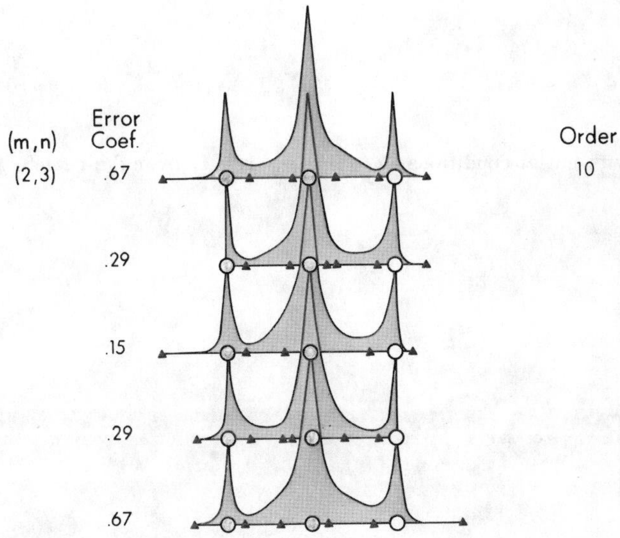

Figure 4. Class II Refiners. (Shaded area contours the magnitude of the coefficients) (3 of 3)

V. NUMERICAL EXAMPLES

The multi-off-grid methods were compared with the same order on-grid method of integration. Figure 5 shows the error resulting from integration of the system:

$$\dot{Y}_1 = Y_2$$

$$\dot{Y}_2 = -(Y_2 t + Y_1)/Y_1^2 t^2$$

with initial conditions $Y_1(1) = Y_2(1) = 1$, over the region $1 \le t \le 11$.

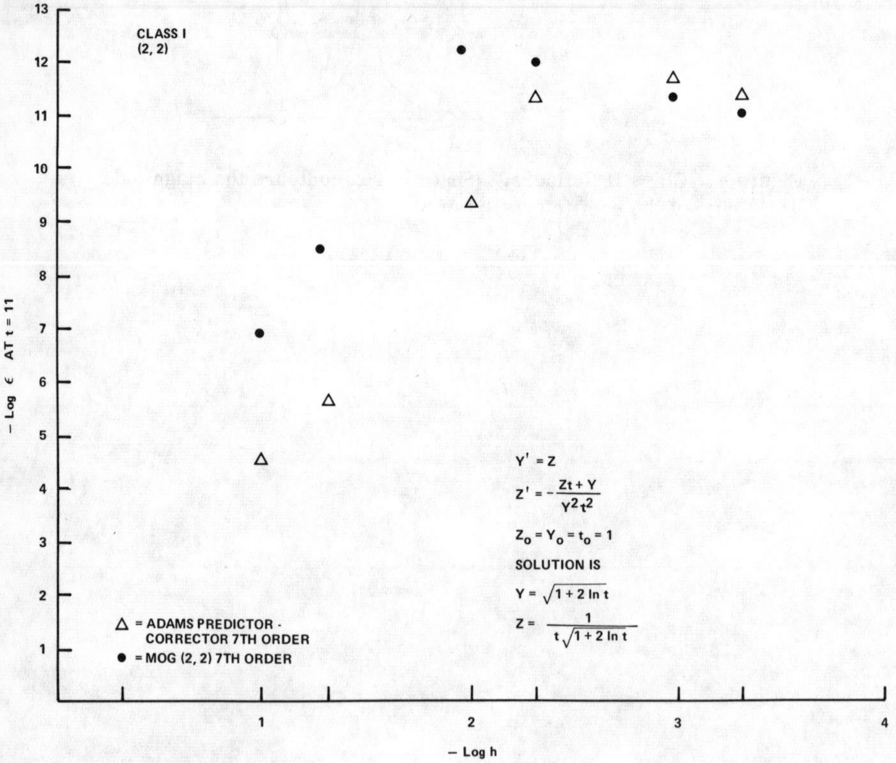

Figure 5. Numerical Example of a 7th Order Class I MOG Integrator Compared to a 7th Order Adams Predictor Corrector Integrator

In this graph, we plot the negative log of the absolute error, ϵ , i.e., (number of significant digits) as a function of the negative log of the step size, h , which is proportional to the log of the total number of function calls. Absolute error was determined by comparison with the closed form integral evaluated analytically. As one can see, the multi-off-grid error at larger step sizes, h, (smaller (- log h)) is about three orders of magnitude smaller than the corresponding on-grid error.

Figure 6 shows similar results for a class II simple harmonic oscillator problem integrated with a 10th order multi-off-grid method compared to the corresponding Cowell on-grid integration technique for $0 \leq t \leq 50$. Again, significant reduction in the error is apparent at larger step sizes, for the multi-off-grid integrator.

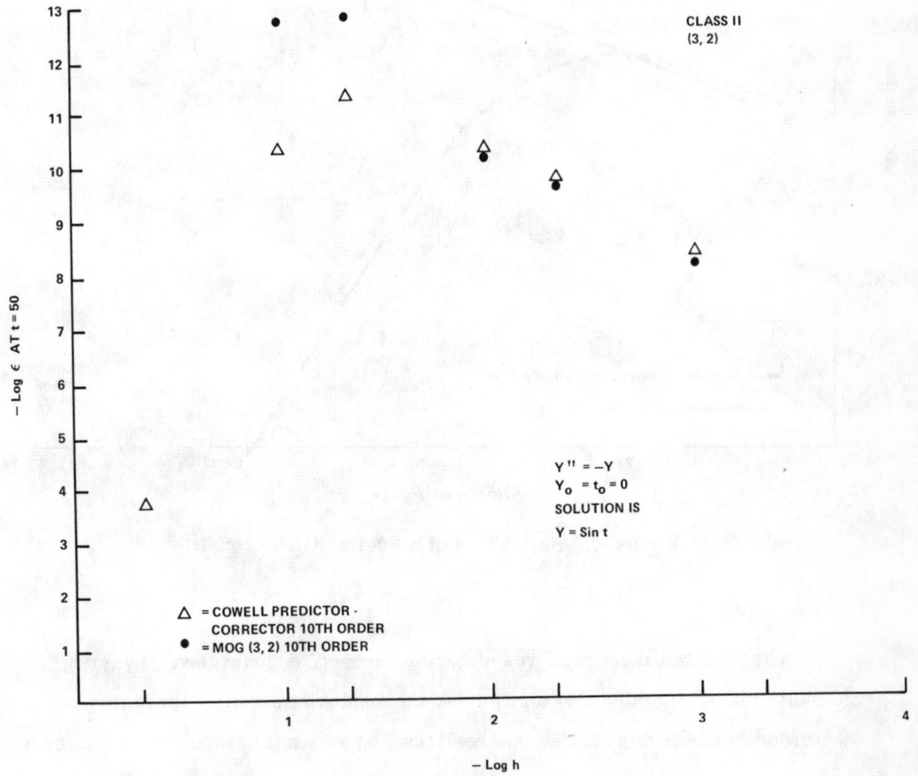

Figure 6. Numerical Example of a 10th Order Class II MOG Integrator Compared to a 10th Order Cowell Integrator

Finally, Figure 7 illustrates the results of the integration of a GEOS satellite orbit using standard Cowell on-grid coefficients vs. the multi-off-grid refiner methods. Two-body dynamics were assumed so that an analytic solution was available for comparison. The orbit was integrated for 30 days. At a step size of 100 seconds, the (2, 3) multi-off-grid method incurred an error of 7 cm, while the corresponding Cowell 10th order method incurred an error of 14, 100 cm.

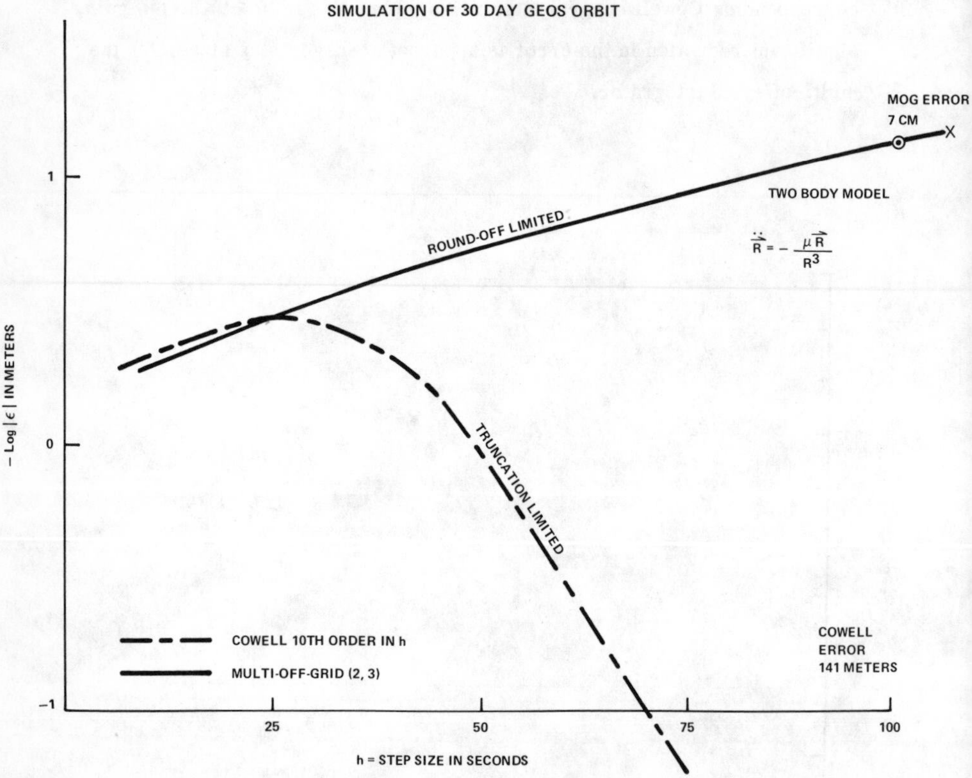

Figure 7. Simulation of a 30-Day GEOS Orbit

These initial numerical results are extremely encouraging. By going to multi-off-grid methods to circumvent the Dahlquist stability limitation on the order of on-grid methods, one realizes a significant advance in the state of integration technology. It is believed that this advance will lead to more accurate and more economical numerical integrators.

REFERENCES

1. Dahlquist, G., "Convergence and Stability in the Numerical Integration of Ordinary Differential Equations." Math. Scand. 4 (1956), 33-53

2. Beaudet, P. R., "Development of Multi-Off-Grid (MOG) Multistep Integration Techniques for Orbital Applications, Volume I Concept," CSC Report 5035-19100-01TR for NASA-GSFC under Contract NAS 5-11790

3. Beaudet, P. R., "Development of Multi-Off-Grid (MOG) Multistep Integration Techniques for Orbital Applications Volume II Numerical Evaluation," CSC Report 5035-19100-01TR for NASA-GSFC under Contract NAS 5-11790

4. Beaudet, P. R., "Development of Optimum Coefficients for Multi-Step Integration Processes," CSC Report 5035-27000-01TR for NASA-GSFC under Contract NAS 5-11790

5. Beaudet, P. R. and Moore, W. H., "Generalized Interpolation Procedures," CSC Report 5035-27000-03TR for NASA-GSFC under Contract NAS 5-11790

6. Henrici, P., Discrete Variable Methods in Ordinary Differential Equations, New York, John Wiley and Sons, 1962

7. Gragg, William B. and Stetter, Hans J., "Generalized Multistep Predictor-Corrector Methods," J. ACM, Vol. 11, No. 2 (April, 1964) pp. 188-209

8. Hansen, Eldon, "Cyclic Composite Multistep Predictor - Corrector Methods," 24th Natl. Confer. ACM (1969), pp. 135-139

9. Donelson, J. III, and Hansen, E., "Cyclic Composite Multistep Predictor-Corrector Methods," Submitted for Publication (1970), 40 pp.

10. Butcher, J. C., "A Modified Multistep Method for the Numerical Integration of Ordinary Differential Equations," J. ACM, Vol. 12, No. 1 (Jan. 1965), pp. 124-135

11. Dyer, James, "Generalized Multistep Methods in Satellite Orbit Computations," J. ACM, Vol. 15, No. 4 (Oct. 1968), pp. 712-719

12. Chester, L. and Pierce, S., "The Application of Generalized, Cyclic, and Modified Numerical Integration Algorithms to Problems of Satellite Orbit Computation," SDC Report TM-4717/000/00 for NASA under Contract NAS 5-11785

13. Velez, C. E., "Local Error Control and Its Effects on the Optimization of Orbital Integration," NASA TND-4542, Goddard Space Flight Center, Greenbelt, Maryland

COMPARISON OF NUMERICAL INTEGRATION TECHNIQUES FOR ORBITAL APPLICATIONS†

By

Hays Moore
Computer Sciences Corporation
8728 Colesville Road
Silver Spring, Maryland 20910

I. INTRODUCTION

The efficiency of integration routines for orbital applications warrants study because of the large amount of machine time required for orbit determination. Complex force function evaluations are costly and the required integration error tolerance is usually very small. Therefore the amount of machine time spent in the integration routine (integrator overhead) and in the force function evaluation routine is important. This paper gives a brief comparison of the performance of programs for integrating differential equations for orbital applications. The evaluation criteria and the method of testing are described, and the results of the test problem set are included. Integration methods that were chosen for comparison include high-order Runge-Kutta methods; a rational extrapolation method (Bulirsch and Stoer, 1966); a variable step, variable order, multistep method (Krogh, 1969); classical multistep methods of Adams and Cowell; and modified multistep methods. The high-order Runge-Kutta methods used in the comparison include RKF 7(8) and RKF 8(9) (Fehlberg, 1968), and RKS 8-10 (Shanks, 1966).

---

†Work supported by NASA Goddard Space Flight Center, Under Contract No. NAS 5-11790

II. A COMPARATIVE STUDY OF NUMERICAL INTEGRATION ROUTINES

The methods of integration that were chosen for comparison reflect the diverse philosophies concerning the solution of differential equations. The Fehlberg techniques pay the price of extra function evaluations per step to gain the benefit of integration step-size control via a local truncation error estimate. The fixed-step Runge-Kutta Shanks method, on the other hand, attempts to minimize the number of function evaluations per step. The multistep methods further minimize the number of function evaluations, but they are not self starting and are generally limited to poor integration step-size controls. Both Fred T. Krogh's multistep method and the rational extrapolation method of Bulirsch and Stoer possess variable order controls as well as variable step-size controls, but they do so at a cost in integrator overhead. An efficient algorithm for solving a differential equation requires a minimum of both computational time and computer storage for a specified tolerance of accuracy. The required computational time for test problems calculated on the same computer measures relative efficiency and overhead of the algorithm, and the required number of function evaluations reflects the efficiency of the method. The required storage for each method proved very dependent upon coding. Storage required for each method in 8-bit bytes (4 bytes per single precision word):

Bulirsch-Stoer	7,575
Krogh	9,636
RKS 8-10	8,374
RKF 7(8)	13,000
RKF 8(9)	17,362
MOG	5,460
Adams-Cowell	4,866

All methods were programmed in FORTRAN V and comparisons were computed in double precision arithmetic on an IBM 360/75.

III. EVALUATION CRITERIA

The efficiency of a program for solving differential equations is usually measured by the required number of function evaluations for a specific accuracy over a finite interval of integration, assuming that the primary cost of computer time lies in the force function evaluation. However, if integration overhead is the primary cost factor, plots of accuracy versus CPU time can best illustrate efficiency. In the following report, the number of calls to the function evaluation routine was used to measure the cost of the computation because the evaluation of complex force functions associated with the problems of celestial mechanics governs the cost of computation. This method of comparison is traditional and computer-independent. This was done by comparing for a given accuracy over a finite integration interval the average distance between function evaluations. The accuracy at the end of the given integration was measured by either the absolute or relative rms error in the positive vector. The problem set of differential equations was chosen so that a variable step-size control would not be needed; however, the methods which had a good step-size control were not restricted to run in the fixed-step mode. In these cases, a range of accuracies was determined by chaning the prespecified error tolerance of each method.

IV. TEST PROBLEM SET

The selected problems are shown in the Appendix and represent various satellite orbits. They include the coplanar circular problem of two bodies, the time-regularized coplanar elliptic problem of two bodies, and three examples of the perturbed four-body problem. Figure 1 illustrates the efficiency of several integration methods for the coplanar circular problem of two bodies after an integration interval of 8 revolutions. The graph shows the efficiency of Class II multistep methods which use two or more function evaluations per integration step. The abscissa measures the negative common logarithm of the average distance (time) between function evaluations over the interval of integration, and it is indicative of the total number of required function evaluations (which are shown in parentheses). The ordinate measures the negative common logarithm of the rms absolute error in the position vector at the final time of integration, and it is indicative of the number of decimal places of accuracy. The most efficient methods of integration occupy the upper left portion of the graph where maximum accuracy is achieved at a cost of fewer function evaluations. The order of each integration method is shown in brackets with the algorithm of each method following. The number of function evaluations per step is given in lieu of the algorithm for single-step methods. The symbols "P," "E," "C," and "R" denote "predict," "evaluate," "correct," and "refine," the latter referring to the use of off-grid point accelerations. In referring to the class of an integration method, the following convention

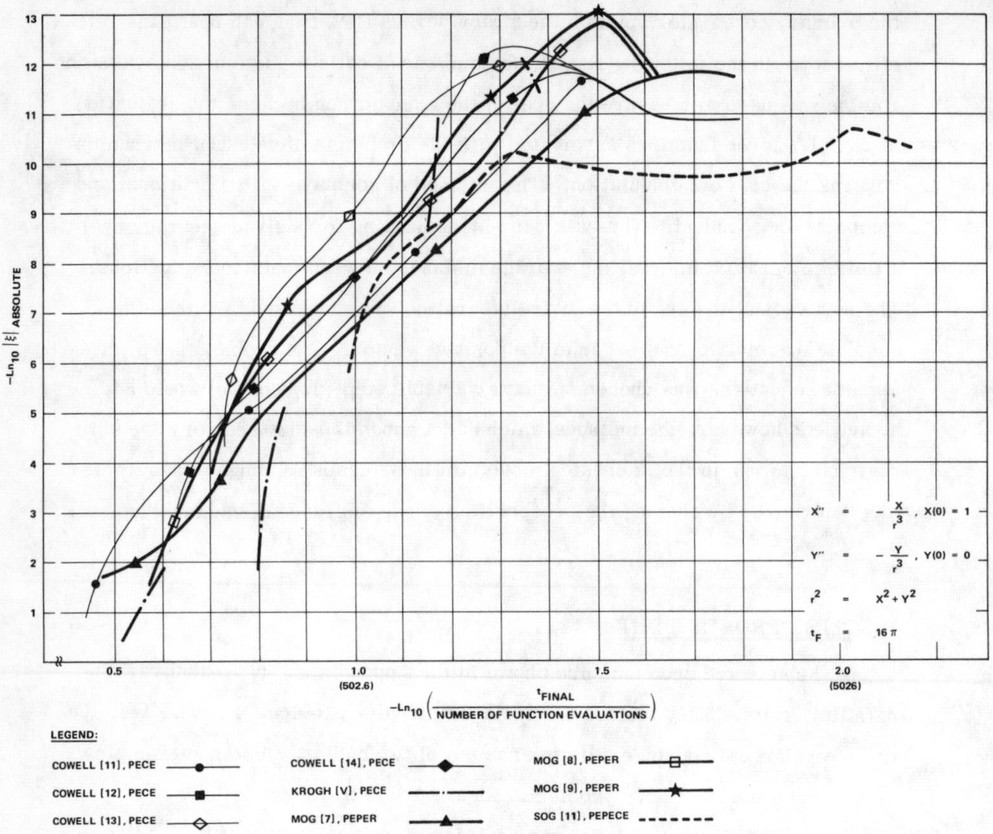

Figure 1. Efficiency Diagram for Coplanar Circular Two-Body Problem for Class II Methods of Integration

will be used: those methods which solve second-order systems of differential equations shall be called Class II methods; those which solve first-order systems shall be called Class I methods.

Figure 1 shows the relationship between the order of integration and numerical stability and accuracy. The accuracy increases with the order of the method, while numerical stability decreases (indicated by a sharp drop in the curve as the step size is increased). Other graphs concerning this problem are shown in the Appendix: Figure 5 illustrates Class II methods with one function evaluation per step, and Figure 6 shows Class I methods.

Figure 2 illustrates the results of the time-regularized coplanar elliptic problem of two bodies as a first-order system of differential equations. The rational extrapolation method of Bulirsch and Stoer proved to be affected by the initial choice of step size, as shown by Fox (1972). But since the revised versions were not available for comparison, several initial step sizes were tried and the best results were plotted.

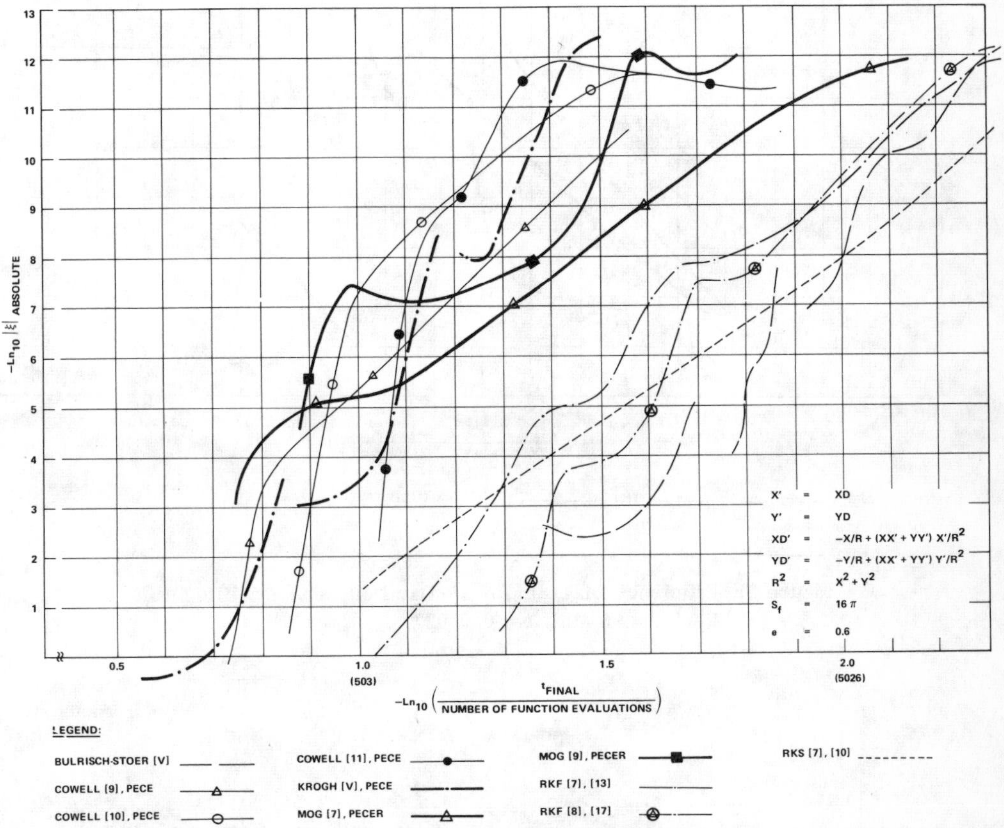

Figure 2. Efficiency Diagram for Time-Regularized Coplanar Elliptic Problem of Two Bodies for Class I Methods of Integration

Figure 3 is an efficiency diagram for a 15-day Delta Pac satellite orbit with period of 96 minutes and small eccentricity. The force model included a 4 x 4 field of the earth's geopotential, and the gravitational attraction of the sun and moon. All methods shown were Class II except the RKF 7.

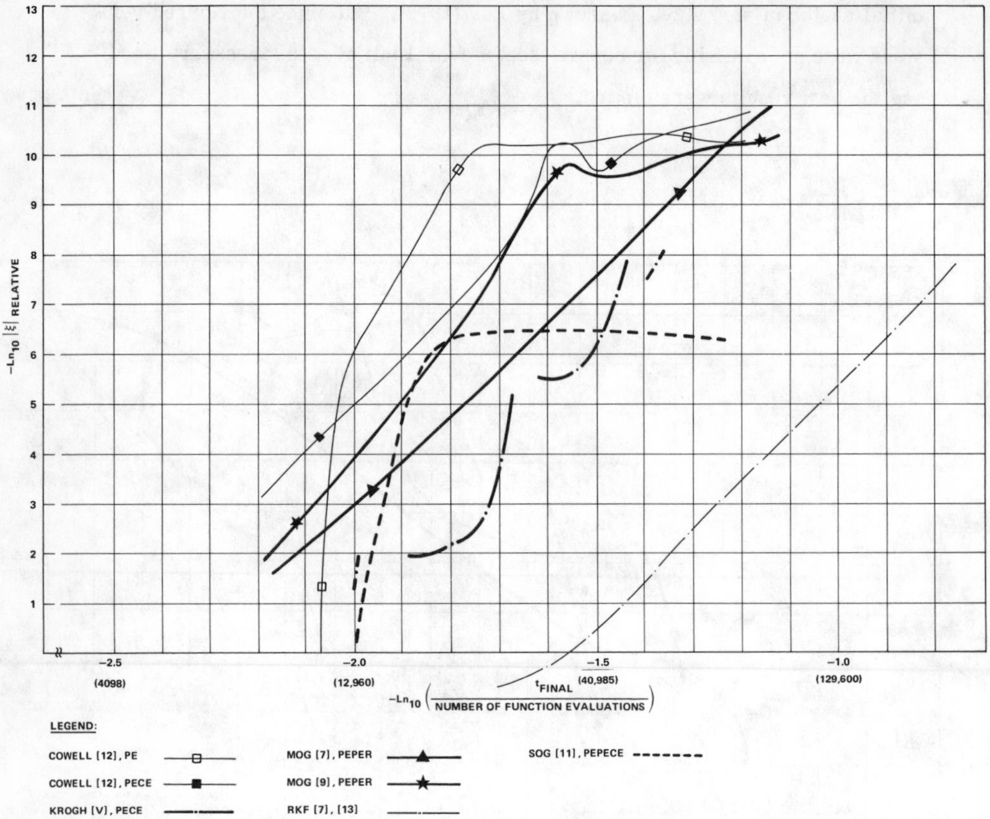

Figure 3. Efficiency Diagram for 15-Day Delta Pac Satellite Orbit
Period = 96 Min, e = 0.00027

To show the importance of the force function evaluation in computer time consumption, Figure 4 illustrates accuracy versus required CPU time for the same problem. Included in this graph is a comparison between a PE algorithm and a PECE* algorithm. The E* denotes a pseudo-evaluate process: the non-two-body effects of the full force function evaluation (denoted by E) are added to the effects of two-body motion. Thus the pseudo-evaluate algorithm evaluates only two-body effects.

Other satellite orbits are shown in the Appendix for different satellite orbits with the same force model.

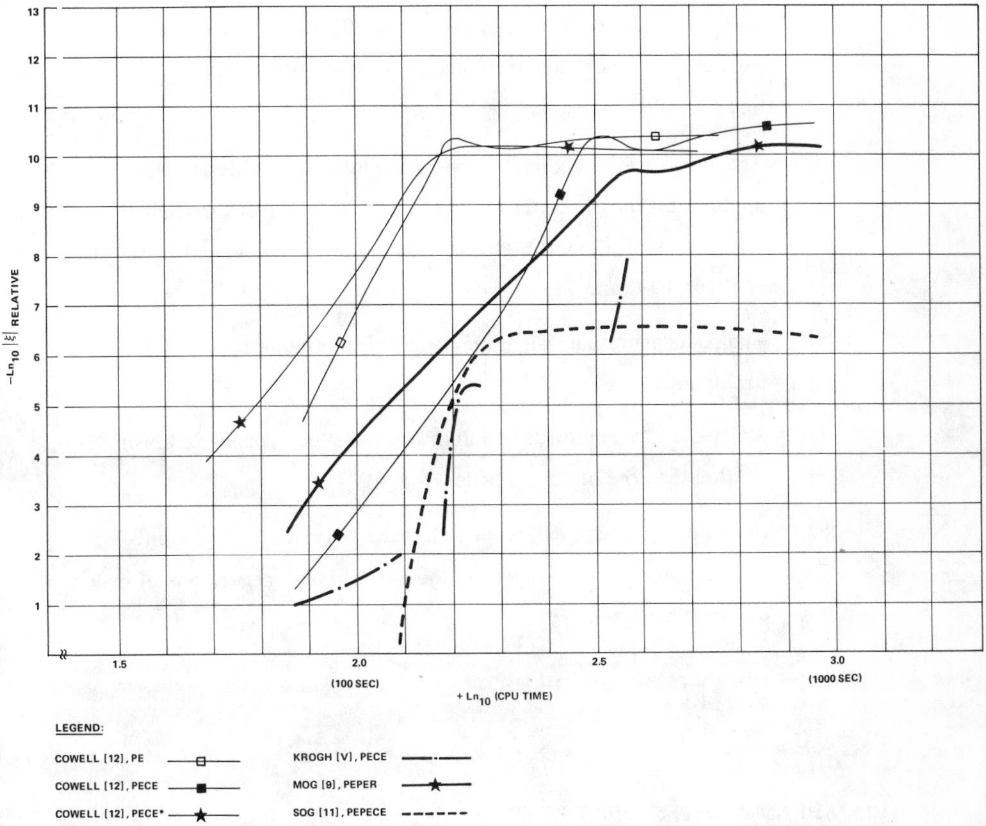

Figure 4. Efficiency Diagram for 15-Day Delta Pac Satellite Orbit

V. DISCUSSION OF TEST RESULTS

In the preceding comparison the generation of starting values for the multistep methods was accomplished using Runge-Kutta integration routines in order to effect a more realistic comparison. The following remarks are presented as a summary of the test results.

1. The rational extrapolation method of Bulirsch and Stoer is very efficient when solving a problem in which the functions vary slowly.

2. When force function evaluations are costly, the algorithms that limited the number of evaluations per integration step were most efficient.

3. Class II methods were more efficient than Class I methods.

4. For the same order of integration, Multi-Off-Grid methods of integration were more accurate and less stable than classical Adams and Cowell techniques.

5. The PE and PEC algorithms were slightly more efficient when applied in Class II methods, but the PECE and PECEC algorithms were better in Class I methods because of the numerical instability of Class I PE and PEC algorithms.

6. The pseudo-evaluate algorithm was very efficient for the satellite problems tested.

7. Multi-Off-Grid methods of integration are less round-off error limited than the other methods.

8. Conjecture: There exists an efficiency "barrier" that the envelopes of all methods approach within the same class, regardless of order or algorithm.

VI. APPENDIX - TEST PROBLEMS

The following problem set was chosen to illustrate the efficiency of numerical integrators applied to the solution of differential equations of orbital nature. The first three problems are variations of the two-body orbit; the next three represent perturbed motion for satellite orbits.

Preliminary Problem Set

Conic equations for elliptical orbit are given by

$$\ddot{\overline{Y}} = -\frac{\mu \overline{Y}}{r^3}$$

Using the transform

$$R = \frac{r}{a}$$

and

$$\tau = \sqrt{\frac{\mu}{a^3}}\ t$$

these equations become

$$\bar{Y}'' = -\frac{\bar{Y}}{R^3}\ ,\ R^2 = Y_1^{\ 2} + Y_2^{\ 2}$$

for coplanar case.

Solution:

$$Y_1 = \cos E - e$$

$$Y_2 = \left(1 - e^2\right)^{1/2} \sin E$$

where E is solution to Kepler's equation $\tau = E - e \sin E$, and e is the eccentricity of the orbit, and τ , the independent variable, is the mean anomaly nt .

$$Y_1' = \frac{1}{1 - e \cos E} (-\sin E) = \frac{-\sin E}{R}$$

$$Y_2' = \frac{1}{1 - e \cos E} \left(1 - e^2\right)^{1/2} \cos E = \frac{\left(1 - e^2\right)^{1/2} \cos E}{R}$$

$$T = 2\pi$$

<u>Test Problem A</u>

Coplanar Circular Two-Body Motion as a System of Second Order Equations.

$$Y_1'' = -Y_1/R^3$$

$$R^2 = Y_1^{\ 2} + Y_2^{\ 2}$$

$$Y_2'' = -Y_2/R^3$$

Initial Conditions:

$$Y_1(0) = 1$$

$$Y_2(0) = 0$$

$$Y_1'(0) = 0$$

$$Y_2'(0) = 1$$

Integration Interval:

$$[0\ ,\ 16\pi]$$

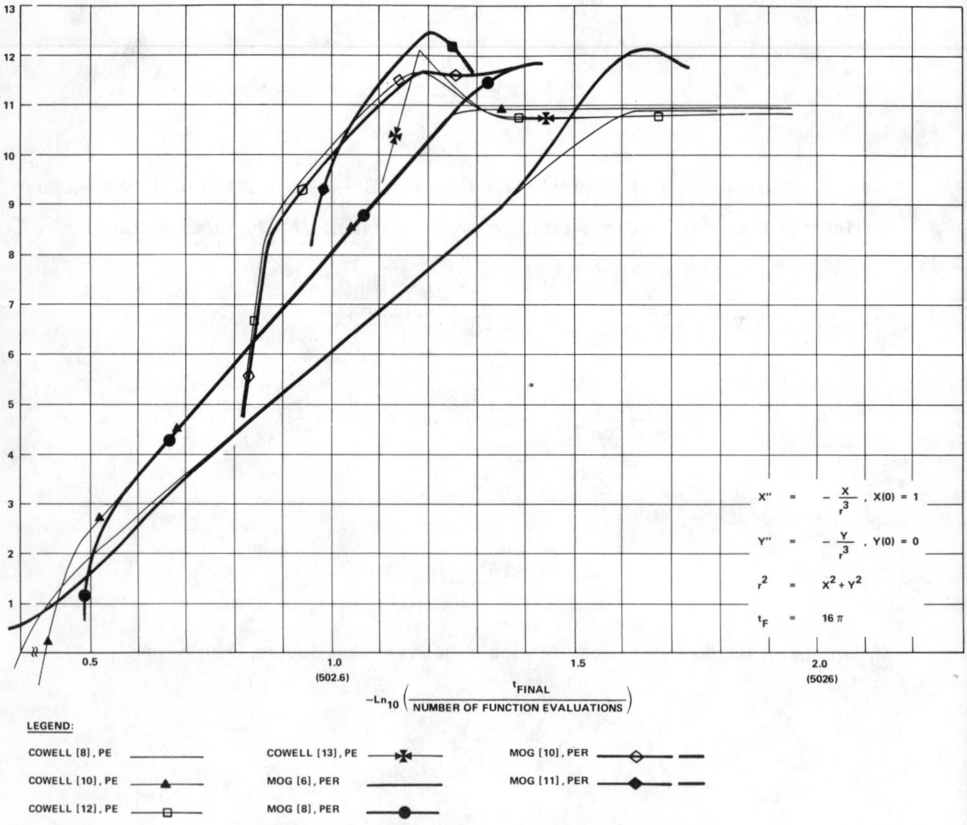

Figure 5. Efficiency Diagram for Coplanar Circular Two-Body Problem for Methods With One Function Evaluation Per Integration Step

Solution:

$$Y_1 = \cos(\tau) = 1 \Big|$$
$$Y_2 = \sin(\tau) = 0 \Big|_{16\pi}$$

Source: Fred T. Krogh

Remarks: Independent variable τ reduces to eccentric anomaly E for circular orbits. Only positions Y_1 and Y_2 were used to determine the rms of the absolute error $|\bar{Y}_{computed} - \bar{Y}_{exact}|$.

Test Problem B

Same as A, as a system of first-order differential equations.

$$Y_1' = Y_3$$
$$Y_2' = Y_4$$
$$Y_3' = -Y_3/R^3$$
$$Y_4' = -Y_4/R^3$$
$$R^2 = Y_1^2 + Y_2^2$$

Initial Conditions:

$$Y_1(0) = 1$$
$$Y_2(0) = 0$$
$$Y_3(0) = 0$$
$$Y_4(0) = 1$$

Integration Interval:

$$[0, 16\pi]$$

Solution:

$$\left. \begin{aligned} Y_1 &= \cos(\tau) &= 1 \\ Y_2 &= \sin(\tau) &= 0 \\ Y_3 &= -\sin(\tau) &= 0 \\ Y_4 &= \cos(\tau) &= 1 \end{aligned} \right|_{16\pi}$$

Remarks: The absolute error was computed.

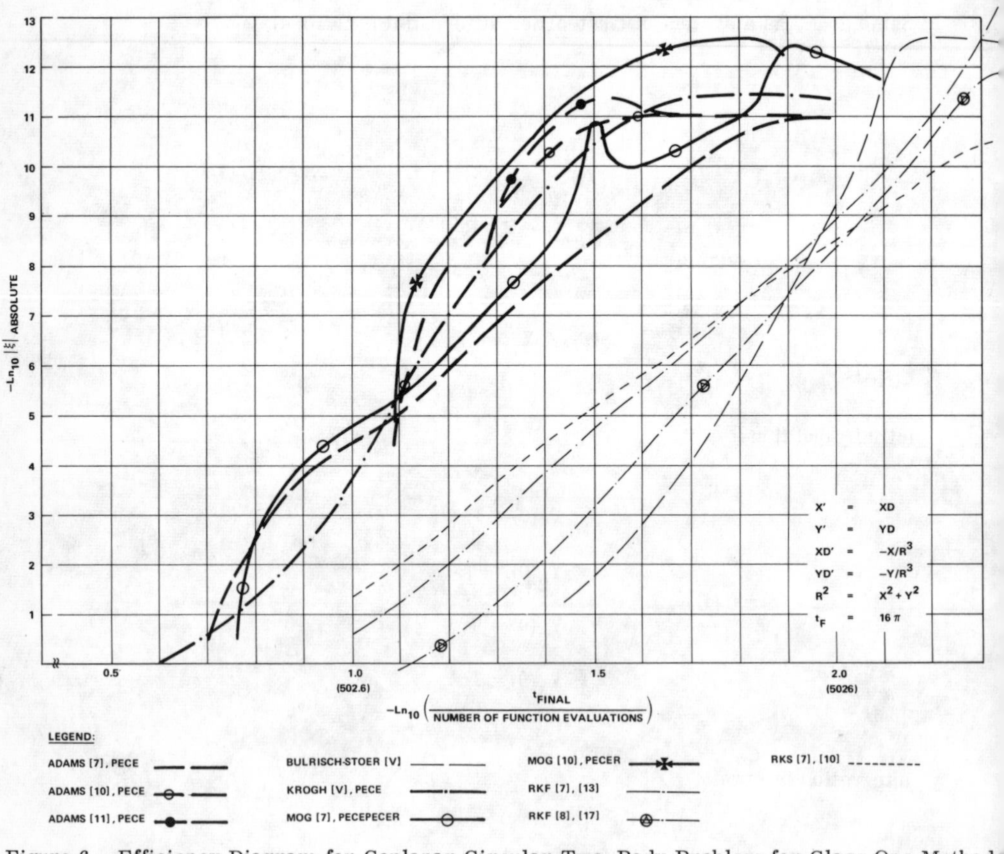

Figure 6. Efficiency Diagram for Coplanar Circular Two-Body Problem for Class One Methods

Test Problem C

Time-Regularized Coplanar Elliptic Two-Body Problem.

The relation $d\tau/ds = R^n$ regularizes the elliptic equations of motion

$$\bar{Y}^{**} = -\bar{Y}/R^{3-2n} + \frac{n\,(\bar{Y}\cdot\bar{Y}^*)\,\bar{Y}^*}{R^2}$$

where $(^*)$ denotes $\left(\frac{d}{ds}\right)$.

If $n = 1$, the first-order differential equations are

$$Y_1^* = Y_3$$

$$Y_2^* = Y_4$$

$$Y_3^* = -\frac{Y_1}{R} + \frac{\left(Y_1 Y_3 + Y_2 Y_4\right)}{R^2} Y_3$$

$$Y_4^* = -\frac{Y_2}{R} + \frac{\left(Y_1 Y_3 + Y_2 Y_4\right)}{R^2} Y_4$$

$$R^2 = Y_1^2 + Y_2^2$$

Initial Conditions:

$$Y_1(0) = \cos\left(E_0\right) - e = 1 - e$$

$$Y_2(0) = \left(1 - e^2\right)^{1/2} \sin E_0 = 0$$

$$Y_3(0) = \left[\left(-\sin E_0\right)/\left(1 - e\cos E_0\right)\right] R_0^{1/2} = 0$$

$$Y_4(0) = \left[\left(\left(1 - e^2\right)^{1/2} \cos E_0\right)/\left(1 - e\cos E_0\right)\right] R_0^{1/2} = \left(1 - e^2\right)^{1/2}$$

For $e = 0.6$

$Y_1(0) = 0.4$

$Y_2(0) = 0$

$Y_3(0) = 0$

$Y_4(0) = (0.0975)^{1/2}$

Integration Interval:

$$[0\ ,\ 16\pi \text{ (8 revolutions)}]$$

Solution:

$$\left.\begin{aligned} Y_1 &= \cos(E) - e = 1 - e \\ Y_2 &= \left(1 - e^2\right)^{1/2} \sin(E) = 0 \end{aligned}\right|_{16\pi}$$

Source: J. Baumgarte

Remarks: Independent variable is the eccentric anomaly E .

Test Problem D

Perturbed Satellite Problem .

Force Model:

Includes a 4 × 4 geopotential and the gravitational attraction of the sun and moon.

Integration Interval:

$$[0\ ,\ 15 \text{ days}]$$

Problem 1:

DELTA-PAC Low Altitude Satellite

period ≐ 2 hours

eccentricity ≐ 0.00027

Problem 2:

ATS-1 Geosynchronous Satellite

period ≐ 24 hrs

eccentricity ≐ 0.0003

Problem 3:

GEOS-B Low Altitude Satellite

period $\doteq$ 2 hrs

eccentricity $\doteq$ 0.07

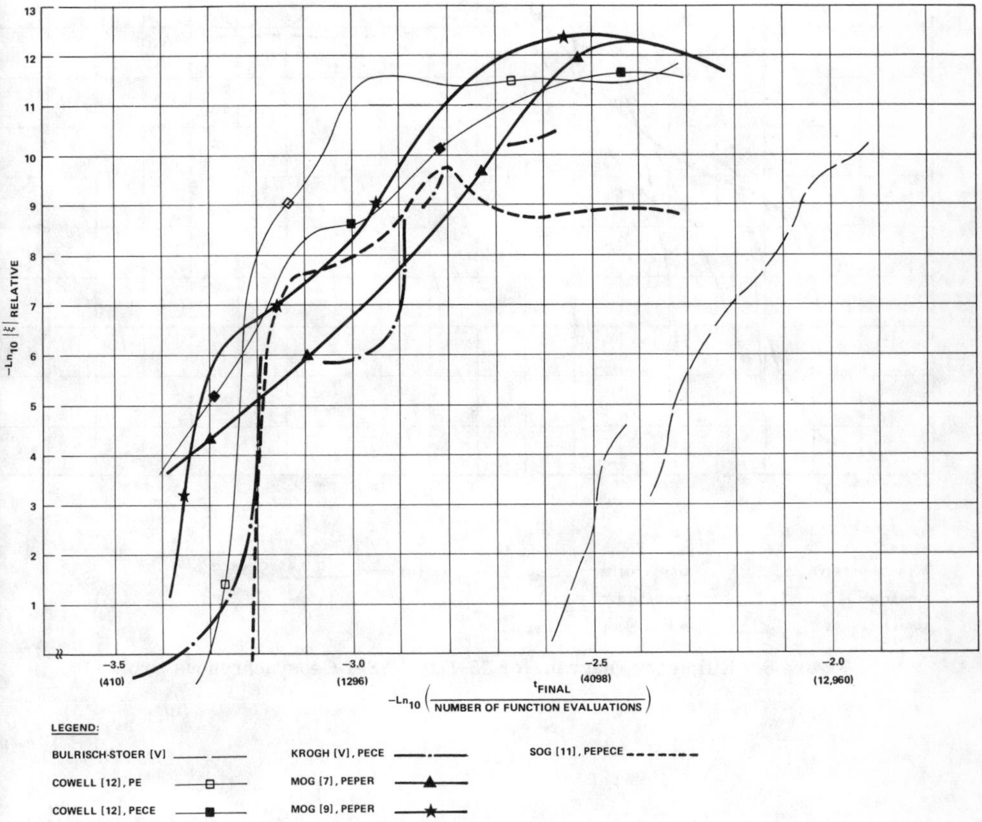

Figure 7. Efficiency Diagram for 15-Day ATS-1 Satellite Orbit
Period = 24 Hrs, e = 0.0003

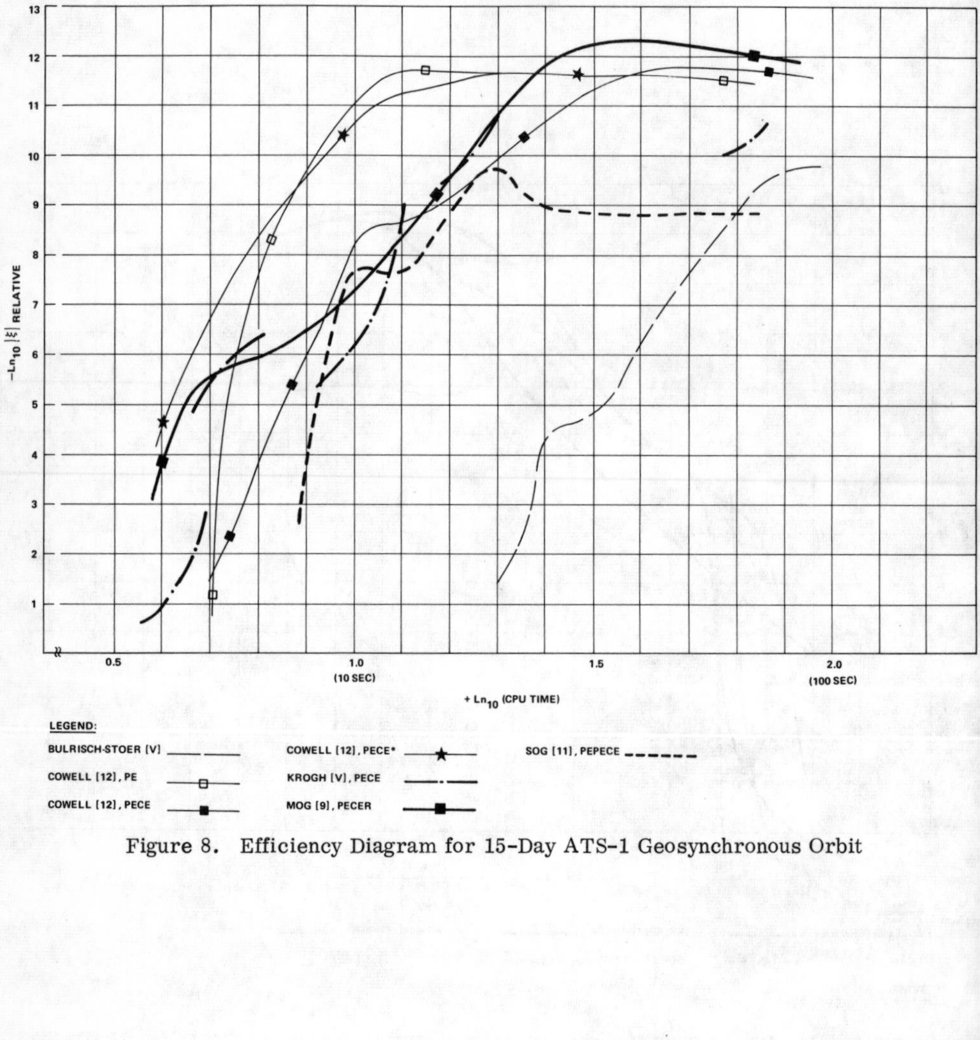

Figure 8. Efficiency Diagram for 15-Day ATS-1 Geosynchronous Orbit

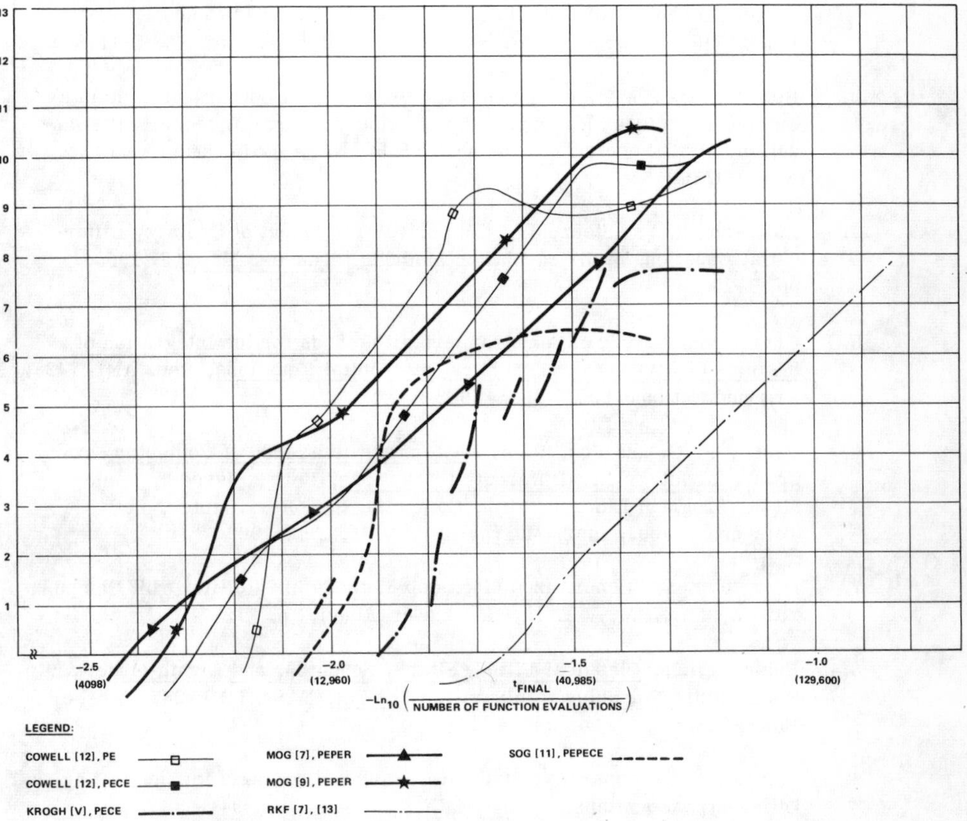

Figure 9. Efficiency Diagram for 15-Day GEOS-B Satellite Orbit
Period = 2 Hrs, e = 0.07

REFERENCES

1. Baumgarte, J., "Numerical Stabilization of the Differential Equations of Keplerian Motion," Celestial Mechanics 5, 1972, pp. 490-501

2. Beaudet, P. R., "Development of Multi-Off-Grid (MOG) Multistep Integration Techniques for Orbital Applications, Volume I Concept," CSC Report 5035-19100-01TR for NASA-GSFC under Contract NAS 5-11790

3. Beaudet, P. R., "Development of Multi-Off-Grid (MOG) Multistep Integration Techniques for Orbital Applications, Volume II Numerical Evaluation," CSC Report 5035-19100-01TR for NASA-GSFC under Contract NAS 5-11790

4. Bulirsch, R. and Stoer, J., "Numerical Treatment of Ordinary Differential Equations by Extrapolation Methods," Numer. Math. 8, 1966, pp. 1-13

5. Clark, N., A Study of Some Numerical Methods for the Integration of Systems of First-Order Ordinary Differential Equations, Rep. ANL-7428, Argonne National Lab., Mar. 1968

6. Crane, P. C. and Fox, P. A., A Comparative Study of Computer Programs for Integrating Differential Equations, Vol. 2, Issue 2, Numer. Math. Program Library Proj., Computing Sci. Res. Center, Bell Telephone Labs., Inc., Murray Hill, N.J., Feb. 1969

7. Dyer, James, "Generalized Multistep Methods in Satellite Orbit Computations," J. ACM, Vol. 15, No. 4 (Oct. 1969), pp. 712-719

8. Fehlberg, E., Classical Fifth-, Sixth-, Seventh-, and Eighth-Order Runge-Kutta Formulas with Stepsize Control, NASA TR R-287, October 1968

9. Fox, P., "A Comparative Study of Computer Programs for Integrating Differential Equations," Num. Math., 15, 1972, pp. 941-948

10. Gragg, W. B. and Stetter, H. J., "Generalized Multistep Predictor-Corrector Methods," J. ACM 11, 2, Apr. 1964, pp. 188-209

11. Krogh, F. T., On Testing a Subroutine for the Numerical Integration of Ordinary Differential Equations, JPL Internal Document, Section 814, Technical Memorandum No. 217, Jet Propulsion Laboratory, Pasadena, Calif., October 1970

12. Krogh, F. T., "A Variable Step Variable Order Multistep Method for the Numerical Solution of Ordinary Differential Equations," Information Processing 68 (Proceedings of the IFIP Congress 1968), North Holland Publishing Co., Amsterdam, 1969, pp. 194-199

NUMERICAL INTEGRATION ASPECTS OF A NUTRIENT UTILIZATION ECOLOGICAL PROBLEM*

by

D.B. Frazho[1], W.F. Powers[2], and R.P. Canale[3]
The University of Michigan

Introduction

The initial motivation for this study was to identify potentially useful interactions among technologists from the traditional fields of aerospace and civil engineering. The particular problem area selected for study was the application of optimal control theory to the modeling and active control of phytoplankton succession in lakes. However, in the early stages of the study it was realized that a careful numerical analysis of certain standard simulation procedures in ecological modeling was in order.

In this paper we shall discuss first the development of the governing differential equations, and then present results of the numerical studies. Details of other aspects of this study, e.g., the application of optimal control to the problems are presented in Ref. 1.

1. Development of the Differential Equations

The derivation of the equations which describe the transient and spatial behavior of phytoplankton and nutrient concentrations in natural waters requires the utilization of mass continuity relations. In general, partial differential equations result which account for the physical transport of the species as well as growth or substrate depletion. However, if it is assumed that no concentration

*This research was supported by The Institute for Environmental Quality, NSF Grant GK-30115, and NOAA Office of Sea Grant #04-3-158-23.

[1] Graduate Student, Aerospace Engineering, The University of Michigan

[2] Associate Professor, Aerospace Engineering, The University of Michigan

[3] Associate Professor, Civil Engineering, The University of Michigan

gradients exist in the natural water body it is possible to describe the system state by a relatively simple set of ordinary differential equations. Such a situation in nature would be approximated by a shallow nonstratified lake with high levels of mixing due to wind and other disturbances. In this case the continuity of species k is defined by Equation 1 below:

$$\frac{d(C_k V)}{dt} = \sum_i A_i J_{i,k} + VS_k \tag{1}$$

where V is the volume of the lake

C_k is the concentration of species k

A_i is the ith area normal to the $J_{i,k}$ flux

$J_{i,k}$ is the ith flux of k

S_k is the summation of internal sources and sinks of k

If the flux of nutrients and organisms to a constant volume lake is the result of feeder streams alone, then Equation 1 can be written as

$$\frac{dC_k}{dt} = \frac{Q}{V} C_{k,I} - \frac{Q}{V} C_k + S_k \tag{2}$$

where $C_{k,I}$ is the weighted average concentration of k in the influent streams,

and Q is the total volumetric flow rate of influent and effluent streams

The lake whose behavior is defined by Equation 2 is illustrated in Figure 1.

The source-sink terms for the phytoplankton include: growth, respiration, excretion, sinking, and uptake by predators. Source-sink mechanisms for algal nutrients such as phosphorus, nitrogen and silica include: uptake by phytoplankton, adsorption, release by phytoplankton, release by organisms in higher trophic levels, and uptake and regeneration by heterotrophic organisms. Although the above factors can be included in lake ecosystem models, it is advantageous for us to consider a somewhat simplified case. Consider a eutrophic lake with dominant phytoplankton assemblages limited to the green and blue-green forms. If the species growth is limited by the supply of nitrogen and phosphorus, the

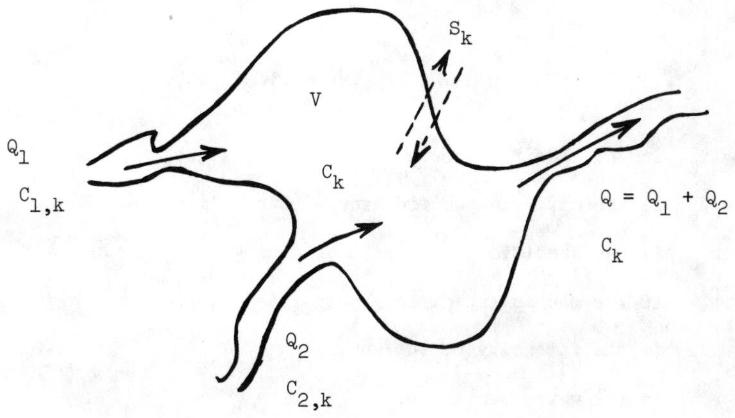

(a)

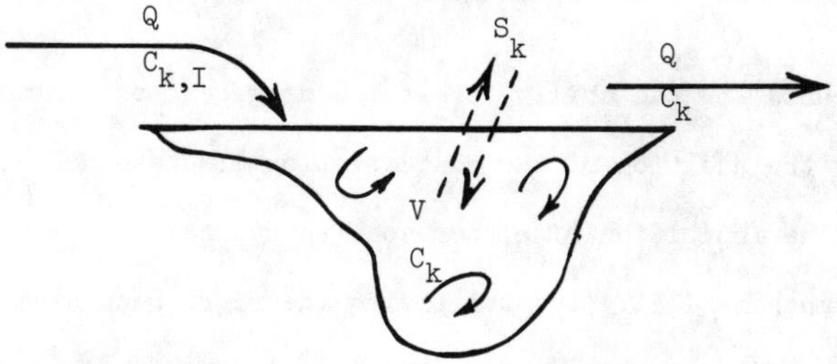

(b)

Figure 1. Homogeneous System in both Longitudinal and Lateral Direction (a) and Vertical Direction (b)

specific growth rate for either species may be written in the form[2]:

$$\mu_G = \hat{\mu}_G(I,T)\left(\frac{P}{K_P+P}\right)\left(\frac{N}{K_n+N}\right) \tag{3}$$

$$\mu_{BG} = \hat{\mu}_{BG}(I,T)\left(\frac{P}{K_P+P}\right)\left(\frac{N+n}{K_n+N+n}\right) \tag{4}$$

where μ_G is the specific growth rate of green algae

$\hat{\mu}_G$ is the maximum specific growth rate of green algae

μ_{BG} is the specific growth rate of blue-green algae

$\hat{\mu}_{BG}$ is the maximum specific growth rate of blue-green algae

I is the intensity of the sunlight

T is the water temperature

P is the concentration of dissolved inorganic phosphorus

N is the concentration of dissolved inorganic nitrogen

n is the concentration of dissolved nitrogen gas

K_P is the half-saturation constant for phosphorus

K_n is the half-saturation constant for nitrogen

Equation 4 accounts for the ability of some blue-green species to fix dissolved nitrogen gas by the addition of the concentration of dissolved nitrogen in the kinetic term. The inhibition of either species by the other species or by itslef can be defined mathematically by mutliplying the right-hand side of Equation 3 or 4 by an inhibition function, u(t). The inhibition function may be considered as a control function, and such a case is considered in Ref. 1. However, it will not be included here since our main interest is in standard digital simulations of the model.

As shown in the above equations the maximum specific growth rates of the green and blue-green algae vary with temperature and sunlight intensity. Problems to be considered in subsequent sections of the paper will be concerned with seasonal variations of plankton and nutrients. Thus it is appropriate to assume that I is constant and that changes in temperature occur on a seasonal basis. Figure 2 shows the assumed functions for $\hat{\mu}_G(T)$ and $\hat{\mu}_{BG}(T)$ with an optimal temperature of 14°C for the greens and 20°C for the blue-greens. Figure 3 shows the assumed T(t) function.

The disappearance of phytoplankton is represented by negative or sink terms in the continuity equations. The various phenomena causing depletion can be collectively characterized by first order kinetic decay terms. The rate coefficient for the green algae is r_G while the decay coefficient for the blue-green algae is r_{BG}.

The transient behavior of the phytoplankton populations is the result of source, sink, and washout effects, and is defined in differential equation form by Equations 5 and 6 below for the case when the input streams are devoid of organisms,

$$\frac{dA_G}{dt} = \hat{\mu}_G(I,T)\left(\frac{P}{K_P+P}\right)\left(\frac{N}{K_n+N}\right)A_G - \frac{1}{\tau}A_G - r_G A_G \tag{5}$$

$$\frac{dA_{BG}}{dt} = \hat{\mu}_{BG}(I,T)\left(\frac{P}{K_P+P}\right)\left(\frac{N+n}{K_n+N+n}\right)A_{BG} - \frac{1}{\tau}A_{BG} - r_{BG}A_{BG} \tag{6}$$

where A_G is the concentration of green algae

A_{BG} is the concentration of blue-green algae

τ is Q/V

If it is assumed that a unit of plankton production is accompanied by a fixed amount of nitrogen and phosphorus depletion, the nutrient transient behavior is defined in differential equation form by Equations 7 and 8.

$$\frac{dP}{dt} = P_I - \frac{P}{\tau} - \frac{1}{Y}\left(\frac{P}{K_P+P}\right)\left[\hat{\mu}_G\left(\frac{N}{K_n+N}\right)A_G + \hat{\mu}_{BG}\left(\frac{N+n}{K_n+N+n}\right)A_{BG}\right] \tag{7}$$

$$\frac{dN}{dt} = N_I - \frac{N}{\tau} - \frac{1}{W}\left(\frac{P}{K_P+P}\right)\left[\hat{\mu}_G\left(\frac{N}{K_n+N}\right)A_G + \hat{\mu}_{BG}\left(\frac{N}{K_n+N}\right)A_{BG}\right] \tag{8}$$

where Y is the yield of either green or blue-green algae per unit of phosphorus uptake

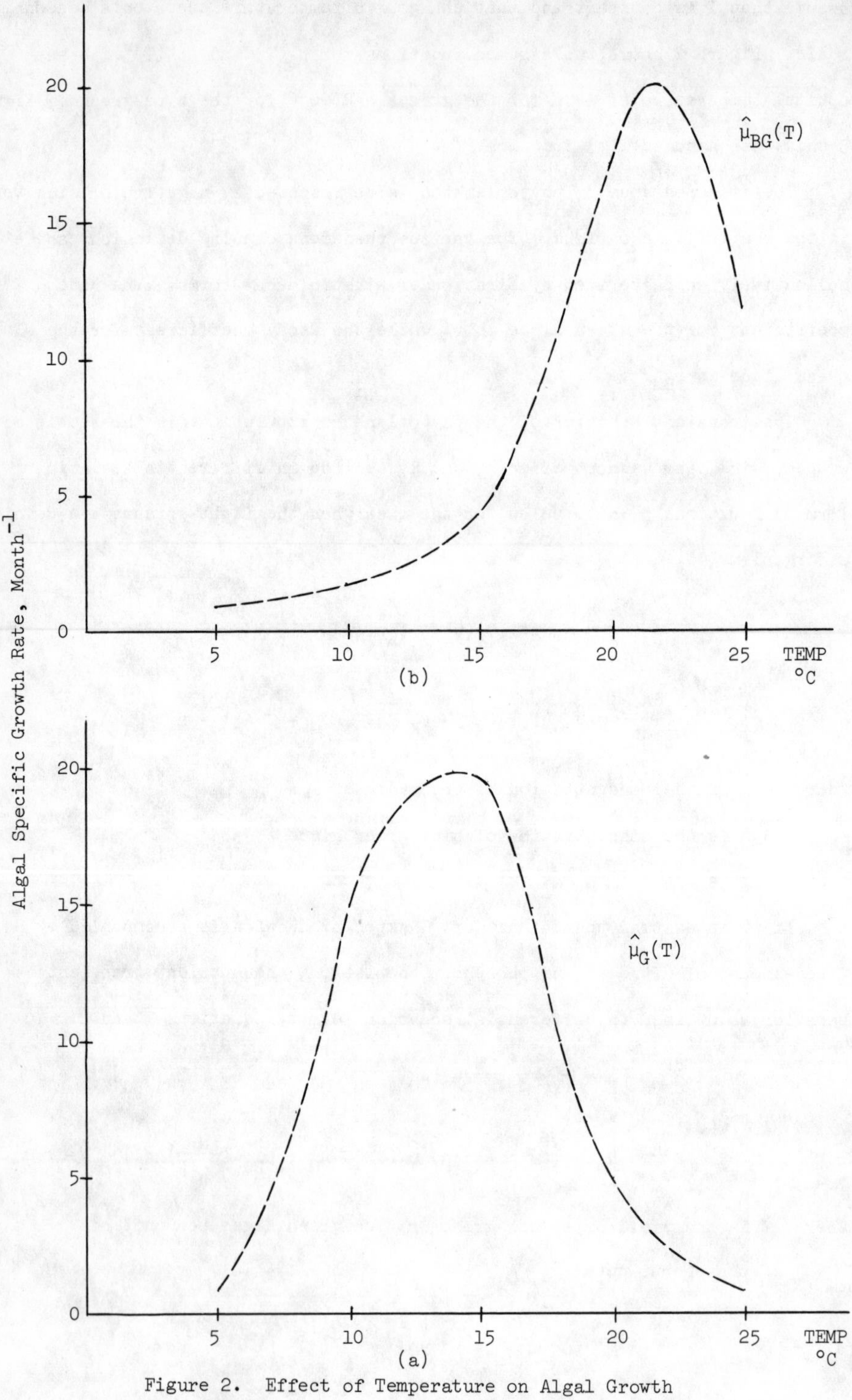

Figure 2. Effect of Temperature on Algal Growth

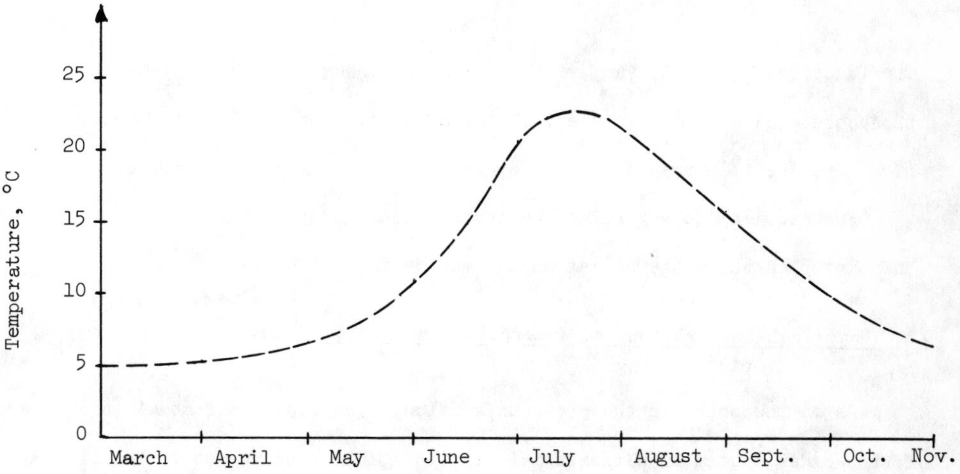

Figure 3. Annual Temperature Cycle

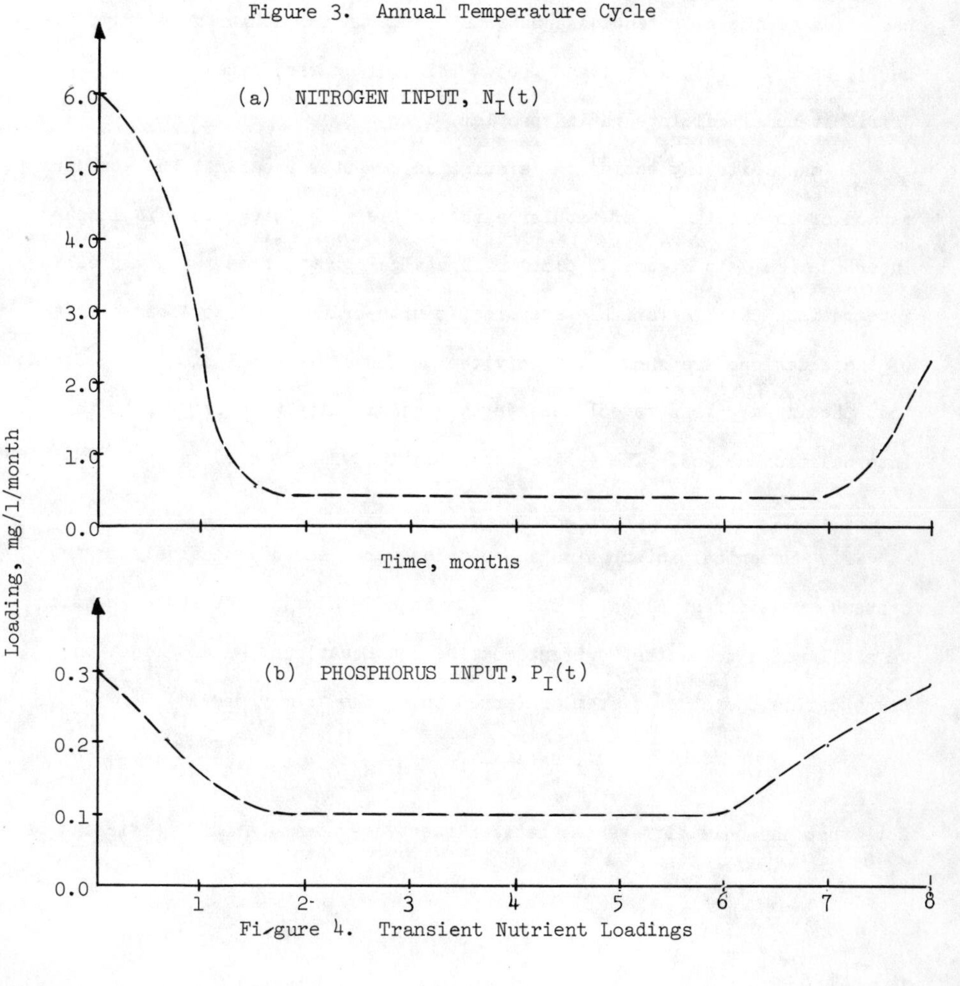

Figure 4. Transient Nutrient Loadings

W is the yield of either green or blue-green algae per unit of nitrogen uptake

The functions P_I and N_I vary with time and represent the loadings of phosphorus and nitrogen in the influent streams. The particular functions used here are illustrated in Figure 4 and are typical of many streams.

Further details regarding the development and application of nutrient and phytoplankton mathematical models may be found in Refs. 3, 4, 5.

2. Numerical Integration and Interpolation Results

As may be noted in the previous section, even though the model is only a fourth-order system of differential equations, numerous inputs to the right-hand side of the differential equations are known in tabular form only, i.e., $\hat{\mu}_G(T)$, $\hat{\mu}_B(T)$, $P_I(t)$, $N_I(t)$, and $T(t)$. This, of course, causes a more difficult numerical integration problem.

In many existing ecological simulation computer programs, the standard method of interpolation of tabular data for use with integrators is linear interpolation. In Figure 5, plots of the algal growth from two numerical integrations (with a variable stepsize, fourth-order predictor-connector) of the equations are shown with only the method of interpolation changed. Note that the integrations reveal considerable sensitivity with respect to the interpolation method. The spline interpolation was employed in the simulations of this study.

Five numerical integration subroutines were tested extensively on two systems of differential equations. Equations 5-8 define one system, which we shall refer to as the "current model", and Equations 4-7 and Equation 9 below define the other system, referred to as the "early model".

$$\frac{dN}{dt} = N_I - \frac{N}{\tau} - \frac{1}{W}\left(\frac{P}{K_P+P}\right)\left[\hat{\mu}_G\left(\frac{N}{K_n+N}\right)A_G + \hat{\mu}_{BG}\left(\frac{N+n}{K_n+N+n}\right)A_{GB}\right] \qquad (9)$$

Note that the only difference between the two systems is the coefficient of

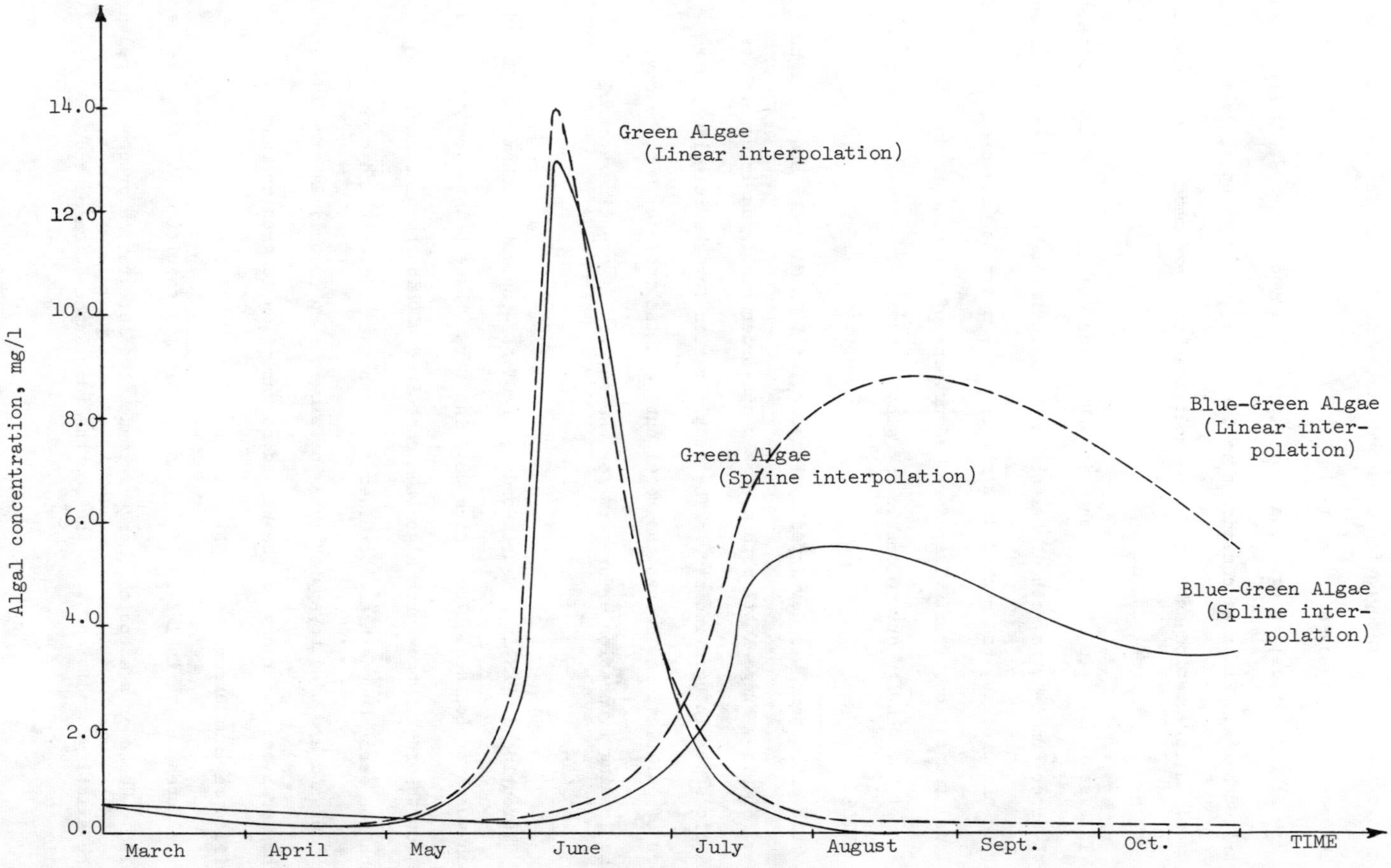

Figure 5. Effect of Interpolation Scheme

A_{BG} in Equations 8 and 9. We shall find that this one small difference has interesting numerical integration consequences. The reasons for the differences in the model are discussed in detail in Ref. 1.

The five numerical integration subroutines employed in the study are summarized below. Since we consider ourselves users, as opposed to developers, of numerical integration subroutines no modifications were made to the routines.

(1) DRKGS[6]

This subroutine is IBM's scientific subroutine package variable stepsize, fourth-order Runge-Kutta method. Although the stepsize is changed easily at any point in the process, the method does not have the means to conveniently estimate the local truncation error. Therefore, adjustment of stepsize is done by comparison of the results due to double and single stepsize.

(2) RK713[7]

This subroutine was developed by Fehlberg of NASA Marshall Space Flight Center. It is a Runge-Kutta method of seventh-order requiring 13 derivative evaluations per integration step. The justification for the relatively high number of calculations is that the resultant formulas include complete coverage of the leading truncation error term for the purpose of stepsize control.

(3) STEPER[8]

Subroutine STEPER is a fourth-order, predictor-corrector scheme developed by Schwausch of Lockhead Electronics Co. The first three points are generated by a fourth-order Runge-Kutta algorithm contained within the subroutine. Truncation error bounds are determined for both the predictor and corrector equations and are used to approximate the single step (local) integration error. If the local error for any of the dependent variables is greater than the specified maximum error the stepsize is halved. If the single step error for all the dependent variables is less than the minimum specified the stepsize is doubled. Whenever a stepsize change is made, the integration process is restarted at the existing conditions with the fourth-order Runge-Kutta method.

(4) DVDQ[9]

Subroutine DVDQ was developed by Krogh of the Jet Propulsion Laboratory. DVDQ is a variable stepsize and variable order predictor-corrector method. The subroutine utilizes Adams type predictor-corrector formulas with orders one through seventeen. Krogh notes that the algorithm was heavily influenced by numerical experimentation. He also points out that although the overhead required by the integrator is considerably greater than most, the integrator requires a near minimum of derivative evaluations.

(5) BULIRSCH-STOER EXTRAPOLATION[10]

The particular Bulirsch-Stoer extrapolation scheme used was Gear's DIFSUB[11] rational function extrapolation version. A discussion of the scheme is presented in Chapter 6 of Ref. 11. This method was not considered until late in the study so that we do not have extensive experience with it. However, on the controlled test problems it performed exceptionally well, as will be discussed later.

In the initial portion of the study, extensive simulations were made with the "early model". However, physically impossible negative nitrogen values for reasonable parameter values resulted. By the process described in Ref. 1, a new model in which nitrogen remains positive was proposed and this is the model of Equations 5-8. Although the two models differ by only one term in one of the four equations, their numerical integration properties are quite different when the standard fourth-order Runge-Kutta and predictor-corrector integrators are employed.

Subroutines 1-5 described above were tested extensively by integrating both systems of differential equations with various inputs and initial conditions. In Table (1) the numerical integration results of simulations for each model are presented, and these results are representative of the many simulations. All subroutines were given the same input data, local error tolerances, etc., and all simulations agreed to roughly five significant figures.

The results presented in Table (1) indicate that RK713 gives the best per-

TABLE 1

Method	Early Model		Current Model	
	CPU Time	# Function Evaluations	CPU Time	# Function Evaluations
RK713*	3.526	1053	3.368	882
DVDQ	5.245	1286	7.830	1392
EXTRAPOLATION+	6.895	2634	4.871	1770
STEPER	10.426	2115	5.542	1257
DRKGS	11.823	2826	5.683	1330

*This RK713 subroutine was supplied to us recently by H.L. Ingram of NASA-MSFC and worked considerably better than an earlier version of RK713.
+Bulirsch and Stoer have recently improved this scheme, but the new subroutine was not employed in this study

NUMERICAL INTEGRATOR COMPARISONS
(See Appendix for parameter values and initial conditions.)

formance with respect to both computer time and number of function evaluations. An earlier version of RK713 did not perform near as well, so DVDQ was used in our optimization studies. Further simulations of the new version of RK713 will be carried out before a switch from DVDQ is made.

Other interesting results from the table are: 1) The Bulirsch-Stoer rational function extrapolation scheme performed well on each example, and with the recent modifications noted by the developers, it may be the best general purpose algorithm with regard to overhead, simplicity of use, and computer time. 2) The classical fourth-order Runge-Kutta and predictor-corrector schemes were the most sensitive to the problem change. Their computation times are competitive with the "current model" but were not with the "early model". 3) As noted previously, we consider ourselves as users and, thus, CPU time is our critical factor. Note that the extrapolation integration scheme usually required a considerably larger number of function evaluations than the other schemes, but was second and third (out of five) with respect to least CPU time.

A final comment about ease of use of the integrators is in order. We have used DVDQ on problems in aerospace, economics, and physiology, and have had little trouble getting the scheme to work. The error specification is straightforward, and the CPU time has on the average been better by a factor of two than the corresponding CPU time for DRKGS. Thus, eventhough on the "current model" DVDQ is the worst of the five integrators, its CPU time is not so excessive that a change will be made without considerable thought.

Conclusions

A model for nutrient utilization by two species of algae has been introduced and various aspects of its digital simulation discussed. After showing a sensitivity to the method of interpolation of the numerous tabular functions, a study of five numerical integration subroutines was conducted. It was found that Fehlberg's RK713 was the best overall performer on two versions of the model while Krogh's DVDQ and Bulirsch-Stoer's extrapolation gave acceptable performance.

The classical fourth-order Runge-Kutta and predictor-corrector routines were more problem dependent, with the CPU time being different by a factor of two for the two models.

Currently we are also analyzing the numerical intergration subroutines on a Space Shuttle reentry problem. An interesting trend is that if the desired number of significant digits is relatively small (e.g., 3 to 4 digits) then RK713 gives the least CPU time. However, as the accuracy is tightened (e.g., to 8 or more digits), DVDQ becomes the best performer with respect to CPU time. In the study of this paper a relatively small number of significant digits (4 to 5 digits) were required, and the results agree with the trend noted above.

APPENDIX

PARAMETER AND INITIAL VALUES

Initial Values*	Parameter Values**	
$A_G = .5$ mg/l	$\tau = 2$ months	$r_G = 2$ month^{-1}
$A_{BG} = .5$ mg/l	$K_p = .03$ mg/l	$r_{BG} = .5$ month^{-1}
$P = .05$ mg/l	$K_n = .25$ mg/l	$Y = 100 \frac{\text{mg algae}}{\text{mg P}}$
$N = 1$ mg/l	$n = .25$ mg/l	$W = 12 \frac{\text{mg algae}}{\text{mg N}}$

*Initial values at t = 0 are assumed to occur on March 1

**The parameter values are order of magnitude estimates only and are not intended to be characteristic of a specific lake or algal population.

REFERENCES

1. Canale, R.P., Frazho, D.B., and Powers, W.F., "Application of Optimal Control Theory in Modeling Nutrient Utilization and Phytoplankton Production in Lakes", University of Michigan, Report AC-101, February 1973

2. Chen, C.W., "Concepts and Utilities of Ecologic Model", Journal of the Sanitary Engineering Division, Proceedings of the American Society of Civil Engineers, Vol. 96, No. SA5, October 1970

3. Riley, G.A., "Theory of Food-Chain Relations in the Ocean", page 438-463 in Hill, M.N. (editor) The Sea, 2, Interscience, New York, 1963

4. Canale, R.P., "A Methodology for Mathematical Modeling of Biological Production", Report to the University of Michigan Sea Grant, 1970

5. DiToro, D.M., O'Connor, D.J., and Thomann, R.V., "A Dynamic Model of the Phytoplankton Population in the Sacramento-San Joaquin Delta", Non-equilibrium Systems in Natural Water Chemistry, Advances in Chemistry Series 106, 1971

6. IBM System/360, Scientific Subroutine Package, (360A-CM-03X) Version III (1968) IBM Technical Publication Department, DRKGS subroutine

7. Fehlberg, E., "Classical Fifth-, Sixth-, Seventh-, and Eighth-Order Runge-Kutta Formulas with Stepsize Control",NASA TR R-287, October 1968

8. Schwausch, O.A., "A Fortran Subroutine for the Numerical Integration of First Order Ordinary Differential Equations Using Either a Fixed or Variable Integration Step Size", Personal Communication, Lockheed Electronics Company, Houston, Texas, 1968

9. Krogh, F.T., "VODQ/SVDQ/DVDQ - Variable Order Integrators for the Numerical Solution of Ordinary Differential Equations", TU Document No. CP-2308, NPO-11643, JPL, Pasadena, Calif., May 1969

10. Bulirsch, R. and Stoer, J., "Numerical Treatment of Ordinary Differential Equations by Extrapolation Methods", Num. Math, Vol. 8, 1966, pp. 1-13

11. Gear, C.W., Numerical Initial Value Problems in Ordinary Differential Equations, Prentice-Hall, 1971

CALCULATION OF PRECISION SATELLITE ORBITS WITH NONSINGULAR ELEMENTS (VOP FORMULATION)

by

C.E. Velez

NASA/Goddard Space Flight Center

Greenbelt, Maryland

and

P.J. Cefola, A.C. Long, and K.S. Nimitz

Computer Sciences Corporation

Silver Spring, Maryland

Abstract

There is numerical evidence to support the conclusion that appropriate "matching" of the formulation of the equations of motion to the numerical integration method can lead to significant improvement in the accuracy and computational efficiency of the orbit generation process. This paper investigates this possibility from the point of view of matching a "Gaussian" variation of parameter (VOP) formulation with Adams type numerical integrators. The performance of the new orbit generators is then compared to the popular classical Cowell/Gauss-Jackson formulation/integrator pair.

Numerical results indicate that the VOP orbit generator can yield significant efficiency advantages. For example, after 28 days of integration of a synchronous orbit, a comparable accuracy (.5m) was attained with a VOP formulation at one-half the number of force model evaluations needed using the Cowell formulation.

It is suspected that the reasons one can do better with the VOP formulation include more favorable truncation error/stability region balancing. For example, it has been reported that lower order integration processes coupled with VOP are "better than" higher order processes. This supports the view that the VOP formulation results in a smaller stability region with the advantage of lower truncation error due to the more slowly varying parameters. An analytical approach to the study of this type of matching of the equation formulation to the numerical integration technique is given.

Introduction

Accurate prediction of the position and velocity of an artifical satellite after several revolutions about a central body is essential to the major functional modes (detailed mission design and targeting, orbit determination and estimation, and geophysical parameter estimation) of an Orbit Determination System. Since orbit determination programs usually require significant computational resources (CPU and core) to operate, the optimization of such programs has received considerable attention. In the near future, the need for optimizing orbit determination programs will be increased due to two related factors:

1. The availability of high data rate, high accuracy sensors (for example, laser devices)
2. The emphasis on earth physics and the use of satellite tracking data for investigations that require very accurate translation of the mathematical model into "computed values."

This report deals with some results obtained in an effort to develop efficient, high precision trajectory computation processes for artificial satellites by optimum selection of the form of the equations of motion of the satellite and the numerical integration method. Experience has shown that evaluation of the complex perturbing accelerations is the major cost factor in earth trajectory prediction programs, and therefore, an optimum method is defined as one which minimizes the number of perturbative acceleration computations for a given accuracy. The development of optimum methods involves a reformulation of the equations of motion to a form which appears to be more suitable for numerical solution, followed by a careful calibration or matching with a numerical integration process. Key elements of this matching process are local truncation error and the stability region of the integrator.

This report investigates this process for the case of Gaussian variation of parameter (VOP) formulations of the equations of motion. The performance of the resulting orbit generators is then compared to the popular classical Cowell/Gauss-Jackson formulation/integrator pair.

The development of the VOP formulations employed in this optimization study follows certain guidelines for reformulation of the equations of motion given by Deprit (Reference 2), who pointed out that the construction of an ephemeris is actually a twofold procedure. First, the curve in phase space originating from the state $(\vec{r}_o, \dot{\vec{r}}_o)$ that the satellite travels must be located - the "track." Second, the displacement of the satellite along this trajectory must be determined - "along track." Both observational and numerical experience show that the second aspect of this process is the larger source of error in the orbit prediction problem.

Integration of the position and velocity state vectors (Cowell method) gives no clue to the source of this difficulty. In contrast, a formulation of the state equations such that a single "fast" variable describes motion along the orbital path with the remaining state parameters describing the slow variation of the orbital or reference plane should result in an improvement in the accuracy of the numerical integration process because of the decoupling of the errors of the along track variable from the motion of the plane which moves slowly and hence, has small associated truncation errors.

For the numerical portion of this study, an orbit generator has been developed which uses a nonsingular VOP formulation in conjunction with an Adams type integrator. This VOP formulation, which is expressed in terms of equinoctial orbital elements (References 3 and 4), partially decouples the motion of the orbital frame from motion within the orbital frame. In addition, motion along the orbit is isolated in one fast variable. The equinoctial formulation of the equations of motion is discussed in Section 1.

For the purpose of the numerical comparison, both the VOP and Cowell orbit generators were optimized with respect to the order of the integrator and the integration algorithm. In both cases, high-order integrators used in a PECE* (Predict-Evaluate-Correct-Partial Evaluate, Reference 10), algorithm were found to be most efficient over a wide range of accuracies. The optimum order of the integrator was determined by numerical stability limits. These results are discussed in Section 3. The numerical stability limits for the Gauss-Jackson and Adams integrators were analytically determined for the PE, PEC, PECE, and $PE(CE)^2$ algorithms by development of the appropriate characteristic polynomials. Stability limits for the PECE* algorithm were determined numerically. These stability regions are discussed in Section 2.

The performance of the VOP orbit generator has been compared to that of the Cowell orbit generator for two distinctly different orbit types. These results are discussed in Section 3.

1. Analytic Formulation

The Cowell Formulation

A Cowell method of orbit prediction requires the numerical solution of the differential equation

$$\ddot{\vec{R}} + \frac{\mu \vec{r}}{|r|^2} = \vec{Q}\ (t, \vec{r}, \dot{\vec{r}}) \tag{1}$$

where $\vec{r}$ = position vector in some inertial system

$|\vec{r}|$ = magnitude of r

$\dot{\vec{r}}$ = velocity vector

$\ddot{\vec{r}}$ = acceleration vector

Q = perturbing acceleration vector

For near-earth trajectories, the function $\vec{Q}$ generally includes accelerations due to the nonsphericity of the earth, aerodynamic drag and lunar-solar gravitational effects. Since such expressions are in general very complex, (see, for example, Reference 11), the efficiency of most computer programs designed to compute precise earth orbits by the Cowell method will be governed by the number of evaluations of $\vec{Q}$ the numerical integration process used to solve (1) requires.

The Variation of Parameter Formulation

The variation of parameter formulation used in this study is basically the Gaussian form

$$\dot{\vec{\alpha}} = \frac{\partial \vec{\alpha}}{\partial t} + \frac{\partial \vec{\alpha}}{\partial \dot{\vec{x}}} \cdot \vec{Q} \tag{2}$$

where $\vec{\alpha}$ is the vector consisting of nonsingular orbital elements, and $\vec{Q}$ is the perturbing acceleration vector given in (1). This form was pointed out by Dallas (Reference 8) to be particularly suitable to high precision calculations because of the general availability of computer routines which calculate the complex perturbing accelerations in rectangular coordinates and because this form allows a simple interchange of orbital elements for varying orbital geometries. Equation (2) can be derived as follows:

Solutions to the unperturbed problem

$$\ddot{\vec{x}} + \frac{\mu \vec{x}}{|x|^2} = 0$$

can be written as

$$\vec{x} = \vec{x}(\vec{\alpha}|t|)$$

$$\dot{\vec{x}} = \frac{\partial x}{\partial \vec{\alpha}} \cdot \dot{\vec{\alpha}}$$

$$\ddot{\vec{x}} = \frac{\partial \dot{\vec{x}}}{\partial \vec{\alpha}} \cdot \dot{\vec{\alpha}}$$

where $\vec{\alpha}$ is any 6 dimensional set of orbit elements for which the determinant

$$\left| \frac{\partial \vec{\alpha}}{\partial \vec{x}} \quad \frac{\partial \vec{\alpha}}{\partial \dot{\vec{x}}} \right| \neq 0$$

and $\dot{\vec{\alpha}} = \frac{d\vec{\alpha}}{dt}$ for two-body motion. Solutions to the perturbed problem can be expressed as osculating solutions to the unperturbed problem

$$\vec{r}(t) = \vec{x}(\vec{\alpha}(t)) \quad .$$

The unperturbed orbit is taken as a reference orbit in the VOP method. The time rate of change of the orbit elements $\vec{\alpha}$ due to the reference motion is separated out such that in the perturbed case

(3) $$\frac{d\vec{\alpha}}{dt} = \left(\frac{d\vec{\alpha}}{dt}\right)_{\text{two-body}} + \left(\frac{d\vec{\alpha}}{dt}\right)_{\text{perturbed}} = \dot{\vec{\alpha}} + \dot{\vec{\alpha}}'.$$

($\vec{\alpha}'$ is commonly called the grave derivative)

Using these definitions

$$\dot{\vec{r}} = \frac{\partial x}{\partial \vec{\alpha}} \cdot \frac{d\vec{\alpha}}{dt} = \dot{\vec{x}} + \frac{\partial \vec{x}}{\partial \vec{\alpha}} \cdot \vec{\alpha}'$$

$$\ddot{\vec{r}} = \frac{\partial \dot{\vec{r}}}{\partial \vec{\alpha}} \quad \frac{d\vec{\alpha}}{dt} = \ddot{\vec{x}} + \frac{\partial \dot{\vec{x}}}{\partial \vec{\alpha}} \cdot \vec{\alpha}' + \frac{\partial}{\partial \vec{\alpha}}\left[\frac{\partial \vec{x}}{\partial \vec{\alpha}} \cdot \vec{\alpha}'\right] \cdot \frac{d\vec{\alpha}}{dt}$$

To make the problem definite, three additional conditions are chosen

(4) $$\frac{\partial \vec{x}}{\partial \vec{\alpha}} \cdot \vec{\alpha}' = 0, \text{ yielding} \quad \begin{aligned} \dot{\vec{r}} &= \dot{\vec{x}} \\ \ddot{\vec{r}} &= \ddot{\vec{x}} + \frac{\partial \dot{\vec{x}}}{\partial \vec{\alpha}} \cdot \vec{\alpha}' \end{aligned} \quad .$$

Substituting into the perturbed problem gives

$$\ddot{\vec{x}} + \frac{\partial \dot{\vec{x}}}{\partial \vec{\alpha}} \cdot \vec{\alpha}' = \vec{Q} - \frac{\mu \vec{x}}{|x|^3}$$

Using the fact that $\vec{x}$, $\dot{\vec{x}}$ are solutions to the unperturbed problem, gives

(5) $$\frac{\partial \dot{\vec{x}}}{\partial \vec{\alpha}} \cdot \vec{\alpha}' = \vec{Q} \quad .$$

Solving the set of equations (4) and (5) results in the equation

$$\vec{\alpha}' = \frac{\partial \vec{\alpha}}{\partial \dot{\vec{x}}} \cdot \vec{Q}$$

which together with equation (3) yields (2), as required.

For this study, a set of nonsingular elements (equinoctial) for nearly circular and equatorial satellites was used because of the interest in this class of satellites from the point of view of long-term high precision calculations. However, it would be expected that the results of this paper would apply to other orbit types using other nonsingular elements. The equinoctial elements are derived in References 3 and 4. These elements do not result in any singularities for small eccentricities.

In order to avoid singularities due to inclination, two sets of elements are used; a direct set to be used for inclinations between 0 and 90 degrees and a retrograde set for inclinations between 90 and 180 degrees.

The equations which follow hold for the direct case if the factor I equals +1 and for the retrograde case if the factor I equals -1.

The equinoctial formulations use the $(\hat{f},\hat{g},\hat{w})$ coordinate system (Figure 1) where

$\hat{f}$ is chosen pointing to the "origin of longitudes."

$\hat{w}$ is chosen in the direction of the angular momentum vector.

$\hat{g}$ is chosen so as to form a right-handed coordinates system.

In terms of the classical elements, (a,e,i,Ω,ω,M) the equinoctial elements are defined as follows:

$$\begin{aligned}
& a \\
& h = \bar{e} \cdot \hat{g} = e \sin(\omega + \Omega I) \\
& k = \bar{e} \cdot \hat{f} = e \cos(\omega + \Omega I) \\
(6) \qquad & p = (\bar{N} \cdot \hat{g}) I = \tan^{I} \tfrac{i}{2} \sin \Omega \\
& g = \bar{N} \cdot \hat{f} = \tan^{I} \tfrac{i}{2} \cos \Omega \\
& \lambda = M + \omega + \Omega I
\end{aligned}$$

In the above equations $\bar{e}$ is a vector in the direction of periapsis with the magnitude of the eccentricity and $\bar{N}$ is a vector pointing to the node with magnitude $\tan^I \frac{i}{2}$.

In terms of the equinoctial elements, the coordinate axes are given by

$$\hat{f} = \frac{1}{1 + p^2 + q^2} \begin{pmatrix} 1 - p^2 + q^2 \\ 2pq \\ -2pI \end{pmatrix}$$

(7)

$$\hat{g} = \frac{1}{1 + p^2 + q^2} \begin{pmatrix} 2pqI \\ (1 + p^2 - q^2) I \\ 2q \end{pmatrix}$$

$$\hat{w} = \frac{1}{1 + p^2 + q^2} \begin{pmatrix} 2p \\ -2q \\ (1 - p^2 - q^2) I \end{pmatrix}$$

The total time derivatives of the equinoctial elements are computed according to Equation 2. The partial derivatives with respect to velocity of the equinoctial elements are given in Table 1.

It is noted that the only nonzero two-body time derivative of the elements is in the fast variable λ,

$$\dot{\lambda}_{\text{two-body}} = n$$

where n is the mean motion. Hence, we can see that for satellite trajectories which are governed by accelerations in which the satellite frequency n is the smallest, the high-order time derivatives of the elements, and hence local truncation error, will be governed by the grave derivatives of n.

2. Numerical Integrators

The integrators used in this study were basically the "summed" form of the popular Adams formulas which are derived from the basic Newtonian polynomials in Reference 11. They can be expressed as follows:

• Class I: $\dot{x} = f(t,x)$

$$\text{(Adams)} \quad x_{n+s} = {}^{I}S_n + \sum_{i=0}^{k} \delta_i(s)\, f_{n-i} + C_k(s)\, h^{k+3} x^{k+3}(\xi)$$

• Class II: $\ddot{x} = f(t,x)$

$$\text{(Stormer/Cowell)} \quad x_{n+s} = {}^{II}S_n + \sum_{i=0}^{k} \beta_i(s)\, f_{n-i} + C_k(s)\, h^{k+5} x^{k+5}(\xi)$$

where predictors are obtained by setting $s = 1$, and correctors with $s = 0$.

Equations of motion in the Cowell formulation were solved using the Class II methods and the VOP equations by the Class I. The parameters which were varied in the matching process were k, effectively the order, and the predictor/corrector algorithm, i.e., PE, PEC, PECE, PECE* methods were tested, where P = predict, E = evaluate, C = correct, and E* = pseudo-evaluate. The pseudo-evaluate methods used are described in Table 2. The matching process basically involved finding the "cheapest" algorithm which yielded the most precise results. It is clear that such a process results in investigating the higher-order methods, which in turn requires an analysis of the numerical stability of the method, which depends on both the predictor/corrector algorithm and the order of each formula. Although not totally rigorous, the question of stability of a predictor/corrector method can be (approximately) reduced to the investigations of the stability of the method as applied to the linear scalar equations

$$\dot{x} = \lambda x \qquad \text{(Class I)}$$

$$\ddot{x} = \lambda^2 x \qquad \text{(Class II)}$$

where for the Cowell orbital equations describing circular motion $\lambda \cong n$, the mean motion, and for the VOP equations

$$\lambda = \text{max. eigenvalue of } \frac{\partial f}{\partial x}$$

over an orbital revolution. This analysis assumes $\partial f/\partial x$ is constant and similar to a diagonal matrix. A more rigorous treatment of this question for an Adams PE method applied to circular motion is given in Reference 12.

Under these assumptions, the stability of the method is determined by the so-called "extraneous" roots of the characteristic polynomial

$$P(\bar{h},z) = 0 \ , \quad \bar{h} = \begin{cases} h\lambda \\ h^2\lambda^2 \end{cases}$$

which generate the solutions of the predictor/corrector difference equation (Reference 9). For Class I PE, PEC, and PECE Adams type methods these polynomials are well-known. The equivalent Class II polynomials were not readily found in the open literature but are easily derived in a fashion similar to the Class I equations. The explicit polynomial expressions are rather lengthy and can be found in Reference 13. Figures 2 and 3 display the $\bar{h}$-plane stability regions obtained from these polynomials. In Figure 2, the Class I PECE stability regions for orders 8 through 12 are displayed, showing the expected reduction in stability as k increases. The figure also shows the sensitivity of the stability region to the algorithms, the PEC being the least stable, and the PECE the most stable. The PECE* stability region was obtained numerically and is only approximate. The two significant results one can obtain from this graph is that in general, (1) stability is proportional to cost and inversely proportional to precision at the same cost, and (2) the PECE* algorithm offers a significant improvement in stability at no substantial increase in cost. As will be seen in the numerical results, this improvement in stability due to the pseudo-evaluate is the key element which makes the PECE* VOP orbit generator competitive or superior to the Cowell. Figure 3 shows similar results for the Class II integrators.

3. Numerical Results

Two orbits were used for the numerical studies. Case I is the orbit of the ATS satellite at near-geosynchronous conditions. The values of the Keplerian elements at epoch for Case I are:

$a = 42167.4$ km $\qquad \Omega = 73.65^o$

$e = .000792$ $\qquad \omega = 149.94^o$

$i = 3.758^o$ $\qquad M = 284.40^o$

Case II is the GEOS-C with a near-circular orbit at 1000 kilometers and high inclination. The values of the Keplerian elements at epoch for Case II are:

$a = 7379.97$ km $\qquad \Omega = 4.0^o$

$e = .005$ $\qquad \omega = 0.0^o$

$i = 55.0^o$ $\qquad M = .015^o$

For this study, the GEOS orbit was integrated for 100 revolutions.

These two orbits were chosen because they are typical of the satellites that are highly stable and for which long arc high precision trajectories are computed. Also, the small eccentricities of these two orbits allow the use of a fixed stepsize, which simplifies the numerical studies.

Complex perturbing acceleration models were used for both cases. For Case I (ATS), a 4 x 4 geopotential model for the earth's gravitational field was used along with the perturbing effects of the moon and sun. For Case II (GEOS-C), a full SAO 15 x 15 model for the earth was employed (Reference 11).

The first step in the numerical analysis was to compute the stepsize constraints imposed by the stability regions computed analytically for the various integrators. The results are displayed in Table 3. For each satellite and formulation, the minimum number of steps per revolution required to maintain stability is given for the PE, PECE, and PECE* methods for orders 9 through 12. These were computed using $\lambda_{max} \cong n$ for the Cowell method and $\lambda \cong$ the maximum eigenvalue of $\partial f/\partial x$ over a revolution for the VOP method.

It is noted that since the stability region is governed by the two-body motion in the Cowell case, the stepsize requirements are independent of type of orbit. On the other hand, the two-body Jacobian of the VOP equations has zero eigenvalues, so that the stability parameter $\bar{h}$ depends on the magnitude of the perturbations. Hence, for example, the integration of the ATS orbit with a twelfth order PE requires

65 steps/revolution, as contrasted to 420 steps/revolution for the GEOS-C case.

The last column in this table shows the minimum number of steps required to achieve 10 meter accuracy for both cases--a requirement imposed by the truncation errors of the methods. This dramatically shows the impact of the PE stability region constraints when applied to the VOP equations. For example, for the case of GEOS, an eleventh order PE method requires a minimum of 150 steps/revolution to maintain the required local accuracy. As a result one would be forced to use a lower order PE method for this case. This result is verified below in the numerical experiments. Table 3 also shows the impact of the pseudo-evaluate technique which in all cases sufficiently stabilized the processes so as to allow the stepsizes used to be constrained by local accuracy.

The next step in the numerical analysis was to actually compute the trajectories for the two cases allowing the parameters k and algorithm to vary.

Based on Reference 8, we decided to experiment first with predict-evaluate (PE) Adams methods of varying orders. In that JPL report, it was concluded that first, a lower order PE method was an optimum integration technique for the VOP equations of those tested and second, VOP/PE offered no significant advantage over Cowell with a high order PECE* integrator. Figure 4 shows numerical results of an integrator of 100 revolutions of the GEOS-C orbit with the VOP integrator. The integrations were carried out with PE Adams processes of orders 7,9, and 11. The cost of a particular run is the number of full perturbing acceleration evaluations. Figure 4 shows that an accuracy requirement must be specified before the optimum order of integration can be determined. However, in a broad sense, it supports the conclusion in Reference 8 that the VOP formulation should be integrated with a lower order PE process than the higher order processes used with the Cowell formulation in modern orbit determination programs. As shown above, results can be easily seen to be attributable to numerical instability (See Table 3).

Figure 5 shows numerical results obtained with the PECE* process for the GEOS-C case. Except for the numerical algorithm, all the conditions are the same as those in Figure 4. Supporting the results of Table 3, the PECE* process provides an improvement in the stability characteristics relative to the PE process. We also performed numerical integrations with the PE, PECE, and PECE* processes for the geosynchronous (ATS) test case; Figure 6 shows the results. Again, the PECE* performs in a superior manner.

A Comparison of VOP with Cowell

Because the main goal is to demonstrate the efficiency of the VOP orbit generation process, Figure 7 compares VOP with the Cowell formulation for the ATS and GEOS-C test cases. For the ATS case, Figure 7 shows a significant reduction in the number of force evaluations required for equivalent accuracy; this is true over a wide range in accuracy requirements. For the GEOS-C (Figure 8) case, there is an apparent penalty involved with the application of the VOP formulation for all but very high accuracy requirements. These results support the conclusion that variational methods about two-body Keplerian motion will be superior whenever the "main" frequency of the motion is the orbital frequency since in a sense the VOP equations remove this from the problem by isolating it into a single "fast" but nearly linear equation. In the geosynchronous orbit, the remaining "main" frequencies are due to lunar/solar and long period zonal harmonic effects and hence, greater than the orbital frequencies. As a result, the equations are more slowly varying than the comparable Cowell equations and hence, result in small truncation errors, allow larger stepsizes, etc. In contrast, in the case of the two-hour satellite, the main frequency is suborbital, in fact, due to J_2, with a frequency of 2n, and this frequency enters into both the Cowell and VOP formulations in the same non-linear manner and therefore, the numerical behavior of both systems, i.e., local errors, are equivalent. Although not available to us at the time of this study, a twelfth-order Class I integrator would have brought the performance of the VOP integrator even closer to the Cowell results for this case.

4. Conclusions

The main result of this report is that in evaluating the value or applicability of various forms of the equations of motion of satellite with the goal of improving the numerical calculations, care must be taken in selecting or matching appropriate numerical integrators. This is particularly important in the light of the many recent proposal modifications of the equations of motion which regularize or stabilize the system (eg. Reference 14).

By appropriate choice of integrator, it was shown that the VOP formulation is competitive or superior to the Cowell formulation for the orbits considered and it would be expected that VOP methods will offer advantages for all orbits for which the main frequency is the mean motion.

Finally, full advantage of the decoupling of the along-track errors into a single fast variable has not yet been achieved. Parameters which achieve full decoupling--the "ideal frame" in reference or numerical methods which handle the along-track problem with more precise integrators are being currently tested in our efforts to achieve this advantage.

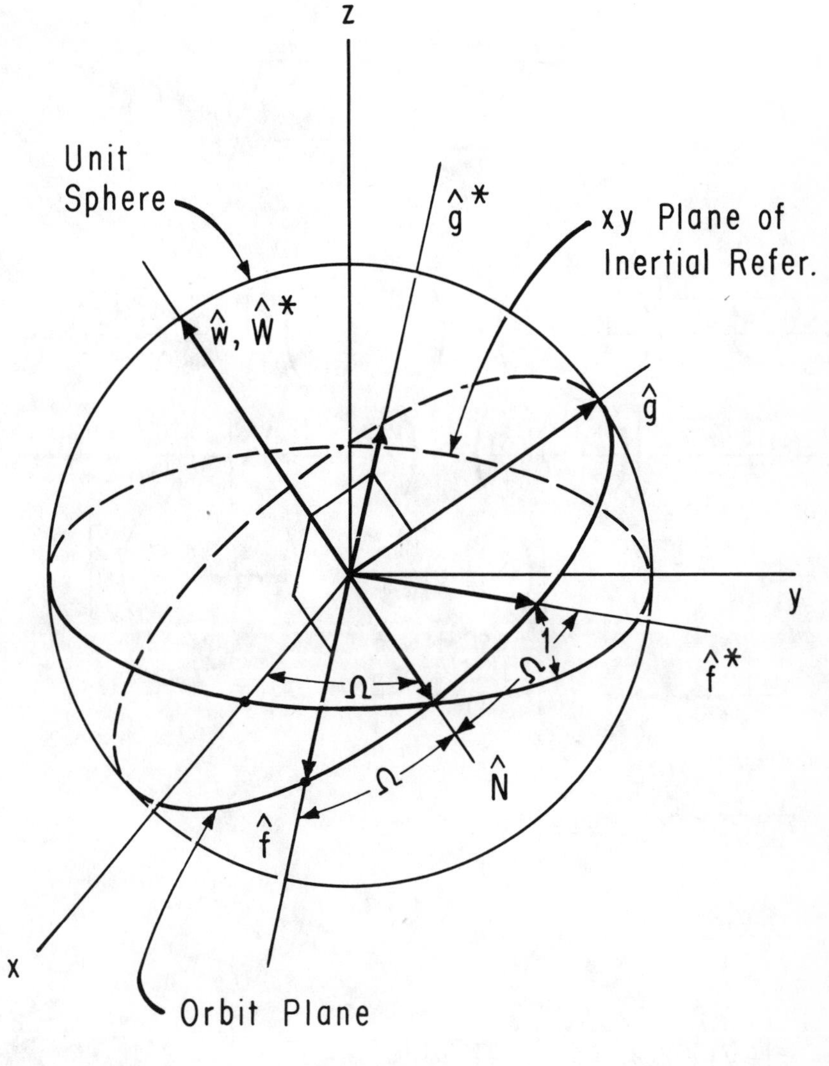

FIGURE 1. The Direct Equinoctial Coordinate Frame $(\hat{f},\hat{g},\hat{w})$
and the Retrograde Equinoctial
Coordinate Frame $(\hat{f}^*,\hat{g}^*,\hat{w}^*)$

TABLE 1

Partial Derivatives of the Equinoctial Elements

With Respect to Velocity

$$\frac{\partial a}{\partial \dot{\underline{x}}} = \frac{2\dot{\underline{r}}}{n^2 a}$$

$$\frac{\partial h}{\partial \dot{\underline{x}}} = \frac{\sqrt{1-h^2-k^2}}{na^2}\left[\left(\frac{\partial X_1}{\partial k} - h\beta\frac{\dot{X}_1}{n}\right)\hat{f} + \left(\frac{\partial Y_1}{\partial k} - h\beta\frac{\dot{Y}_1}{n}\right)\hat{g}\right] + \frac{k\,(qY_1 - pX_1)}{na^2\sqrt{1-h^2-k^2}}\,\hat{w}$$

$$\frac{\partial k}{\partial \dot{\underline{x}}} = -\frac{\sqrt{1-h^2-k^2}}{na^2}\left[\left(\frac{\partial X_1}{\partial h} + k\beta\frac{\dot{X}_1}{n}\right)\hat{f} + \left(\frac{\partial Y_1}{\partial h} + k\beta\frac{\dot{Y}_1}{n}\right)\hat{g}\right] - \frac{h\,(qY_1 - pX_1)}{na^2\sqrt{1-h^2-k^2}}\,\hat{w}$$

$$\frac{\partial \lambda}{\partial \dot{\underline{x}}} = \frac{-2}{na^2}\,\underline{r} + \frac{\sqrt{1-h^2-k^2}}{na^2}\,\beta\left[\left(h\frac{\partial X_1}{\partial h} + k\frac{\partial X_1}{\partial k}\right)\hat{f} + \left(h\frac{\partial Y_1}{\partial h} + k\frac{\partial Y_1}{\partial k}\right)\hat{g}\right]$$

$$+ \frac{1}{na^2\sqrt{1-h^2-k^2}}\,(qY_1 - pX_1)\,\hat{w}$$

$$\frac{\partial p}{\partial \dot{\underline{x}}} = \frac{1+p^2+q^2}{2\,na^2\sqrt{1-h^2-k^2}}\,Y_1\hat{w}$$

$$\frac{\partial q}{\partial \underline{x}} = \frac{(1+p^2+q^2)}{2\,na^2\sqrt{1-h^2-k^2}}\,X_1\hat{w}$$

$$\frac{\partial X_1}{\partial h} = a\left[-(\lambda - F)\left(\beta + \frac{h^2\beta^3}{1-\beta\beta}\right) - \frac{a}{r}\cos F\,(h\beta - \sin F)\right]$$

$$\frac{\partial X_1}{\partial k} = -a\left[(\lambda - F)\frac{hk\beta^3}{1-\beta} + 1 + \frac{a}{r}\sin F\,(\sin F - h\beta)\right]$$

$$\frac{\partial Y_1}{\partial h} = a\left[(\lambda - F)\frac{kh\beta^3}{1-\beta} - 1 + \frac{a}{r}\cos F\,(\cos F - k\beta)\right]$$

$$\frac{\partial Y_1}{\partial k} = a\left[(\lambda - F)\left(\beta + \frac{k^2\beta^3}{1-\beta}\right) + \frac{a}{r}\sin F\,(\cos F - k\beta)\right]$$

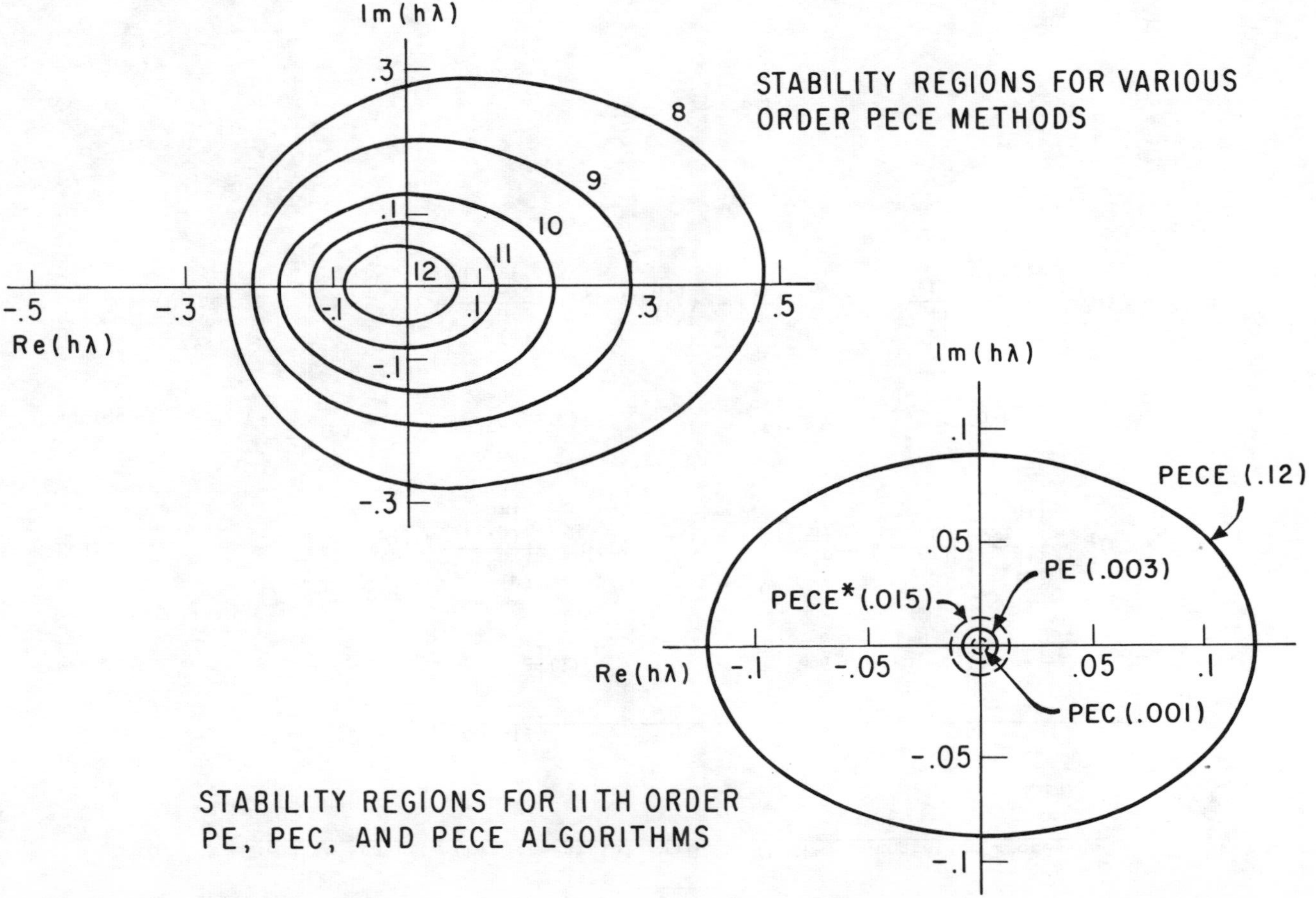

FIGURE 2. CLASS I STABILITY REGIONS

TABLE 2

Predict-Evaluate-Correct-Partial Evaluate (PECE*) Integration Processes

	Cowell	VOP
Predict	$x_{n+1}^{(p)} = II_{S_n} + \sum_{i=0}^{k} \beta_i(1) f_{n-i}$	$a_{n+1}^{(p)} = I_{S_n} + \sum_{i=0}^{k} \delta_i(1) f_{n-i}$
Evaluate	$f(t, x_{n+1}^{(p)}) = \frac{-\mu x_{n+1}^{(p)}}{\lvert r_{n+1}^{(p)}\rvert^3} + \vec{Q}(t, x_{n+1}^{(p)})$	$f(t, a_{n+1}^{(p)}) = \frac{\partial a_{n+1}^{(p)}}{\partial t} + \frac{\partial a_{n+1}^{(p)}}{\partial \dot{\vec{x}}_{n+1}} \cdot \vec{Q}(t, x_{n+1}^{(p)})$
Correct	$x_{n+1}^{(c)} = II_{S_n} + \sum_{i=0}^{k} \beta_i(0) f_{n-i+1}$	$a_{n+1}^{(c)} = I_{S_n} + \sum_{i=0}^{k} \delta_i(0) f_{n-i+1}$
Pseudoevaluate	$f(t, x_{n+1}^{(c)}) = \frac{-\mu x_{n+1}^{(c)}}{\lvert r_{n+1}^{(c)}\rvert^3} + \vec{Q}(t, x_{n+1}^{(p)})$	$f(t, a_{n+1}^{(c)}) = \frac{\partial a_{n+1}^{(c)}}{\partial t} + \frac{\partial a_{n+1}^{(c)}}{\partial \dot{\vec{x}}_{n+1}} \cdot \vec{Q}(t, x_{n+1}^{(p)})$

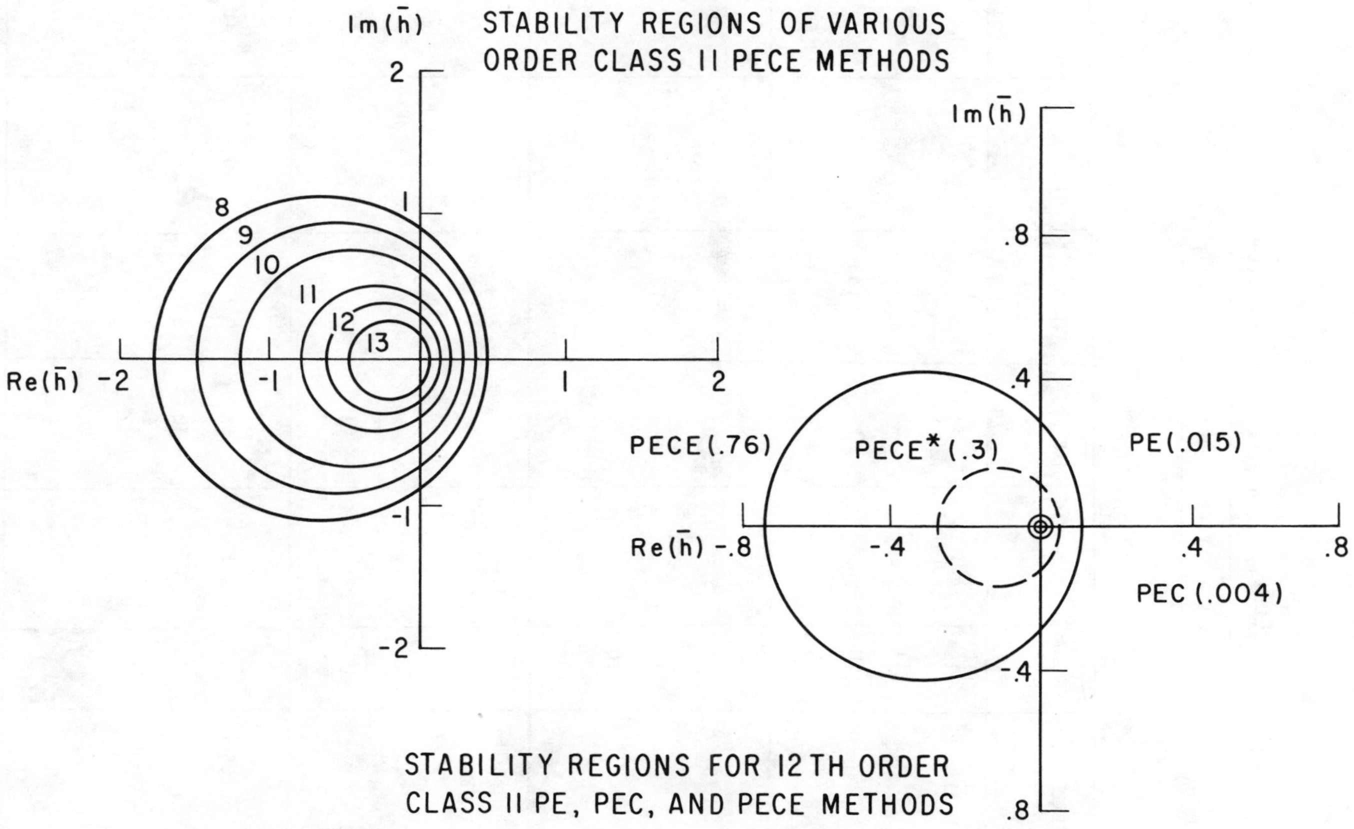

FIGURE 3. CLASS II STABILITY REGIONS

TABLE 3

Comparison of Accuracy and Stability

Characteristics of VOP and Cowell

METHOD	SATELLITE	ORDER	MAX	MIN N_{STAB} PE	MIN N_{STAB} PECE	MIN N_{STAB} PECE*	MIN N_{ACC} (10 m)
VOP	ATS	12	$.74 \times 10^{-6}$	65	.7		
		11		21	.5	4.5	17
		10		11	.43		
		9		4.5	.33		
COWELL	ATS	12	$.5 \times 10^{-8}$	63	7		30
		11		50	6.5		
		10		36	5.5		
		9		26	5		
VOP	GEOS-C	12	$.71 \times 10^{-4}$	420	5		
		11		150	3.3	28	50
		10		75	3		
		9		30	2.4		66
COWELL	GEOS-C	12	1×10^{-6}	68	7	10	40
		11		50	6.5		
		10		36	5.5		
		9		26	5		

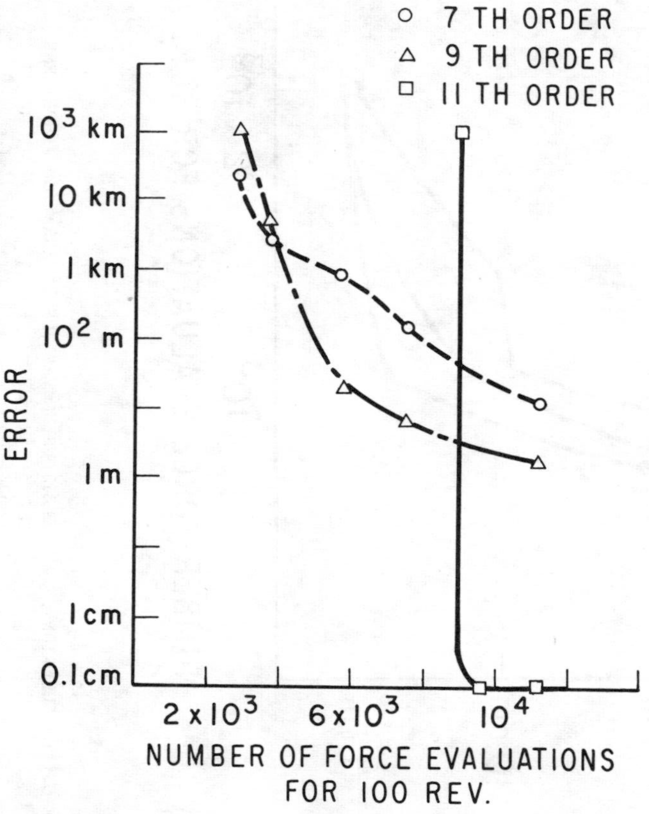

FIGURE 4. CALCULATION OF GEOSC-C ORBIT USING PE ADAMS PROCESSES

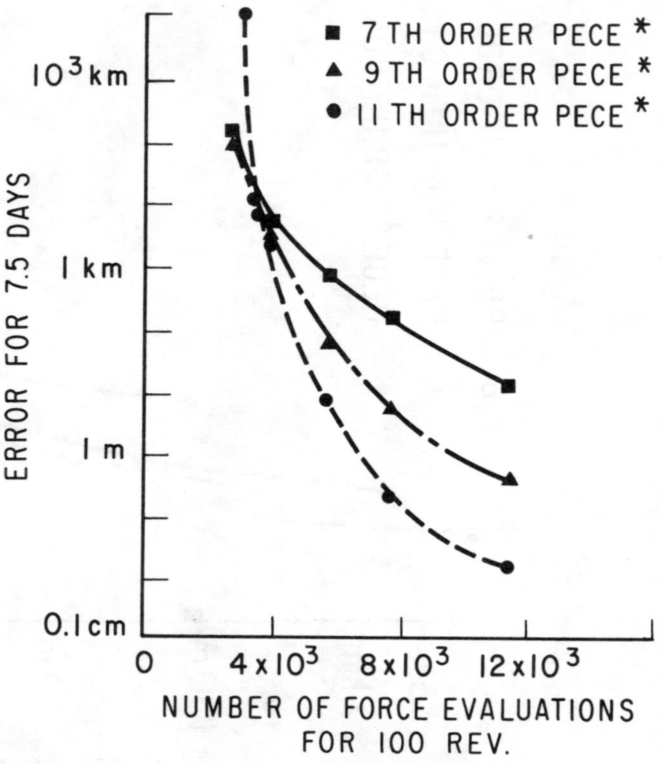

FIGURE 5. CALCULATION OF GEOSC-C ORBIT USING VARIOUS ORDER PECE* INTEGRATORS (VOP FORMULATION)

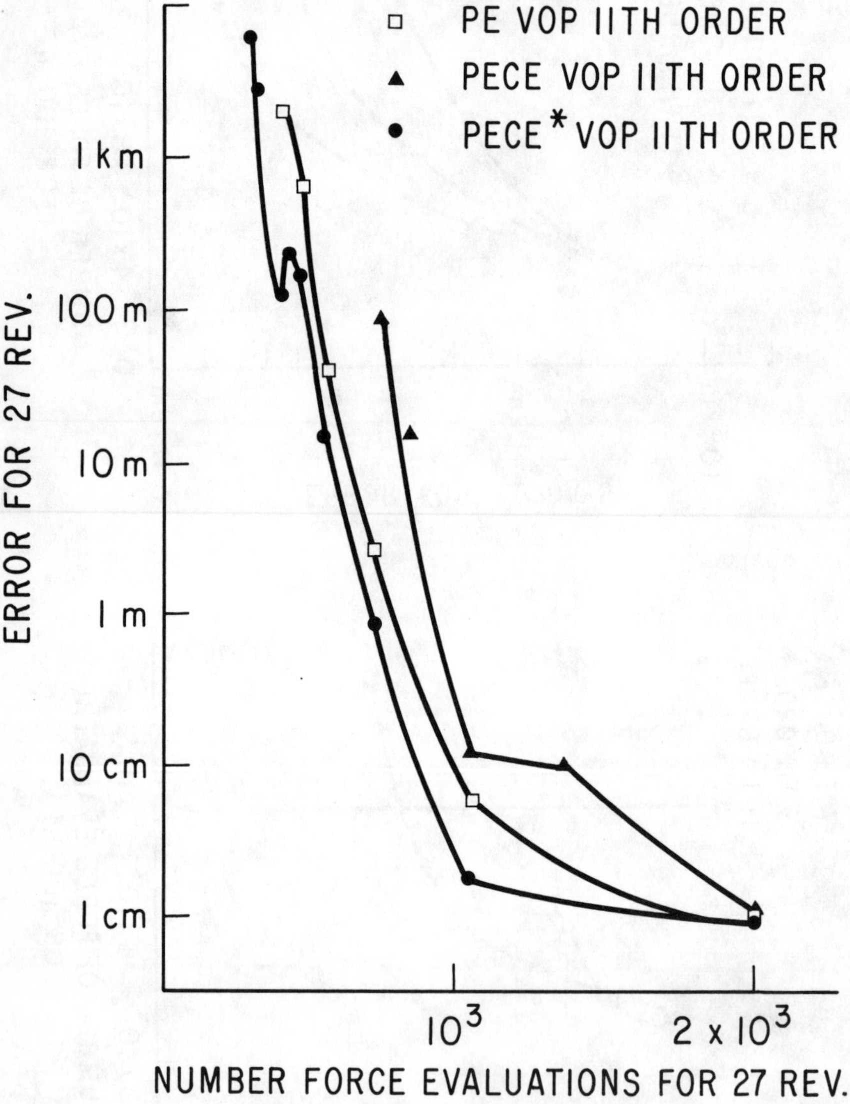

FIGURE 6. CALCULATION OF ATS ORBIT USING VARIOUS INTEGRATION ALGORITHMS

(VOP-λ FORMULATION)

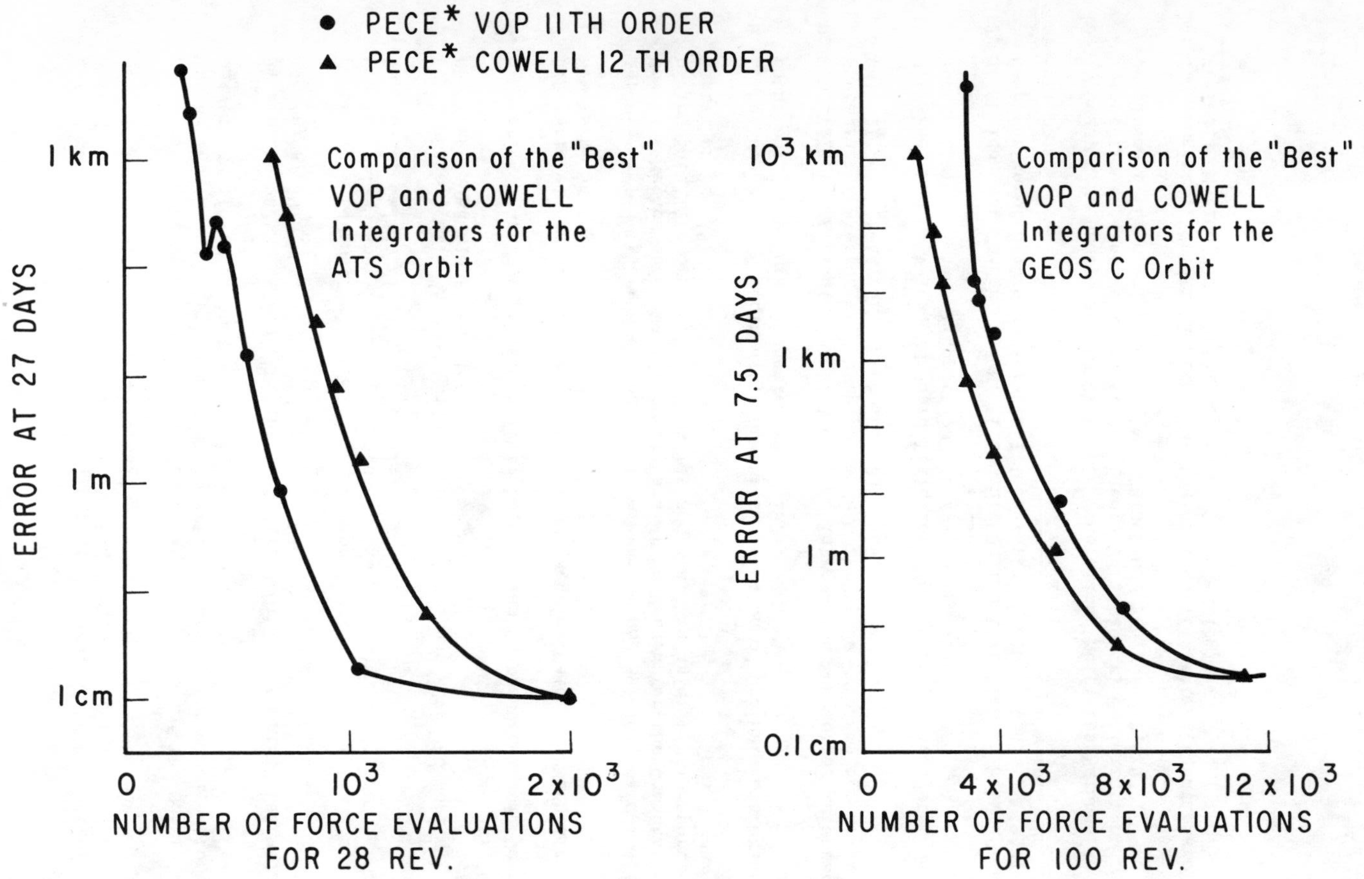

FIGURE 7. COMPARISON OF VOP AND COWELL INTEGRATORS

References

1. Jet Propulsion Laboratory, Technical Report 32-1526, "A Comparison of Cowell's Method and a Variation-of-Parameters Method for the Computation of Precision Satellite Orbits," Vol. V., S.S. Dallas and E.A. Rinderle, October 1971.

2. A. Deprit, "The Theory of Special Perturbations--Motion of the Orbital Plane," (notes from lectures given at the Summer Institute in Dynamical Astronomy, University of Texas, Austin, Texas, May 1970).

3. R. Broucke and P. Cefola, "On the Equinoctial Orbit Elements," Celestial Mechanics, 1972, Vol. 5, No. 3, pp. 303-310.

4. P. Cefola, "Equinoctial Orbit Elements--Application to Artificial Satellite Orbits," AIAA Paper 72-937, presented at the AIAA/AAS Astrodynamics Conference, Palo Alto, California, September 11-12, 1972.

5. R.M.L. Baker, Jr., "Astrodynamics, Applications, and Advanced Topics," New York, Academic Press, 1961.

6. S. Herrick, "Astrodynamics," Vol. 1, London: Van Nostrand Reinhold Company, 1971.

7. E.L. Stiefel and G. Scheifele, "Linear and Regular Celestial Mechanics," New York, Springer-Verlag, 1971.

8. Jet Propulsion Laboratory, Technical Memorandum No. 392-101, "A Comparison of Cowell's Method and a Variation-of-Parameters Method for the Computation of Precision Satellite Orbits," S.S. Dallas and E.A. Rinderle, Final Report, September 20, 1972.

9. L. Lapidus and J.H. Seinfeld, "Numerical Solution of Ordinary Differential Equations," New York, Academic Press, 1971.

10. T.D. Talbot, "A Predict-Partial Correct Integration Scheme for Integrating Spacecraft Trajectories," Deep Space Network, Space Programs Summary 37-59, Vol. II, pp. 78-84, JPL, September 30, 1969.

11. W. Wagner and C. Velez, "Goddard Trajectory Determination Subsystem Mathematical Specifications," NASA X-Report 552-72-244, March 1972.

12. J. Sheldon, B. Zondek, and M. Friedman, "On the Time-Step to be Used for the Computation of Orbits by Numerical Integration," MTOAC, Vol. II, 1957, pp. 181-189.

13. J. Dyer, R.F. Haney, and L. Chesler, "Generalized Multistep Methods in Orbit Computation: Studies in Existence Theory, Efficiency, Optimization," System Development Corporation Report TM-4888/000/00, February 28, 1972.

14. J. Baumgarte, "Numerical Stabilization of the Differential Equations of Keplerian Motion," Celestial Mechanics 5, 490-501, 1972.

EXAMPLES OF TRANSFORMATIONS IMPROVING THE NUMERICAL ACCURACY OF THE INTEGRATION OF DIFFERENTIAL EQUATIONS

by

J. Baumgarte and E. Stiefel

Swiss Federal Institute of Technology

Zurich, Switzerland

Abstract

Methods for transforming differential equations are developed in such a way that the transformed equation is better suited for numerical integration. In order to obtain theoretical insight into such mechanisms we discuss mostly differential equations which are also solvable in closed analytical form. The methods under consideration are aimed at application in cases where small perturbing terms are added to these equations.

1. Objective of the Study.

In order to solve a differential equation (d.eq.)

$$\ddot{x} = f(t,x,\dot{x}) \tag{1}$$

for an unknown function x(t) we are used to subject it - as it stands - to a well programmed numerical integration routine. An automatic step regulator may be built in for taking care of regions where rapid variations of the solution occur. Such events are likely in the vicinity of singularities of the d.eq. at hand.

It may be more efficient, however, to transform the d.eq. before putting electronic machinery into operation. The goal of such transformations is to obtain a new d.eq. which is better adapted to numerical integration since the solutions vary more smoothly or since by the transformation singularities are removed.

In this paper three methods of transforming d.eq. are discussed, namely

1. Analytic step adaption.

2. Stabilization.

3. Regularization

Before introducing them some remarks are in order. The d.eq.(1) is called regular at the point $(t_o, x_o, \dot{x}_o)$ provided f as well as $\frac{\partial f}{\partial x}$ and $\frac{\partial f}{\partial \dot{x}}$ are finite and continuous functions at any point (t,x) inside an open domain containing (t_o, x_o). If this condition is not satisfied, the point is called singular. One should distinguish carefully between such singularities of a d.eq. and the singularities of its solutions. The linear d.eq., for instance,

$$(1-\cos t)\,\ddot{x} - \sin t\cdot\dot{x} + x = 0 \tag{2}$$

is singular at the point t=0, x=0 since the denominator in the expression

$$f = \frac{\dot{x}\sin t - x}{1-\cos t}$$

vanishes. The general solution of the d.eq. is

$$x = c_1(1-\cos t) + c_2 \sin t$$

and this function is everywhere regular, showing that singularities of differential equations do not necessarily imply singularities of the solutions. A suitable transformation for removing the singularities from the d.eq. (2) is the introduction of the new dependent variable y by putting

$$x = \sin\frac{t}{2}\cdot y.$$

The transformed equation is the harmonic oscillation

$$\ddot{y} + \frac{1}{4}\,y = 0.$$

2. Analytic step adaption.

This is performed by introducing a new independent variable s by virtue of a differential relation

$$dt = \mu\,(x)\;ds \tag{3}$$

and by appropriate choice of the scaling function $\mu(x)$. In a region near a singularity one should choose μ small such that a constant step size ds produces tiny steps with respect to the old variable t. Our differential equation is transformed into the new one

$$x'' = g(t,x,x'), \tag{4}$$

where the accent denotes differentiation with respect to s. This d.eq. together with

$$t' = \mu(x) \tag{5}$$

is a differential system for determining the functions x(s) and t(s). The scaling function $\mu(x)$ must satisfy the requirement to be $\geqslant 0$ on the solution x(s) in order to ensure a correspondance between s and t which is uniquely determined in both directions. The value $\mu=0$ is only permitted at isolated values of s.

Singularities are mostly produced by vanishing denominators and such a denominator is a good candidate for the scaling function μ. Let us consider, for instance, Keplerian motion. The gravitational force behaves like $\frac{1}{r^2}$ in magnitude, where r is the distance of the moving particle from the central mass. Thus we may try

$$\mu = r^2 \quad , \quad dt = r^2\, ds \; . \tag{6}$$

The new "fictitious time" is then essentially the true anomaly. The proposals

$$dt = r\, ds \; , \quad dt = r^{3/2}\, ds \tag{7}$$

are also found in the literature.

Let us now restrict ourselves to the d.eq. of analytic dynamics; in this field guiding rules for the choice of μ are available. We begin with the motion of a particle which is subjected to conservative forces. Let x be its position vector and U(x) the acting potential. The equations of motion and of energy are

$$\ddot{x} = -\frac{\partial U}{\partial x} \, , \tag{8}$$

$$\frac{1}{2}\dot{x}^2 + U = E \, , \tag{9}$$

where E is the total energy. Large values of U should be compensated by our step-size adaption, thus the choice $\mu = \frac{1}{U}$ may be recommended. As J. Baumgarte observed recently, the integration of the additional d.eq. (5) can be avoided by adopting a slightly modified scaling function μ. For reaching that goal the variation of the momentum of inertia

$$J = \frac{1}{2}|x|^2 \tag{10}$$

is discussed. It follows that

$$\dot{J} = (x,\dot{x}) \, , \quad \ddot{J} = (x,\dot{x})^{\cdot} = (x,\ddot{x}) + \dot{x}^2 .$$

$(x,\dot{x})$ denotes the scalar product of the vectors x and $\dot{x}$. By substituting eqs. (8)

(9) into this, one obtains

$$\ddot{J} = (x,\dot{x})^{\cdot} = -\left(x, \frac{\partial U}{\partial x}\right) + 2(E-U).$$

By using the abbreviation

$$(11) \qquad W(x) = 2\,U(x) + \left(x, \frac{\partial U}{\partial x}\right)$$

we have

$$(12) \qquad (x,\dot{x})^{\cdot} = 2\,E - W\ ,$$

and by integrating

$$(13) \qquad (x,\dot{x}) = 2\,E\,t + C - \int W\,dt.$$

C is the constant of integration. Now we propose to choose the scaling function

$$(14) \qquad \mu\,(x) = \frac{\lambda}{W(x)}\ , \quad \text{thus } dt = \frac{\lambda}{W}\,ds\ ,$$

where λ is an arbitrary constant. It follows

$$(15) \qquad (x,\dot{x}) = 2\,E\,t + C - \lambda s$$

or

$$(16) \qquad t = \frac{\lambda s - C}{2E} + \frac{(x,\dot{x})}{2E}.$$

In practice this explicit relation is used for computing t as a function of the new independent variable s. No numerical integration of the d.eq. (14) is needed and thus <u>the total differential order</u> of the problem at hand <u>is not raised</u> by the introduction of s.

<u>Euler's theorem, examples.</u> In the sequel we are repeatedly faced with the special case where the potential is a homogeneous function of the state vector x, hence

$$(17) \qquad U(\alpha x) = \alpha^n U\,(x).$$

α is an arbitrary scalar and n the degree of homogeneity. From Euler's theorem there follows

$$(18) \qquad \left(x, \frac{\partial U}{\partial x}\right) = n\,U(x), \quad \text{thus } W = (2+n)U.$$

As follows from eq. (14) the scaling function is essentially the reciprocal of the potential. This was our first guess.

<u>Example 1.</u> Keplerian motion.

$$(19) \qquad U = -\frac{K^2}{r}\ ,$$

K^2 = gravitational parameter, r = distance from the origin. Degree of homogeneity: $n = -1$, $W = U = -\frac{K^2}{r}$. By adopting $\lambda = -K^2$ we obtain the law of time transformation

(20) $$dt = r\,ds$$

which was already mentioned in formula (7). For elliptic motion it is more practical to use the negative total energy h = (-E). Eq. (16) is thus transformed into

(21) $$t = \frac{K^2 s + C}{2h} - \frac{(x,\dot{x})}{2h} = \frac{K^2 s + C}{2h} - \frac{(x,x')}{2hr} .$$

The equations of motion are

(22) $$\ddot{x} = -\frac{K^2}{r^3}\,x$$

and with respect to the independent variable s

$$x'' - \frac{r'}{r}\,x' + \frac{K^2}{r}\,x = 0$$

or

(23) $$x'' - \frac{1}{r^2}\,(x,x')x' + \frac{K^2}{r}\,x = 0 .$$

The energy equation (9) is transformed into

(24) $$\frac{1}{2r^2}\,x'^2 - \frac{K^2}{r} + h = 0.$$

Remark. The variable s is essentially the eccentric anomaly. This is discussed in detail in the book [2]. Since the coordinates of the particle are regular functions of the eccentric anomaly, we are faced once more to the fact, that a singular d.eq. (23) has only regular solutions.

Example 2. Zonal harmonics of the earth potential.

These potentials have the structure

(25) $$K^2 V_n, \quad V_n = c_n \frac{P_n(\cos\theta)}{r^{n+1}}, \quad n = 0,1,2,\ldots$$

c_n is a constant and P_n the n-th Legendre polynomial. θ is the geographical colatitude. V_n is homogeneous of degree -(n+1). Let us take into account the central potential $-\frac{K^2}{r}$ and, in addition, the dominant term $K^2 V_2$ of the oblateness potential. From Euler's theorem it follows

$$\left(x, \frac{\partial V_2}{\partial x}\right) = -3V_2 .$$

With $\lambda = -K^2$ the results

$$W = -\frac{K^2}{r} + K^2\left[2\,V_2 + \left(x, \frac{\partial V_2}{\partial x}\right)\right] = -K^2\left(\frac{1}{r} + V_2\right) ,$$

(26) $$dt = \frac{ds}{\frac{1}{r} + V_2} ,$$

is obtained. The transformed equations of motion are

(27) $$\left(\frac{1}{r} + V_2\right)\left[\left(\frac{1}{r} + V_2\right)x'\right]' + \frac{K^2}{r^3}\, x = 0 \ ,$$

and the formular (16) for the computation of time is reformulated as

(28) $$t = \frac{K^2 s + C}{2h} - \frac{\frac{1}{r} + V_2}{2h}\,(x,x') \quad .$$

<u>The time-element</u>. The basic equation (16) may be split up into

(29) $$\tau = \frac{\lambda\, s - C}{2E}$$

and

(30) $$t = \tau + \frac{(x,\dot{x})}{2E} \quad .$$

τ is called the "time-element" since it is a linear function of the independent variable s. (As in celestial mechanics the name of "element" is given to any quantity which is constant during the motion or varies linearly).

We proceed now to the discussion of a <u>perturbed motion</u>. Let us assume that, in addition to the potential U, a perturbing force $P(t,x,\dot{x})$ is acting such that the equations of motion are

(31) $$\ddot{x} = -\frac{\partial U}{\partial x} + P \quad .$$

The total energy

(32) $$E = \frac{1}{2}\dot{x}^2 + U$$

is no longer constant but varies according to the law of energy

(33) $$\dot{E} = (P,\dot{x}) \quad .$$

Similarly the time-element, defined by eq. (30), does no longer vary linearly but obeys a d.eq. which is obtained as follows:

$$\tau = t - \frac{(x,\dot{x})}{2E}$$

$$\dot{\tau} = 1 - \frac{(x,\ddot{x})}{2E} - \frac{\dot{x}^2}{2E} + \frac{\dot{E}}{2E^2}\,(x,\dot{x})$$

$$= 1 + \frac{1}{2E}\left(x,\frac{\partial U}{\partial x}\right) - \frac{1}{2E}\,(x,P) - 1 + \frac{2U}{2E} + \frac{\dot{E}}{2E^2}\,(x,\dot{x})$$

(34) $$\dot{\tau} = \frac{W}{2E} - \frac{(x,P)}{2E} + \frac{\dot{E}}{2E^2}\,(x,\dot{x}) \quad .$$

The relations (31)(32)(11) were used. With respect to s it follows

(35) $$\tau' = \frac{\lambda}{2E} - \frac{\lambda}{W}\,\frac{(x,P)}{2E} + \frac{W}{\lambda}\,\frac{E'}{2E^2}\,(x,x')$$

where

$$E' = (P,x'). \tag{36}$$

The element E is almost constant and τ varies almost linearly provided the perturbing force is small.

Example 3. Problem of three bodies. Let m_i be the masses and r_{ij} the mutual distances of the bodies. The gravitational potential is

$$U = -k^2 \sum_{i<j} \frac{m_i m_j}{r_{ij}} . \tag{37}$$

The 9 components of the three position vectors x_1, x_2, x_3 may be interpreted as a 9-dimensional vector x. As a function of x the potential is homogenous of degree (-1). By making use of Euler's theorem it is seen that

$$\sum_{i=1}^{3} \left(x_i, \frac{\partial U}{\partial x_i}\right) = -U . \tag{38}$$

Let us discuss again the moment of inertia

$$J = \frac{1}{2} \sum m_i x_i^2$$

$$\dot{J} = \sum m_i (x_i, \dot{x}_i), \qquad \ddot{J} = \sum m_i (x_i, \dot{x}_i)^{\cdot} = \sum (x_i, m_i \ddot{x}_i) + \sum m_i \dot{x}_i^2$$

$$\ddot{J} = \sum m_i (x_i, \dot{x}_i)^{\cdot} = - \sum \left(x_i, \frac{\partial U}{\partial x_i}\right) + 2(E-U) = 2E - U. \tag{39}$$

This relation is known as Jacobi's equation. By integration it follows

$$\sum m_i (x_i, \dot{x}_i) = 2E \cdot t + C - \int U \, dt . \tag{40}$$

For the purpose of step-size adaption the new argument s is introduced by

$$dt = - \frac{1}{U} ds \tag{41}$$

and thus the formula

$$t = - \frac{s+C}{2E} + \frac{1}{2E} \sum_{i=1}^{3} m_i (x_i, \dot{x}_i) \tag{42}$$

for computing the time is obtained. By relation (41) a device for obtaining a slow motion picture is switched in whenever close approaches of the bodies occur.

The generalization to n>3 bodies is trivial.

Poincaré's transformation. This is a canonical version of the technique for introducing a new independent variable s. Let x,p be the vectors of position and momentum of a canonical system. The Hamiltonian H(x,p) is assumed to depend not explicitly on the time t, such that the problem is conservative. The Hamiltonian is constant during the motion, this is to say

(43) $$H[x(t),p(t)] = E,$$

where E is the total energy. The equations of motion are

(44) $$\dot{x} = \frac{\partial H}{\partial p}, \qquad \dot{p} = -\frac{\partial H}{\partial x}.$$

By virtue of eq. (3)

(45) $$dt = \mu(x)\, ds$$

the new argument s is introduced, where μ is arbitrary and not necessarily determined according to the rule (14). By combining this with eqs. (44) the new differential system

(46) $$x' = \mu \frac{\partial H}{\partial p}, \qquad p' = -\mu \frac{\partial H}{\partial x}$$

is obtained which is no longer of canonical type. The canonical form can be re-established by introducing the new Hamiltonian

(47) $$H^* = \mu(H-E) .$$

The canonical equations corresponding to H* are

(48) $$x' = \frac{\partial H^*}{\partial p} = \mu \frac{\partial H}{\partial p}, \qquad p' = -\frac{\partial H^*}{\partial x} = -\mu \frac{\partial H}{\partial x} - \frac{\partial \mu}{\partial x}(H-E) .$$

From the analytical point of view this set is equivalent to eqs. (46) since the the parenthesis vanishes by virtue of eq. (43). But from the numerical point of view one should distinguish carefully between (46) and (48). In eq. (48) the computational procedure is modified by adding the term

(49) $$-\frac{\partial \mu}{\partial x}(H-E) .$$

This term vanishes provided the correct solution is inserted. During the automatic numerical integration it may not vanish due to truncation and round-off errors. Such a term will be called a control term. For integrating the system (48) the numerical value of E must be communicated to the computer. In contrast, this value is not needed for integrating (46).

Next we discuss the particular case of a particle moving in space.

(50) $$H = \frac{1}{2} p^2 + U(x),$$

where U is the conservative potential

(51) $$H^* = \mu\left(\frac{p^2}{2} + U - E\right)$$

$$x' = \mu p, \quad p' = -\mu\,\frac{\partial U}{\partial x} - \left(\frac{p^2}{2} + U - E\right)\frac{\partial \mu}{\partial x}\,,$$

or, by eliminating p:

(52) $$\left(\frac{x'}{\mu}\right)' = -\mu\,\frac{\partial U}{\partial x} - \left(\frac{x'^2}{2\mu^2} + U - E\right)\frac{\partial \mu}{\partial x}\,.$$

Example 4. Keplerian motion with control term.

This is the special case

$$U = -\frac{K^2}{r}\,, \qquad \mu = r\,,$$

thus

(53) $$\left(\frac{x'}{r}\right)' = -\frac{K^2}{r^2}\,x - \left(\frac{x'^2}{2r^2} - \frac{K}{r} + h\right)\frac{x}{r}$$

It is seen that the gravitational parameter K cancels out and is replaced by the total energy (-h). The final equation

(54) $$x'' - \frac{r'}{r}\,x' + \left[\frac{x'^2}{2r^2} + h\right]x = 0$$

or

(55) $$x'' - \frac{(x,x')}{r^2}\,x' + \left[\frac{x'^2}{2r^2} + h\right]x = 0$$

was discussed in detail in the publication [19]. The generalization of Poincaré's transformation to non-conservative systems was suggested by G. Scheifele [2] and successfully applied by him [21] and U. Krichgraber [22]. In these more general theories the scaling function μ is allowed to depend not only on the state variables x but also on the momenta, on t, on E and on s.

3. Stabilization. At this state of the discussion we are faced with the following three types of d.eq. for Keplerian motion:

(56) Eq. (22) $$\ddot{x} = -\frac{K^2}{r^3}\,x$$

(57) Eq. (23) $$x'' - \frac{1}{r^2}\,(x,x')x' + \frac{K^2}{r}\,x = 0$$

(58) Eq. (55) $$x'' - \frac{1}{r^2}\,(x,x')x' + \left[\frac{x'^2}{2r^2} + h\right]x = 0$$

In eqs. (57)(58) the independent variable is s, defined by dt = r ds. Eq. (58) is obtained by adding to (57) the control term

(59) $$\left[\frac{x'^2}{2r^2} - \frac{K^2}{r} + h\right]x\,.$$

The bracket is the balance of energy. The unmanipulated Newtonian equation (56) is unstable in the sense of Ljapunow. This is seen by discussing two circular motions on circles of radii r and $(r+\Delta r)$, where Δr is small. According to Kepler's third law the revolution time depends on the radius. On our two circles the difference of the revolution times is (up to terms of higher order)

$$\frac{3\pi}{K}\sqrt{r}\,\Delta r.$$

Hence the particle of radius r runs away of the particle of radius $(r+\Delta r)$. Eq. (57) is also unstable, the difference of revolution times((with respect to the t-time) is

$$\frac{\pi}{K}\sqrt{r}\,\Delta r\quad .$$

The step-size adaption reduced therefore the unstability by a factor 3.

In order to discuss the stability of eq. (58) let us forget about the history of that equation. Hence it is considered as a d.eq. of second order for a state vector x as function of the independent variable s. The letter h is considered as a numerical value appearing in the d.eq. and given in advance. This value has no connection whatsoever with the initial conditions. The equation may be solved under arbitrary initial conditions. Of course the solution is not necessarily a Keplerian orbit. This is the numerical situation and this is the set of information available to the computer. A circular solution is still possible. Let ϕ be the polar angle. Then the angular velocity turns out to be

$$\phi' = \sqrt{2h}$$

and it is thus independent of the radius, this is to say of the initial conditions. This situation suggests stability. In the article [19] it is proved, indeed, that eq. (58), restricted to elliptic motion, is stable in the sense of Ljapunow, this is to say with respect to variations of the inital conditions. For purposes of numerical integration this fact is welcome since the results of the n-th step of integration are the initial conditions of the (n+1) -th step.

To sum up it is seen that Poincaré's control term achieves stabilization.

Canonical approach. We proceed to discussing numerical stability by making use of the canonical theory as an instrument. Let again be H(x,p) a given Hamiltonian.

x and p are n-vectors with components x_i, p_j respectively, (i=1,2,...,n). Let us assume that a canonical introduction of elements was performed. More precisely we assume that a transformation to <u>action-and angle variables</u> α_i, β_j is feasible. This means that the canonical transformation

(60) $$x_i = x_i(\alpha_i, \beta_j) , \qquad p_j = p_j(\alpha_i, \beta_j)$$

is 2π-periodic with respect to each variable $\beta_1, \ldots, \beta_n$ and that the transformed Hamiltonian depends only on the action-type variables:

(61) $$H = H(\alpha_\rho) , \qquad \rho = 1,2,\ldots,m \leq n$$

The transformed equations of motion are

(62) $$\dot{\alpha}_i = 0, \qquad \alpha_i = \text{const} \qquad (i = 1,2,\ldots,n).$$

(63) For $\rho \leqslant m$: $$\dot{\beta}_\rho = \frac{\partial H}{\partial \alpha_\rho} , \; \beta_\rho = \frac{\partial H}{\partial \alpha_\rho} t + c_\rho .$$

(64) For $j > m$: $$\dot{\beta}_j = 0 , \qquad \beta_j = \text{const} .$$

The c_ρ are constants of integration.

As follows for $\rho \leq m$ from eq. (63) the elements β_ρ are linear functions of the independent variable t with angular velocities

(65) $$\omega_\rho = \frac{\partial H}{\partial \alpha_\rho} \qquad \rho = 1,2,\ldots,m.$$

These quantities will also be called <u>fundamental frequencies</u> and $m \leq n$ is the number of fundamental frequencies.

Let us assume now that at least one of the fundamental frequencies - for instance ω_1 - depends on the α_ρ, this is to say on the initial conditions. Hence a modification of the initial conditions does modify this frequency into a new value $(\omega_1 + \Delta\omega_1)$. This implies a change of the function $\beta_1(t)$ by the amount $(\Delta\omega_1)t$. This means <u>linear instability</u>. There results the necessary and sufficient condition for stability;

<u>The fundamental frequencies do not depend on the initial conditions</u>.

A good example is furnished by the Keplerian motion. The Hamiltonian attached to the Newtonian equation (56) is

$$H = \frac{1}{2} p^2 - \frac{K^2}{r} .$$

Delaunay's elements [2] build a set of action-angle variables. They are: ℓ = mean anomaly, g = angular distance node to pericentre, Ω = longitude of node. The conjugated action variables L, G, I determine axis, eccentricity and inclination of the orbit. In terms of these elements the Hamiltonian is

$$(66) \qquad H = -\frac{K^4}{2L^2} \quad .$$

There is one fundamental frequency (m=1)

$$(67) \qquad \omega_1 = \frac{\partial H}{\partial L} = \frac{K^4}{L^3}$$

which depends via L on the initial conditions. Thus the d.eq. (56) is unstable.

In contrast the Hamiltonian pertinent to eq. (58) is by virtue of Poincaré's construction (51)

$$(68) \qquad H^* = r\left(\frac{p^2}{2} - \frac{K^2}{r} + h\right) \quad .$$

By making use of Jacobi's integration method appropriate action-angle-variables can be constructed [21] which reduce the Hamiltonian to the expression

$$(69) \qquad H^* = \sqrt{2h}\ \alpha_1 \quad .$$

The only fundamental frequency

$$(70) \qquad \omega_1 = \frac{\partial H}{\partial \alpha_1} = \sqrt{2\,h}$$

is determined by the Hamiltonian (68) under consideration, and it is thus independent of the initial conditions. Equation (58) is stable.

The gain of accuracy by the transition from (56) to (58) is very spectacular for highly accentric orbits. For an eccentricity e = 0.8, for instance, the gain is about 6 decimals (2 revolutions). Above all the step adaption is responsible for this result, the control term is probably only efficient after many revolutions.

The linearized stability of eqs. (56) (57) (58) was discussed by H.R. Schwarz in [23].

Dissipative stabilization. The foregoing theory is good enough for discussing the stability of a d.eq. under consideration. But no guiding rule for achieving stabilization is established. We describe therefore another line of approach which was adopted in the publication [20]. Let us begin with the general d.eq. (1) of second order

(71) $$\ddot{x} = f(t,x,\dot{x})$$

and let us assume that we know a <u>first integral</u> $N(t,x,\dot{x})$. We adopt the following definition of this notion. The rate of change of any function N on a solution of eq. (71) is

(72) $$\dot{N} = \frac{\partial N}{\partial t} + \frac{\partial N}{\partial x}\dot{x} + \frac{\partial N}{\partial \dot{x}} f \quad .$$

N is a first integral provided the expression on the right-hand side vanishes identically with respect to the three independent variables $t,x,\dot{x}$. Furthermore N is assumed to vanish initially. As an example we quote the law of energy

$$N = \frac{1}{2}\dot{x}^2 + U - E$$

of the conservative problem (8). Other examples are integrals produced by outer constraints. The equation

$$\ddot{x} = -\frac{\dot{x}^2}{x^2}x$$

for a vector x in space has for example as a first integral the scalar product $(x,\dot{x}) = 0$, and by integration to constraint $x^2 = \text{const}$ is obtained. (Motion on a sphere).

From the computational point of view the relation $N(t) = 0$ may be used as a check. It is our objective to go much further by modifying the d.eq. at hand in such a way that an erroneous value of N (due to computational errors) is readjusted to its required value zero. This is achieved by adding the control term $(-Nq)$, where q is a vector to be chosen later on. The modified equation is thus

(73) $$\ddot{x} = f(t,x,\dot{x}) - Nq.$$

As in the case of Poincaré's control term this modification is meaningless from the analytical point of view since N vanishes on the exact solution. There results the differential relation

(74) $$\dot{N} = \left(\frac{\partial N}{\partial t} + \frac{\partial N}{\partial x}\dot{x} + \frac{\partial N}{\partial \dot{x}} f\right) - N\frac{\partial N}{\partial \dot{x}} q \quad .$$

By virtue of our definition of a first integral, the paranthesis vanishes, hence

(75) $$\dot{N} = - N \frac{\partial N}{\partial \dot{x}} \; q \; .$$

The right-hand side should be read as a scalar product. Multiplication with N yields

$$\frac{1}{2} (N^2)^{\cdot} = - N^2 \frac{\partial N}{\partial \dot{x}} q \; ,$$

thus N decreases in absolute value provided

$$\frac{\partial N}{\partial \dot{x}} q > 0 \; .$$

This requirement is satisfied by choosing

(76) $$q = \lambda \frac{\partial N}{\partial \dot{x}} \; ,$$

where λ is a positive function vanishing only in isolated situations. Eq. (75) is thus reformulated as

(77) $$\dot{N} = - \lambda N \left(\frac{\partial N}{\partial \dot{x}}\right)^2$$

and the modified eq. (73) becomes

(78) $$\ddot{x} = f(t,x,\dot{x}) - \lambda N \frac{\partial N}{\partial \dot{x}} \; .$$

A good choice is

(79) $$\lambda = \frac{\gamma}{\left(\frac{\partial N}{\partial \dot{x}}\right)^2}$$

where γ is a constant > 0. In this case N is adjusted exponentially as follows from

$$N(t) = N(0) \, e^{-\gamma t} \; .$$

In practice one expects that this <u>technique of readjustment</u> reduces a numerical error of N, which occurs at the n-th step, during the subsequent computation.

Let us apply this to the d.eq. (8) and its first integral

$$N = \frac{1}{2} \dot{x}^2 + U - E \; .$$

$$\frac{\partial N}{\partial \dot{x}} = \dot{x} \; , \qquad \lambda = \frac{\gamma}{\dot{x}^2} \; .$$

Modified equation:

(80) $$\ddot{x} = - \frac{\partial U}{\partial x} - \frac{\gamma}{\dot{x}^2} \left(\frac{1}{2} \dot{x}^2 + U - E \right) \dot{x} \; .$$

The weight vector $\dot{x}$ of the control term shows clearly that this technique of readjustment is dissipative in character. In contrast Poincaré's control term (52) is conservative since it stems from the Hamiltonian (51). [24] is dedicated to generalizations of the dissipative theory.

Example 5. Keplerian motion, readjustment of the energy balance.

$$\ddot{x} = -\frac{K^2}{r^3}x - \frac{\gamma}{\dot{x}^2}\left(\frac{1}{2}\dot{x}^2 - \frac{K^2}{r} + h\right)\dot{x} \tag{81}$$

In [23] H.R. Schwarz proved that this equation is stable for elliptic motion in the sense of linearized stability. This statement is correct regardless of the value of $\gamma > 0$. In this connection we mention the fact that the choice of γ is restricted by the step-size of the numerical integration. A large value of γ - that is a fast adjustment - is only permitted provided the step is suitably small (cf. section 5).

In contrast to Poincaré's control term the technique under consideration is not linked to analytic step adaption. Of course these two techniques are applicable simultaneously. [20,23]. In the publication [24] the method of readjustment is generalized to the case of several first integrals.

Similar methods were proposed in [18] and [25].

Perturbations. For sake of completness we list the generalizations of the stabilized equations (58)(81) when a perturbing force P occurs in addition to the central attraction.

First set. Independent variable s.

$$x'' - \frac{1}{r^2}(x,x')x' + \left[\frac{x'^2}{2r^2} + h\right]x = r^2 P \quad . \tag{82}$$

$$h' = -(P,x'), \qquad \tau' = \frac{1}{2h}\left[K^2 + \frac{(x,x')}{rh}\;(P,x') + r(P,x)\right] \tag{83}$$

$$t = \tau - \frac{(x,x')}{2hr} \tag{84}$$

Second set. Independent variable t.

$$\ddot{x} + \frac{K^2}{r^3}x + \frac{\gamma}{\dot{x}^2}\left(\frac{1}{2}\;\dot{x}^2 - \frac{K^2}{r} + h\right)\dot{x} = P \quad , \; \dot{h} = -\,(P,\dot{x}) \tag{85}$$

The first set is recommended for highly eccentric orbits, the second set is appropriate for computing many revolutions of an orbit of modest accentricity.

Canonical discussion. In the beginning of this section it was seen that a control term may improve the stability of a differential equation. Consequently we ask the question whether the readjustment technique is able to stabilize and how many first integrals are needed for achieving stabilization. The answer is given by the foregoing discussion of action-angle variables. Clearly the action-type variables $\alpha_1, \alpha_2, \ldots, \alpha_m$ are first integrals since they are constant during the motion. If deteriorated by numerical erros these variables must be readjusted in order to keep the fundamental frequencies

$$\omega_\rho = \frac{\partial H}{\partial \alpha_\rho}, \quad \rho = 1,2,\ldots,m \leq n$$

on their correct values. Hence the number of first integrals needed for stabilization is at least equal to the number of fundamental frequencies.

It is in order to mention the following exceptional case. It may happen that one variable, for instance α_1, does not influence the fundamental frequencies, consequently

$$\frac{\alpha\omega_\rho}{\partial \alpha_1} = 0, \quad \frac{\partial^2 H}{\partial \alpha_1 \partial \alpha_\rho} = 0, \quad \frac{\partial \omega_1}{\partial \alpha_\rho} = 0, \quad \rho = 1,2,\ldots,m.$$

Thus ω_1 is a constant known in advance. We draw the conclusion that the number m of needed first integrals is reduced by the number of fundamental frequencies which are constants independent on the inital conditions.

These results are illustrated by the Keplerian motion. In case of the Newtonian equations (56) the number m of fundamental frequencies was found to be = 1 and thus only one first integral (the energy balance) must be adjusted. This was performed in eq. (81).

As far as the version (58) and the Hamiltonian (68)(69) is concerned we are faced to the exceptional case since the only frequency (70) is an "a priori" constant independent of the initial conditions. No adjustment at all is needed.

Example 6. Gyroscope. (Figure)

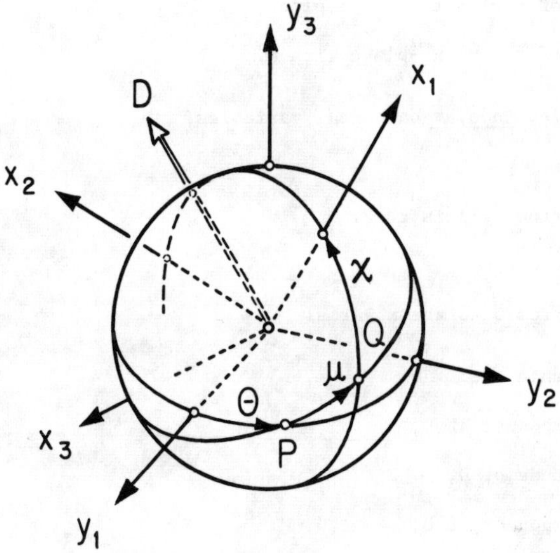

GYROSCOPE

Let be

x_1, x_2, x_3 the frame of the principle axis of inertia,

A, B, C corresponding moments of inertia,

y_1, y_2, y_3 inertial frame fixed in space,

D = vector of angular momentum,

Π = plane perpendicular to D.

Let be on the unit-sphere

P the node of Π with the y_1, y_2-plane,

Q the node of the x_1, x_2-plane with Π.

We use the so called Andoyer canonical variables.

They are:

Θ = angle from y_1-axis to P.

μ = angle P Q.

χ = angle from Q to the x_1-axis.

The conjugate momenta are:

p_Θ = y_3-component of D,

p_μ = magnitude of D,

p_χ = x_3-component of D,

and the Hamiltonian is

$$H = \frac{1}{2}\left(p_\mu^2 - p_\chi^2\right)\left(\frac{\sin^2\chi}{A} + \frac{\cos^2\chi}{B}\right) + \frac{p_\chi^2}{2C} + U , \tag{86}$$

where U is the potential of the exterior forces. In the particular case of a symmetrical top in free motion this reduces to

$$H = \frac{1}{2A}\left(p_\mu^2 - p_\chi^2\right) + \frac{p_\chi^2}{2C} \tag{87}$$

by virtue of A = B and U = 0. Since H depends only on the action variables p_χ, p_μ the transformation into elements is achieved. From

$$\dot{\mu} = \frac{1}{A} p_\mu, \qquad \dot{\chi} = \left(\frac{1}{C} - \frac{1}{A}\right) p_\chi$$

it follows that there are m = 2 fundamental frequencies

$$\omega_\mu = \frac{1}{A} p_\mu \qquad \omega_\chi = \left(\frac{1}{C} - \frac{1}{A}\right) p_\chi \tag{88}$$

Both depend on the initial conditions provided $A \neq C$, and thus the ecceptional case does not occur. During a stabilized numerical integration the first integrals p_μ, p_χ (magnitude of the angular momentum and its x_3-component) should be readjusted. This is unimportant for the free motion of a symmetrical top since a closed analytical solution is available. But it is recommended for handling perturbation as is, for instance, a small difference between A and B or a small exterior force.

4. Regularization. Preliminaries. Unfortunately the theory of transforming a singular differential equation into a regular one is not far developed. Keeping this situation in mind we restrict ourselves to the discussion whether the instruments obtained during the foregoing investigations are fit for achieving regularization.

Let us consider, for instance, Poincaré's transformation (51) which, for purposes of analytic step regulation, introduces the new independent variable s by means of

$$(89) \qquad dt = \mu(x) \; ds$$

and the corresponding Hamiltonian

$$(90) \qquad H^* = \mu\left(\frac{p^2}{2} - E\right) + (\mu U) \quad .$$

It is likely that by appropriate choice of the scaling function $\mu(x)$ a singularity of U can be compensated in order to obtain a regular expression (μU) which leads to a regular Hamiltonian H^*. The corresponding equations of motion

$$(91) \qquad x' = \mu p, \qquad p' = -\left(\frac{p^2}{2} - E\right)\frac{\partial \mu}{\partial x} - \frac{\partial}{\partial x}(\mu U)$$

are regular provided the expressions

$$(92) \qquad \frac{\partial \mu}{\partial x} \quad , \qquad \frac{\partial}{\partial x}(\mu U)$$

are regular together with (μU).

Let us test this proposal for Keplerian motion

$$U = -\frac{K^2}{r}$$

by adopting as a guess

$$(93) \qquad \mu = r^\alpha \; , \qquad dt = r^\alpha \, ds \; , \qquad \mu U = -K^2 r^{\alpha-1} \; ,$$

(94) $$\frac{\partial\mu}{\partial x} = \alpha\, r^{\alpha-2}x\,, \quad \frac{\partial}{\partial x}(\mu U) = -K^2(\alpha-1)r^{\alpha-3}x\,.$$

An easy discussion shows that the foregoing requirements imply $\alpha \geq 3$. But such a choice is not allowed for the following reason. Regularization should permit us to compute an orbit which reaches the singularity at a finite value of the independent variable s, else regularizing is not really of interest. In the book [2, p. 78] it is demonstrated that this desire implies $\alpha < \frac{3}{2}$. Consequently our analytic step adaption does not regularize 3-dimensional Kepler motion.

It is of interest to discuss the one-dimensional case (rectilinear motion). By virtue of $r = x$ the relations (93)(94) reduce to

$$\mu\, U = -K^2 x^{\alpha-1}\,, \quad \frac{\partial\mu}{\partial x} = \alpha\, x^{\alpha-1}\,, \quad \frac{\partial}{\partial x}(\mu U) = -K^2(\alpha-1)x^{\alpha-2}$$

and consequently the value $\alpha = 1$ is singled out by our various conditions. The equations of motion (91) become

(95) $$x' = x\,p\,, \quad p' = -\left(\frac{p^2}{2} + h\right)\,,$$

where $E = -h$. This is a regular set of first-order d.eq.. The resulting second-order equation

$$x'' - \frac{x'^2}{2x} + h\,x = 0$$

is still singular. Only by substituting the energy equation

$$\frac{x'^2}{2x^2} = \frac{K^2}{x} - h$$

the regular oscillator equation

(96) $$x'' + 2\,hx = K^2$$

is obtained.

Coordinate transformations. Whereas analytic step adaption means transformation of the independent variable we proceed now to transformations of the dependent variables which build the state vector x. For the sake of simplicity we restrict ourselves to two-dimensional motion and to conformal transformations. Let x_1, x_2 be the original and u_1, u_2 the transformed coordinates. The mapping

(97) $$x_1 = x_1(u_1,u_2)\,, \qquad x_2 = x_2(u_1,u_2)$$

satisfies the Cauchy-Riemann d.eq.

$$\frac{\partial x_1}{\partial u_1} = \frac{\partial x_2}{\partial u_2} , \qquad \frac{\partial x_1}{\partial u_2} = - \frac{\partial x_2}{\partial u_1} . \tag{98}$$

Let be p_1, p_2 the momenta attached to x_1, x_2 and q_1, q_2 those attached to u_1, u_2. The transformation (97) can be supplemented by a suitable transformation of the momenta such that a canonical transformation is obtained [2, p.196]. The result is

$$q_1 = \frac{\partial x_1}{\partial u_1} p_1 + \frac{\partial x_2}{\partial u_1} p_2 , \qquad q_2 = \frac{\partial x_1}{\partial u_2} p_1 + \frac{\partial x_2}{\partial u_2} p_2 . \tag{99}$$

Or, by virtue of eq. (98)

$$q_2 = - \frac{\partial x_2}{\partial u_1} p_1 + \frac{\partial x_1}{\partial u_1} p_2 ,$$

hence

$$q_1^2 + q_2^2 = (p_1^2 + p_2^2)\Delta , \quad \Delta = \left(\frac{\partial x_1}{\partial u_1}\right)^2 + \left(\frac{\partial x_2}{\partial u_2}\right)^2 . \tag{100}$$

Reformulation in complex notation yields

$$x_1 + ix_2 = z , \qquad u_1 + iu_2 = w ,$$

$$z = f(w), \qquad \Delta = \left|\frac{df}{dw}\right|^2 . \tag{101}$$

According to eqs. (50)(100) the transformed Hamiltonian is

$$H = \frac{q_1^2 + q_2^2}{2\,\Delta} + U . \tag{102}$$

As was expected the singularities of the potential U remain unchanged. In order to manipulate them we add an analytic step size adaption by choosing

$$\mu = \Delta, \qquad dt = \Delta\, ds. \tag{103}$$

After the corresponding Poincaré transformation (47) the new Hamiltonian is

$$H^* = \frac{q_1^2 + q_2^2}{2} + \Delta(U - E) \tag{104}$$

and the new equations of motion are

$$u_1'' = - \frac{\partial}{\partial u_1} [\Delta\cdot(U-E)] , \qquad u_2'' = - \frac{\partial}{\partial u_2} [\Delta\cdot(U-E)] . \tag{105}$$

Consequently the result of the combined transformations is to replace the given potential U by the potential $\Delta(U-E)$. In order to compensate a pole of U at the origin the mapping function f(w) should have an unconformal point there, this is to say $f'(0) = 0$. The simplest analytic function of this behaviour is

$$f(w) = w^n, \qquad n = \text{integer} \geq 2.$$

With $\rho = |w|$ it is found

(106) $$r = \rho^n, \qquad \Delta = n^2 \rho^{2n-2} = n^2 r^{2-2/n}$$

The exponent α defined above is thus

(107) $$\alpha = 2 - \frac{2}{n}.$$

For Keplerian motion:

$$U = -\frac{K^2}{r} = -K^2\rho^{-n}, \qquad \Delta(U-E) = -K^2n^2\rho^{n-2} - n^2E\rho^{2n-2}.$$

The condition for a regular potential is therefore

$$n \geq 2$$

and the earlier condition for reaching the singularity at a finite value of s

$$\alpha = 2 - \frac{2}{n} < \frac{3}{2}.$$

Thus n=2 or n=3. The latter value leads to singular d.eq. n=2 leads to the transformation of Levi-Civita:

(108) $$z = w^2, \quad x_1 = u_1^2 - u_2^2, \quad x_2 = 2u_1 u_2, \quad r = u_1^2 + u_2^2$$

$$\Delta(U-E) = -4K^2 + 4h(u_1^2+u_2^2).$$

Equations of motion:

(109) $$u_1'' + 8\,h\,u_1 = 0, \qquad u_2'' + 8\,h\,u_2 = 0.$$

These d.eq. of harmonic oscillation are not only regular but also stable in the sense of Ljapunow provided h is considered as a fixed quantity which is available in the storage of the computing machine. A variation of initial conditions produces, indeed, modifications of the solutions which remain bounded.

The foregoing discussion of regularization by conformal mapping is due to G.D. Birkhoff [26] who applied it to the restricted problem of three bodies. The discussion raises the feeling that the possibilities of regularization are rather restricted.

KS-transformation. This name was given to the 3-dimensional generalization [4,5] of Levi-Civita's transformation (108). Four generalized coordinates u_1, u_2, u_3, u_4 were needed for representing the rectangular coordinates x_1, x_2, x_3:

$$x_1 = u_1^2 - u_2^2 - u_3^2 + u_4^2 \ , \ x_2 = 2(u_1 u_2 - u_3 u_4), \quad x_3 = 2(u_1 u_3 + u_2 u_4)$$

(110)

$$r = u_1^2 + u_2^2 + u_3^2 + u_4^2 \ .$$

These formulae together with the time transformation

$$dt = r\, ds \tag{111}$$

lead to the transformed differential equations

$$u_j'' + \frac{h}{2} u_j = 0, \qquad j = 1,2,3,4 \tag{112}$$

where h is the negative total energy. The three-dimensional Keplerian motion is therefore equivalent to a four-dimensional harmonic oscillator. The d.eq. (112) are regular as well as stable. The computation of the physical time t as function of s is, of course, not performed by integrating the eq. (111) numerically but by making use of the time element, which was described in section 2.

One may ask the question whether it is necessary to introduce four variables u_j for regularizing a motion with three degrees of freedom. For Kepler motion the answer is affirmative by virtue of the following theorem of B. Gagliardi [27]. Let us consider the following family of admissible transformations. The cartesian coordinates x_1, x_2, x_3 are transformed into three new coordinates u_1, u_2, u_3 by

$$x_i = x_i(u_j) \ , \qquad i,j = 1,2,3 \ .$$

A time transformation of the type

$$\frac{dt}{ds} = \mu(u_i, \dot{u}_i, s)$$

is admitted which introduces the new independent variable s. Velocity is transformed by the formula

$$\frac{dx_i}{dt} = \frac{1}{\mu} \sum_{(\rho)} \frac{\partial x_i}{\partial u_\rho} u_\rho' \ , \qquad u_\rho' = \frac{du_\rho}{ds} \ .$$

Gagliardi proved that such a set of transformation is not able to regularize perturbed Kepler motion. (Perturbing forces are assumed to remain regular if the moving particle collides with the central mass).

From the computational point of view much advantage may be taken by integrating the linear d.eq. (112) instead of the nonlinear d.eq. (56). The monography [2] is dedicated to this guiding idea. We mention in this connection only the fact that the well-known difference formulae for numerical integration can be modified in such a way that the d.eq. (112) of harmonic oscillation is integrated without truncation error. [2,14,15,16]. In [17] also the method of power expansion was adapted to this particular d.eq.

Such techniques reduce the numerical errors during the integration of a perturbed Kepler motion, where a perturbing term appears on the right-hand side of eq. (112).

The linearity of the KS-d.eq. of the Kepler motion facilitates also the discussion of boundary-value and optimization problems [29].

Many other regularizing techniques were proposed. For perturbed Kepler motion we mention [6,7]. Much work was done for regularizing the restricted [11,12] and the general [8,9, 10] problem of three bodies. Finally we draw the attention of the reader to the bibliography [13].

Elements. As in section 2 we give this name to any quantity which is constant or varies linearly during the unperturbed motion. It must be emphasized that all regularization techniques use such elements. In the KS-equations (112), for instance, there appears the energy-element h and the time-element is used for computing t from s. Eight more elements α_j, β_j are obtained by solving the d.eq. (112),

(113) $$u_j = \alpha_j \cos \omega s + \beta_j \sin \omega s \ , \qquad \omega = \sqrt{\frac{h}{2}} \ .$$

The elements α_j, β_j, h vary slowly if our particle is subjected to small perturbing forces; the time-element varies almost linearly.

In classical celestial mechanics the elements of Keplerian motion were defined to be

a = semi-major axis, e = eccentricity, J = inclination,

Ω = longitude of node, ω = argument of pericentre,

M = mean anomaly.

In unperturbed motion M varies linearly whereas the remaining 5 elements are constant.

At this stage it is of interest to discuss once more the event of singularities. The singularity of collision with the central mass is inherent in the Keplerian problem and is thus of physical nature. The KS-elements α_j, β_j, h (+ time-element) are well defined in regular manner for any Keplerian orbit including a collision orbit. In contrast the classical elements do not only break down in case of collision but new singularities are produced by the geometrical nature of these elements. For instance, Ω is undetermined provided $J=0$ and a similar event occurs with ω, e. This type of singularity may be called topological singularity and it should not be confused with physical singularity. For the sake of smooth numerical behavior one should avoid, if possible, topological singularities. Consequently the KS-elements are recommended.

Topological singularities are less dangerous if analytical methods are applied. Nevertheless some effort was made recently [28] to take advantage from the covariance properties of analytical perturbation theories for translating results - obtained with singular elements - to regular elements or to the rectangular coordinates of the particle.

As an example of topological singularities we mention the Andoyer variables (section 3) or the Eulerian angles which are used for determining the position of a spinning rigid body. In contrast the following quaternion variables are regular. (Notations of section 3, "gyroscope").

$\omega_1, \omega_2, \omega_3$ components of the rotation vector in the x-frame,

u_1, u_2, u_3, u_4 components of the quaternion u.

A,B,C principal moments of inertia.

Differential system for free rotation

(114) $$\dot{u} = B(\omega)u \ .$$

(115) $$A\dot{\omega}_1 = (B-C)\omega_2\omega_3 \ , \quad B\dot{\omega}_2 = (C-A)\omega_3\omega_1 \ , \quad C\dot{\omega}_3 = (A-B)\omega_1\omega_2 .$$

$B(\omega)$ is the matrix

(115) $$B(\omega) = \begin{pmatrix} 0 & \omega_3 & -\omega_2 & \omega_1 \\ -\omega_3 & 0 & \omega_1 & \omega_2 \\ \omega_2 & -\omega_1 & 0 & \omega_3 \\ -\omega_1 & -\omega_2 & -\omega_3 & 0 \end{pmatrix}$$

In contrast to the 6 Andoyer variables we are faced here with 7 variables u_j, ω_k. Again the question is important whether such an increase of the number of elements is necessary for obtaining a regular set of elements. Again Gagliardi has shown that redundant elements are necessary for achieving regularity for some particular problems of mechanics. (Gyroscope, Kepler motion).

5. Numerical aspects. Analytical versus numerical step-size adaption. The former as defined in section 2. The transformation (3)

(116) $$dt = \mu(x) \, ds$$

was proposed. In practice it is almost equivalent to order to the automatic computer a variable step-size in t according to the rule

(117) $$h_n = \mu(x_n)k \ ,$$

h_n = step-size to be applied for the n-th step.

k = constant.

x_n = computed values at the beginning of the n-th step.

The given d.eq. (1) is integrated; no transformed equation (4) is established. This numerical step-size control should not be confused with the automatic step-size control which is customary in computing centers. The former uses a scaling function μ chosen in advance whereas the latter adapts the step-size according to the experiences made during the preceeding steps. We made the experience that analytic control is somewhat more efficient than numerical control. Consequently analytic control is recommended provided the transformed d.eq. are not too complicated.

Practical version of the readjustment technique. In section 3 a control term was added to the d.eq. at hand in order to readjust a first integral $N(t,x,\dot{x})$. The thus augmented d.eq. (78) was

$$(118) \qquad \ddot{x} = f(t,x,\dot{x}) - \lambda N \frac{\partial N}{\partial \dot{x}} .$$

In eq. (79) we recommended the choice

$$(119) \qquad \lambda = \frac{\gamma}{\left(\frac{\partial N}{\partial \dot{x}}\right)^2} .$$

By combining this with eq. (77) it turns out that N obeys the d.eq. of readjustment

$$(120) \qquad \dot{N} = - \gamma N.$$

The question of a suitable choice of the constant $\gamma > 0$ was left open. During his numerical experiments G. Gose observed that he obtained best results by choosing $\gamma = \frac{1}{h}$, where h = step-length, and by holding N fixed on the right-hand side of eq. (118) during one step of integration. More precisely this fixed value is chosen as the value N_o of the function N at the beginning of the step. Eq. (120) is thus replaced by

$$(121) \qquad \dot{N} = - \frac{1}{h} N_o ,$$

whence it follows

$$N = - \frac{1}{h} N_o (t-t_o) + N_o ,$$

where t_o is the value of the time at the beginning of the step. At the end $(t=t_o+h)$ one obtains $N = 0$. Thus the first integral is readjusted, in principle, to its nominal value zero during one single step. In order to be explicit, we list the computational program when the trapezoidal rule is used for integrating the system

$$\dot{x} = \xi , \quad \dot{\xi} = f(t,x,\xi) - \frac{1}{h} \frac{N}{\left(\frac{\partial N}{\partial \xi}\right)^2} \frac{\partial N}{\partial \xi} ,$$

which is obtained by splitting up eq. (118) into a system of first order. Then the rule is

$$x_1 = x_o + \frac{h}{2}(\xi_o + \xi_1)$$

$$\xi_1 = \xi_o + \frac{h}{2}(f_o + f_1) - \frac{N_o}{2}\left[\frac{\left(\frac{\partial N}{\partial \xi}\right)_o}{\left(\frac{\partial N}{\partial \xi}\right)_o^2} + \frac{\left(\frac{\xi N}{\partial \xi}\right)_1}{\left(\frac{\partial N}{\partial \xi}\right)_1^2}\right].$$

Experiments. Dissipative stabilization of Kepler motion. Eq. (81) was integrated by making use of the practical version of readjustment as was described above. The notations are explained in eq. (71) - (81) and (118) - (121).

Geometrical data: Cartesian x_1,x_2-frame, plane motion.

Keplerian ellipse, apocentre on the positive x_1-axis.

Semi-major axis a = 1, eccentricity e, K = 1. Computation of $49\frac{1}{2}$ revolutions.

Numerical data: Runge-Kutta method of 4-th order, 100 steps per revolution.

First experiment: Constant step-size, eccentricity e = 0.1.

Errors after $49\frac{1}{2}$ revolutions.

Without readjustment ($\gamma = 0$ in eq. (81)):

$$\Delta x_1 = 1.0 \cdot 10^{-5}, \quad \Delta x_2 = 3.0 \cdot 10^{-3}, \quad N = 5.7 \cdot 10^{-6}.$$

With readjustment:

$$\Delta x_1 = 9.0 \cdot 10^{-7}, \quad \Delta x_2 = 6.3 \cdot 10^{-5}, \quad N = 2.6 \cdot 10^{-9}.$$

Second experiment: Numerical step-size control as was defined above. Step-size

$$\Delta t = r \cdot \frac{2\pi}{100},$$

Eccentricity e = 0.7.

Errors after $49\frac{1}{2}$ revolutions.

Without readjustment:

$$\Delta x_1 = 1.3 \cdot 10^{-2}, \quad \Delta x_2 = 2.4 \cdot 10^{-1}, \quad N = 2.1 \cdot 10^{-4}.$$

With readjustment:

$$\Delta x_1 = 2.6 \cdot 10^{-5}, \quad \Delta x_2 = 7.1 \cdot 10^{-6}, \quad N = 6.9 \cdot 10^{-9}.$$

Remarks. Stabilization is less efficient after an integer number of revolutions.

More examples are contained in [2].

References

[1] V. Szebehely: Theory of orbits. The restricted problem of three bodies, Academic press 1967.

[2] E.L. Stiefel, G. Scheifele: Linear and regular celestial mechanics, Springer 1971.

[3] C.L. Siegel, J.K. Moser: Lectures on celestial mechanics, Springer 1971.

[4] P. Kustaanheimo, E. Stiefel: Perturbation theory of Kepler motion based on spinor regularization, J.reine und angew. Math. 218, 1965, p. 204-219.

[5] E. Stiefel, M. Rössler, J. Waldvogel, C.A. Burdet: Methods of regularization for computing orbits in celestial mechanics, NASA contractor report CR-769, 1967.

[6] H. Sperling: Computation of Keplerian conic sections, ARS journal, 1961, p. 660-661.

[7] C.A. Burdet: Regularization of the two-body problem, ZAMP 18, 1967, p. 434-438.

Theory of Kepler motion: The general perturbed two-body problem, ZAMP 19, 1968, p. 345-368.

[8] K.F. Sundman: Mémoire sur le problème des trois corps, Acta math. 36, 1913, p. 105-179.

[9] G. Lemaître: Regularization dans le problème des trois corps, Acad. Roy. Belg. Bull.d.Sci. (5) 40, 1954, p. 759-767.

[10] J. Waldvogel: A new regularization of the planar problem of three bodies, Cel. Mech., vol. 6, No. 2, 1972.

[11] R. F. Arenstorf: New regularization of the restricted problem of three bodies, Astr. J. 68, No. 8, 1963., p. 548-555.

[12] J. Waldvogel: Die Verallgemeinerung der Birkhoff-Regularisierung auf das räumliche Dreikörperproblem, Bulletin astr. série 3, t. II, fasc. 2, 1967.

[13] H. J. Sperling: Bibliography on the singularities of the equations of motion of celestial mechanics, Second edition, Internal note, Marshall Space Flight Center, Febr. 1970.

[14] E. Stiefel, D.G. Bettis, Stabilization of Cowell's method. Numer. Math. 13, 1969, p. 154-175.

[15] D.G. Bettis, Stabilization of finite difference methods of numerical integration, Cel. Mech., vol. 2, 1970, p. 282-295.

[16] D.G. Bettis: Numerical integration of products of Fourier and ordinary polynomials, Numer. Math. 14, 1970, p. 421-434.

[17] G. Scheifele: On numerical integration of perturbed linear oscillating systems, ZAMP, vol. 22, fasc. 1, 1971, p. 186-210.

[18] P.E. Nacozy: The use of integrals in numerical integrations of the N-body problem, Astrophysics and space science 14, 1971, p. 40-51.

[19] J. Baumgarte: Numerical stabilization of the differential equations of Keplerian motion, Cel. Mech., vol. 5, No. 4, p. 490-501.

[20] J. Baumgarte: Stabilization of constraints and integrals of motion in dynamical systems, Comp. meth. in applied mech. and engineering 1, 1972, p. 1-16.

[21] G. Scheifele: On nonclassical canonical systems, Cel. Mech. 2, p. 296, 1970.

G. Scheifele, E. Stiefel: Canonical satellite theory based on independent variables different from time, Report to ESRO, ESOC contract 219/70/AR, 1972.

[22] U. Kirchgraber: An analytical perturbation theory based on polar coordinates in the four-dimensional KS-space, Thesis ETH Zürich, 1972.

[23] H.R. Schwarz: Stability of Kepler motion, Comp. meth. in appl. mech. and engineering 1. p. 279-299.

[24] J. Baumgarte: Asymptotische Stabilisierung von Integralen bei gewöhnlichen Differentialgleichungen 1. Ordnung. To appear.

[25] E. Hochfeld (ed): Stabilization of Computer Circuits, U. of Chicago, Nov. 1957. Available from US Govt., Printing Office, Washington D.C. as WADC TR 57-425 (AD 155740).

[26] G.D. Birkhoff: The restricted problem of three bodies, Rendiconti Palermo 39, 1915, p. 314.

[27] B. Gagliardi: Ueber die Dimensionserhöhung bei der Regularisierung des Keplerproblems, Thesis ETH Zürich, 1972

[28] U. Kirchgraber: The transformational behaviour of perturbation theories. To appear in Cel. Mech. , vol. 7, No. 4.

[29] K. Kocher: Eine lineare Theorie der Optimierungsprobleme bei Raketen mit kleinem Schub (Low thrust), Thesis ETH Zürich, 1971.

The authors are indebted to Mr. D. Bettis for organizing the conference, to Mr. N. Sigrist for his suggestions concerning the canonical line of approach and to Mr. G. Gose for numerical advice and experiments. Our thanks go to the Swiss Committee on Space Research for sponsoring the collaboration of the two authors.

COMPUTATION OF SOLAR PERTURBATIONS WITH POISSON SERIES

by

R. Broucke

Jet Propulsion Laboratory, Pasadena, California

and University of California, Los Angeles

Abstract

A method is described to compute solar perturbations of a natural or artificial satellite, with the aid of computerized series expansions. The method has been implemented on the Univac - 1108 computer. A worked example is described in detail in the present article.

Introduction

The purpose of this project is to compute first-order perturbations of natural or artificial satellites, by integrating the equations of motion on a computer with automatic Poisson Series Expansions. We prepared software for the treatment of solar perturbations, planetary perturbations, mutual satellite and oblateness perturbations. The basic methods of solution will be illustrated here with the solar perturbation problem only.

A basic feature of our method of solution is that we use the classical variation of parameters formulation rather than rectangular coordinates. However our variation of parameters formulation uses the three rectangular components of the disturbing force rather than the classical disturbing function. There will thus be no problem here to expand the disturbing function in series. However the problem of

*The present research was carried out at the Jet Propulsion Laboratory, California Institute of Technology, under contract NAS7-100, sponsored by the National Aeronautics and Space Administration.

expanding the three disturbing force components is of about the same order of difficulty. Our methods for performing this expansion will be explanined in detail below. Another characteristic of our variation of parameter formulation is that six rather unusual variables are used, in order to avoid singularities at the zero eccentricity and zero (or 90°) inclination, the equinoctial elements (reference 6).

The present work is in fact an extension of our previous researches on planetary perturbations. Our previous work on planetary orbits is characterized by the fact that the two orbit elements ω (argument of periapsis) and Ω (longitude of ascending node) are considered as constants. The numerical values of these angles are substituted in the equations before any expansion is made. On the other hand, in satellite theory, and thus in our present work, the angles ω and Ω are kept in a litteral form in the Fourier series until the expansions and integrations have been made. These angles are then replaced by linear functions of time (rather than constants). The presence of a larger number of litteral variables in the series makes it much more difficult to investigate satellite orbits than planetary orbits.

The most outstanding original aspect of the present work is probably the use of the computer to perform all the fastidious algebra on Fourier and Poisson Series, (References 2,3,5,7). The scope of the work is limited because of the fact that the method is only of first-order. However we think that this work is justified by the fact that the programs are efficient. It takes about two minutes of Univac 1108 computer time to compute the complete solar perturbations of the example given in the last sections of this article, including the verification by numerical integration.

We believe that our present solution could be used for all the natural satelites of the solar system except the moon, the seven outer satellites of Jupiter and the last satellite of Saturn (Phoebe), the limitation being due to the first-order of the theory. Other difficulties that may be encountered in the application of the theory are related to resonances and long-period terms. The results that are presented here have been verified by numerical integration of a few revolutions of the satellite. More extensive verifications over a large number of revolutions are now in progress.

In what follows we will first give a short theoretical exposition of the method (sections 2,3,4, and 5). In the last sections some numerical results will be described, in relation to an Earth-satellite.

2. Equations of Motion

The basic equations of motion of a satellite perturbed by the sun have the general form (reference 1):

$$\ddot{\vec{x}} = -k^2(m_0 + m_1)\frac{\vec{x}}{r^3} - k^2 m_p \left[\frac{\vec{x} - \vec{x}_p}{\Delta^3} - \frac{\vec{x}_p}{r_p^3}\right] \tag{1}$$

The symbols m_0, m_1, m_p respectively represent the masses of the central planet, the perturbed satellite and the perturbing body. The rectangular coordinates of the satellite are $\vec{x}$ while the same quantities for the perturbing body are $\vec{x}_p$. The distances Δ, m, m_p are related to the coordinates $\vec{x}$, $\vec{x}_p$ by the equations

$$\Delta^2 = (x_p - x)^2 + (y_p - y)^2 + (z_p - z)^2 \tag{2a}$$

$$r^2 = x^2 + y^2 + z^2 \tag{2b}$$

$$r_p^2 = x_p^2 + y_p^2 + z_p^2 \tag{2c}$$

Although the above equations of motion are relatively simple, it is not indicated to integrate them directly in rectangular coordinates. The reason for this is that the coordinates are Fast variables. The coordinates change rapidly and by large amounts during a single revolution of the satellite. It is thus more efficient to replace the above three second-order differential equations by six first-order equations with six slowly varying variables. It is standard to use the six classical orbit elements of the orbit as principal variables. However we will rather use six other slow variables which are functions of the classical elements. The classical elements a, e, i, M_o, ω, Ω are too well known to be defined here. The elements that are always used in our work are the so-called equinoctial elements p_α (α= 1,2,...,6). They will also be designated by the letters a, h, k, λ_0, p, q and they are related to the six classical elements by (reference 6):

$$a = a \tag{3a}$$

$$h = e \sin (\omega + \Omega) \tag{3b}$$

$$k = e \cos (\omega + \Omega) \tag{3c}$$

$$\lambda_0 = M_0 + \omega + \Omega \, , \tag{3d}$$

$$p = \tan \frac{1}{2} \sin \Omega \, , \tag{3e}$$

$$q = \tan \frac{1}{2} \cos \Omega \, . \tag{3f}$$

These equinoctial variables have a remarkable advantage over the classical elements, because the corresponding equations of motion have no singularities when i = 0, i = $0^\circ.0$ or i = $90^\circ.0$. These equations of motion may be written in the form (references 1 and 4):

$$\frac{dp_\alpha}{dt} = \sum_{i=1}^{3} \frac{\partial p_\alpha}{\partial \dot{x}_i} X_i = \frac{\partial p_\alpha}{\partial \dot{\vec{x}}} \cdot \vec{X} \; ; \; (\alpha = 1,2, \ldots 6). \tag{4}$$

We generally use greek indices (running from 1 to 6) for the elements and latin indices (running from 1 to 3) for the three coordinates x, y, z or velocity components. The vector $\vec{X}$ = (X, Y, Z) represents the disturbing acceleration, which, according to equation (1), can be written

$$\vec{X} = -k^2 m_p \left[\frac{\vec{x} - \vec{x}_p}{\Delta^3} - \frac{\vec{x}_p}{r_p^3} \right] . \tag{5}$$

It should be pointed out that the equations of motion (1) and (4) are completely equivalent. In other words no approximation whatsoever has been made in deriving equations (4) from equations (1). The partial derivatives of the equinoctial elements p_α with respect to the velocity components $\dot{X}_i$ have previously been described in detail, (references 4 and 6).

3. The Basic Method of Solution

We solve the equations of motion (4) by first representing all the quantities present in the right-side by Fourier Series or Poisson Series in several variables. In the present work, the Poisson Series contain only one polynomial variable (time t). The arguments of the sines and cosines in these series are all assumed to be linear functions of time. Again in the present application, we have restricted ourselves to a maximum of five angles, (this restriction is related to a practical argument only; the 36-bit wordlength of a Univac 1108 computer). More precisely the five angles that are used here are (see references 2,3,7):

M_1 = Undisturbed mean anomaly of the satellite around the central planet.

M_2 = Mean anomaly of the planet in a Keplerian ellipse around the sun.

M_3 = Mean anomaly of a perturbing planet in a Keplerian ellipse around the sun; (this angle will not be used in the present case of solar perturbations).

M_4 = angle ω ($= \omega_0 + \dot{\omega}t$) of the satellite's orbit.

M_5 = angle Ω ($= \Omega_0 + \dot{\Omega}t$) of the satellite's orbit.

The integration process starts by assuming that all the orbit elements present in the right-side of (4) are constants (including ω and Ω). These right-sides are then simple Poisson-series which can be obtained with the use of the Bessel expansions of the two-body problem and with the iteration methods explained below. These Poisson series can then be integrated term by term and a first-order solution is obtained. From this first-order solution the secular rates $\dot{\omega}$ and $\dot{\Omega}$ become known (together with the perturbed mean motion n), and the integration of all the Fourier series can be performed again (in a slightly improved way). The result is thus a little better than a pure first-order solution. In fact the operations are all performed in a single integration (or in a single pass through the computer programs!)

The first-order solution can thus be written as follows

(6) $$p_\alpha(t) = p_\alpha(0) + \int_0^t \sum_{i=1}^{3} \frac{\partial p_\alpha}{\partial \dot{x}_i} X_i \, dt$$

As can be seen from the theory of equinoctial elements (see reference 6), the partial derivatives of the two-body problem are Poisson series of the form

(7) $$\frac{\partial p_\alpha}{\partial \dot{x}_i} = \sum_{j_1,j_4,j_5} (A_{\alpha i} + B_{\alpha i} t) \binom{\sin}{\cos} (J_1 M_1 + j_4 M_4 + j_5 M_5),$$

where j_1 is non-negative and where j_4 and j_5 are less than 3 in absolute value. The coefficients $B_{\alpha i}$ can be non-zero only for the partial derivatives $\partial\lambda_o/\partial\dot{x}_i$. In the next two sections we describe methods of obtaining series expansions for $\vec{X}$.

4. Expansion of $1/\Delta$ with Newton Iterations

The most serious difficulty in the series manipulations solution is in the constructions of the Fourier series for $1/\Delta$ (or $1/\Delta^3$) present in equation (5). This difficulty essentially comes from the fact that only three elementary operations (addition, subtraction and multiplication) can easily be performed on Poisson series. There are no direct rules for the division (or inverse) of such series. For this reason we use a Newton iteration scheme for this purpose (reference 5).

First of all, the series expansions of the coordinates $\vec{x}$ and $\vec{x}_p$ can be obtained with the use of Bessel functions. The unperturbed coordinates $\vec{x}$ of the satellite are pure Fourier series of the form:

(8) $$x_i = \sum_{j_1,j_4,j_5} A_i \binom{\cos}{\sin} (j_1 M_1 + j_4 M_4 + j_5 M_5) \quad .$$

The integer j_1 is non-negative while j_4 and j_5 are less than 2 in absolute value. The coordinates $\vec{x}_p$ of the perturbing body (here the sun around the central planet) are similar series containing only one angle however (the angles ω and Ω of the sun are considered constant and are thus not carried in a litteral form in the series):

(9) $$x_{pi} = \sum_{j_2} B_i \binom{\cos}{\sin} (j_2 M_2) \quad .$$

We can now obtain the series for $A = \Delta^2$ (see equation 2a) by addition, subtraction and multiplication of Poisson series. The problem is then to obtain a series expansion for the quantity

(10) $$y = \frac{1}{\Delta} = A^{-1/2}$$

This can be done in an iterative way. The iteration formula that is needed is obtained by solving the equation

(11) $$F(y) \equiv y^{-2} - A = 0$$

with the classical Newton iteration formula:

(12) $$y_{n+1} = y_n - \frac{f(y_n)}{f'(y_n)} = \frac{1}{2} y_n \left[3 - A y_n^2 \right] \quad .$$

We see that this iteration formula contains no division (of series) operations. On the other hand there are three series multiplications at each iteration. Experiments have shown that in most natural satellite situations (including the moon) only about four iterations with the previous formula are needed. This is essentially due to the quadratic convergence property of Newton's method. The series $1/\Delta^3$ can be obtained from the series $1/\Delta$ by just two supplementary series multiplications.

Finally, the disturbing acceleration vector (see equation 5) is obtained by elementary operations on Poisson series; (The series $1/r_p$ is obtained with a well-known Bessel expansion).

5. Construction of $1/\Delta^3$ with an Encke-type Expansion

Because of the importance and difficulty of the construction of the series $1/\Delta^3$, a second method has been devised. This allows us to perform a very useful verification of this part of the series manipulations. We again start from the expression for Δ^2 as given in equation 2a.

For natural satellites the distance Δ is very nearly equal to r_p and this fact allows us to perform a remarkably converging series expansion similar to the expansion used in Encke's numerical integration method.

$$\Delta^2 = r_p^2 + r^2 - 2(xx_p + yy_p + zz_p)$$

(13)
$$= r_p^2 \left[1 + (\frac{r}{r_p})^2 - 2\left(\frac{x}{r_p} \cdot \frac{x_p}{r_p} + \frac{y}{r_p} \cdot \frac{y_p}{r_p} + \frac{z}{r_p} \cdot \frac{z_p}{r_p} \right) \right]$$

$$= r_p^2 \; [1 - \varepsilon],$$

where the (small) series ε is defined by:

(14)
$$\varepsilon = 2(\frac{x}{r_p} \cdot \frac{x_p}{r_p} + \frac{y}{r_p} \cdot \frac{y_p}{r_p} + \frac{z}{r_p} \cdot \frac{z_p}{r_p}) - (\frac{r}{r_p})^2 \quad .$$

Now we have the following expression for $1/\Delta^3$:

(15)
$$\frac{1}{\Delta^3} = \frac{1}{r_p^3} [1 - \varepsilon]^{-3/2}$$

and by using the binomial expansion we get the following elegant result:

(16)
$$\frac{1}{\Delta^3} = \frac{1}{r_p^3} \left[1 + \frac{3}{2}\varepsilon + \frac{3.5}{2.4} \varepsilon^2 + \frac{3.5.7}{2.4.6} \varepsilon + \dots \right]$$

$$= \frac{1}{r_p^3} \; [1 + G(\varepsilon)]$$

where the (small) series G (ε) is defined by

$$G(\varepsilon) = \frac{3}{2}\varepsilon + \frac{3.5}{2.4}\varepsilon^2 + \frac{3.5.7}{2.4.6}\varepsilon^3 + \dots \tag{17}$$

Using this result, the perturbation component for the third body effects may be written in the form:

$$\begin{aligned}\vec{X} &= k^2 m_p \left[\frac{\vec{x}_p - \vec{x}}{r_p^3}(1 + G(\varepsilon)) - \frac{\vec{x}_p}{r_p^3}\right] \\ &= \frac{k^2 m_p}{r_p^3}\left[(\vec{x}_p - \vec{x})\, G(\varepsilon) - \vec{x}\right] ,\end{aligned} \tag{18}$$

The principal advantage of the above formula is in the fact that it avoids the subtraction of nearly equal large numbers $1/\Delta^3$ and $1/r_p^3$, (or series). This problem results in a serious loss of precision if the Newton iteration method and equation (5) are used.

6. Numerical Application to an Earth-Satellite

The orbital elements of this hypothetical earth satellite are:

$$\begin{cases} a = 0.000045\text{AU} = 6731.905 \text{ Km}, \\ e = .02, \\ i = 10^\circ.0, \\ M_o = \omega = \Omega = 0. \end{cases}$$

The following values have also been used for some important constants

$$\begin{cases} m_1 = 10^{-24}, \quad m_o = 1/328900.1 \quad , m_p = 1, \\ k^2 = 0.000295912208, \\ n = 5693^\circ.1588/\text{day}. \end{cases}$$

Before calculations are started, all input data are reduced to canonical units. All lengths are changed by a factor .000045, so that the new unperturbed semi-major axis is one canonical unit of length. Time is changed in such a way that the unperturbed mean motion n is one radian per time unit. Finally all masses are changed, so that the central planet (the earth) has a mass of one unit.

Next, we make the classical series expansions of the two-body coordinates and partial derivatives. These series are easy to obtain and are not reproduced here.

It should also be mentioned here that there are a few very important input parameters used by the computer programs for series manipulations. These are the so-called epsilon's determining the truncation limit in the Poisson series expansions. In the present programs the limit is taken to be the product of two numbers, εL, where L is the largest coefficient of the series (in absolute value) and ε a small input number. For instance if L is 100 and ε is 10^{-10}, then the smallest coefficient that will be carried in the series is 10^{-8}. In other words ε determines the number of digits of the coefficients of the Poisson series. The example that is described below has been worked with different values of ε at different stages of the work:

1. In the construction of the partial derivatives, $\varepsilon = 10^{-14}$.
2. In the construction of the disturbing components, $\varepsilon = 10^{-16}$.
3. In the integration of the series, $\varepsilon = 10^{-15}$.
4. In all other multiplications of Poisson series, $\varepsilon = 10^{-10}$.

In general the different values of the epsilon's are determined experimentally.

7. Expansion of the Perturbation Components.

The first non-negligible operation consists in finding the accurate series expansion for the three perturbation components X, Y, Z, according to the methods described previously in sections 4 and 5. The Keplerian series for the coordinates (x, y, z, x_p, y_p, z_p) are obtained by classical Bessel Expansions. Then all the other series are derived by computerized series manipulations.

One of the most important series, Δ^2 (see equation 2a) is given below:

$$\begin{aligned}\Delta^2 = 493827161 \\ +44098 \cos (M-M'+\omega+\Omega) \\ -1323 \cos (M'-\omega-\Omega) \\ +441 \cos (2M-M'+\omega+\Omega) \\ +337 \cos (M+M'+\omega-\Omega) \\ -10 \cos (M'+\omega-\Omega) \\ +6 \cos (3M-M'+\omega+\Omega) \\ +3 \cos (2M+M'+\omega-\Omega) \\ +2 \cos (M+M'-\omega-\Omega).\end{aligned}$$

As was said above, the units are canonical units. M' represents the mean anomaly of the earth around the sun, while M, ω, Ω refer to the satellite. These angles are all in radians. The actual series used had 26 rather than 9 terms as shown above. The small terms have not been copied here in order to preserve space. The smallest term in the series had a magnitude of 10^{-7}.

Next the series $1/\Delta$ has to be obtained, with the Newton iteration method explained previously (see equation 12). The initial approximation for the iterations is the series for $1/r_p$. This turns out to be a very close guess in the present case and actually only 3 iterations are needed. During these three iterations the terms changed approximately by 10^{-8}, 10^{-12}, and 10^{-18}. The resulting series for $1/\Delta$ has exactly 50 terms ranging from 10^{-5} to 10^{-20}. The 9 most important terms are given below:

$$\begin{aligned}1/\Delta = 4.50000000D-5 \\ -2.0092D-9 \cos (M - M' + \omega + \Omega) \\ +6.03D-11 \cos (M' - \omega - \Omega) \\ -2.01D-11 \cos (M-M' + \omega + \Omega) \\ -1.54D-11 \cos (M + M' + \omega - \Omega) \\ +4.\ D-13 \cos (M' + \omega - \Omega)\end{aligned}$$

$$-3.\ D\text{-}13 \cos\ (3M-M' + \omega + \Omega)$$

$$-1.\ D\text{-}13 \cos\ (M + M' - \omega - \Omega)$$

$$-1.\ D\text{-}13 \cos\ (3M - M' + \omega + \Omega)$$

In the above series, the usual FORTRAN symbol D is used to designate powers of 10.

In order to obtain an independent verification, the Encke-type of expansion (see section 5) has also been used in order to obtain the series expansions for the disturbing accelerations. We will only give here the series ε (equation 14) and $G(\varepsilon)$ (equation 17). In the present case, terms up to the fourth power in ε have been added in order to construct the series $G(\varepsilon)$. The nine most important terms of the series ε are as follows

$$\begin{aligned} \varepsilon = & -8.9298D\text{-}5 \ \cos(M-M'+\omega+\Omega) \\ & +2.679D\text{-}6 \ \cos(M'-\omega-\Omega') \\ & -8.93D\text{-}7 \ \cos(2M-M'+\omega+\Omega) \\ & -6.83D\text{-}7 \ \cos(M-M'+\omega-\Omega) \\ & +2.1D\text{-}8 \ \cos(M'+\omega-\Omega) \\ & +1.3D\text{-}8 \ \cos(3M-M'+\omega+\Omega) \\ & -7.D\text{-}9 \ \cos(2M+M'+\omega-\Omega) \\ & -4.D\text{-}9 \ \cos(M+M'-\omega-\Omega) \\ & -2.D\text{-}9 \end{aligned}$$

The actual series ε that has been used for the integration had 31 terms, the smallest being about 10^{-19}. On the other hand the series $G(\varepsilon)$ had 104 terms ranging from 10^{-4} to 10^{-20} in magnitude. The 10 largest terms are given below.

$$\begin{aligned} G(\varepsilon) = & -1.33948D\text{-}4 \ \cos(M-M'+\omega+\Omega) \\ & +4.019D\text{-}6 \ \cos(M'-\omega-\Omega) \\ & -1.339D\text{-}6 \ \cos(2M-M'+\omega+\Omega) \\ & -1.025D\text{-}6 \ \cos(M-M'+\omega-\Omega) \end{aligned}$$

$$+3.1D-8 \quad \cos(M'+\omega-\Omega)$$
$$-2.0D-8 \quad \cos(3M-M'+\omega+\Omega)$$
$$-1.0D-8 \quad \cos)2M+M'+\omega-\Omega)$$
$$+7.D-9 \quad \cos(2M-2M'+2\omega+2\Omega)$$
$$-7.D-9 \quad \cos(M+M'-\omega-\Omega)$$
$$+4.D-9$$

With the above results, it is now possible to obtain the series expansions of the disturbing acceleration $\vec{X}$ in two different ways, with equation (5) or with equation (18). The result is identical in both cases and this gives a worthwhile verification. The 3 components of $\vec{X}$ are:

$$\begin{aligned} X = &- 44 \cos(\omega + \Omega) \\ &- 133 \cos(2M' - \omega - \Omega) \\ &- 1 \cos(2M' + \omega - \Omega) \\ &+4460 \cos(M - 2M' + \omega + \Omega) \\ &+ 34 \cos(M + 2M' + \omega - \Omega) \\ &+ 44 \cos(2M - 2M' + \omega + \Omega) \\ &+ 14 \cos(2M + \omega + \Omega) \\ &+ 11 \cos(M + \omega - \Omega) \\ &+1486 \cos(M + \omega - \Omega). \end{aligned}$$

$$\begin{aligned} Y = &- 44 \sin(\omega + \Omega) \\ &- 133 \sin(2M' - \omega - \Omega) \\ &- 1 \sin(2M' + \omega - \Omega) \\ &-4460 \sin(M - 2M' + \omega + \Omega) \\ &+ 34 \sin(M + 2M' + \omega - \Omega) \\ &- 44 \sin(2M - 2M' + \omega + \Omega) \\ &+ 14 \sin(2M + \omega + \Omega) \\ &- 11 \sin(M + \omega - \Omega) \\ &+1486 \sin(M + \omega + \Omega). \end{aligned}$$

and

$$\begin{aligned} Z = &\ 15 \sin\omega \\ &-520 \sin(M + \omega) \\ &- 5 \sin(2M + \omega). \end{aligned}$$

In order to facilitate the writings, the coefficients in the above three series have all been multiplied by a constant factor 10^{11}. Except for this factor, the three series are in canonical units. The magnitude of the components is in the vicinity of 10^{-7}. A numerical verification also showed that the above series represent the expression (5) with a precision of 10^{-15}. Of course the actual series contain a few more small terms than what is shown above. We also see a remarkable symmetry between the X and Y - components. This is essentially due to the special choice of initial conditions for the present particular problem.

We also observe a well known classical property in the series for X, Y, and Z: There are three kinds of terms; long-period terms, solar terms and short-period terms The terms on the first line of X and Y are long-period terms, while on the second and third line we have terms with a period of half a year. All the other terms are short-period terms. The series Z contains one long-period term and two short-period terms. Of course the three perturbation components X, Y, Z are expected to be pure periodic functions and no constant term is present in the corresponding series. For the same reason no secular or mixed secular terms are present.

8. Series Expansion of the Derivatives of Orbit Elements

The next important operation is to construct the series for the derivatives of the six orbit elements, according to the equations (4). It is seen that three series multiplications have to be performed for each of the six orbit elements. This is thus a non-negligible part of the total amount of work involved in this problem. Also in order to have another verification of the work, two different programs have been made, using different orbit elements: classical or equinoctial elements, as was explained in section two. The two integrations give the same series for the semi-major axis, but all other five orbit elements are different. However we can easily convert the classical element series to equinoctial series, by taking derivatives of (3).

Many important features of the equinoctial elements become then apparent. For instance, whenever there is a secular term in ω and Ω (i.e. a constant term in $\dot{\omega}$ and $\dot{\Omega}$), we see that there will be long period terms in $\dot{h}$, $\dot{k}$, and $\dot{p}$, $\dot{q}$. The coefficients of $\cos(\omega + \Omega)$ in $\dot{h}$ and $\sin(\omega + \Omega)$ in $\dot{k}$ are $\pm e(\dot{\omega} + \dot{\Omega})$, while the coefficients of $\cos \Omega$ in $\dot{p}$ and $\sin \Omega$ in $\dot{q}$ are $\pm \tan \frac{i}{2} \dot{\Omega}$. The secular perturbation rates in the orbit elements M_o, ω, Ω can thus be extracted from some important long-period terms in the equinoctial orbit elements. In any way the formulas (3) and their derivatives can be used to convert classical series to equinoctial series, and this is a useful verification of both calculations. On the other hand it should be kept in mind that for small e and i, the classical element series have an inherent lack of precision due to the presence of singularities in the classical variation of parameter equations.

The series for the quantities $\dot{a}$, $\dot{e}$, $\dot{i}$, $\dot{\Omega}$, $\dot{h}$, $\dot{k}$, $\dot{\lambda}_0$, $\dot{p}$ and $\dot{q}$ are now given below. The series for $\dot{M}_o$ and $\dot{\omega}$ are omitted because of their length. The different quantities are in canonical units and radians, but they have all been multiplied by a constant factor 10^{11}. Again, only the largest terms of each series are reproduced here, in order to conserve space.

$$\begin{aligned}\dot{a} = \; & 265 \sin(M-2M'+2\omega+2\Omega) \\ & +57 \sin(M) \\ & -8846 \sin(2M-2M'+2\omega+2\Omega) \\ & -135 \sin(2M+2\omega) \\ & -265 \sin(3M-2M'+2\omega+2\Omega)\end{aligned}$$

$$\begin{aligned}\dot{e} = \; & -221 \sin(2M'-2\omega-2\Omega) \\ & +33 \sin(M-2M'+2\Omega) \\ & -6639 \sin(M-2M'+2\omega+2\Omega) \\ & +1430 \sin(M) \\ & -101 \sin(M+2\omega) \\ & +33 \sin(M+2M'-2\Omega) \\ & +44 \sin(2M-2M'+2\omega+2\Omega) \\ & +14 \sin(2M) \\ & -2209 \sin(3M-2M'+2\omega+2\Omega) \\ & -33 \sin(3M+2\omega) \\ & -88 \sin(4M-2M'+2\omega+2\Omega).\end{aligned}$$

$$\dot{i} = -390 \sin(2M'-2\Omega) - 23 \sin(M-2M'+2\omega+2\Omega) + 23 \sin(M+2\omega) + 387 \sin(2M-2M'-2\omega+2\Omega) - 384 \sin(2M+2\omega)$$

$$\dot{\Omega} = -2215 + 2215 \cos(2M'-2\Omega) - 44 \cos(M-2M'+2\Omega) + 133 \cos(M-2M'+2\omega+2\Omega) + 88 \cos(M) - 132 \cos(M+2\omega) - 44 \cos(M+2M'-2\Omega) - 2228 \cos(2M-2M'+2\omega+2\Omega) + 2211 \cos(2M+2\omega) + 17 \cos(2M+2M'+2\omega-2\Omega) - 44 \cos(3M-2M'+2\omega+2\Omega) + 44 \cos(3M+2\omega).$$

$$\dot{h} = 42 \cos(\omega+\Omega) + 221 \cos(2M'-\omega-\Omega) - 6637 \cos(M-2M'+\omega+\Omega) - 33 \cos(M-2M'+\omega+3\Omega) - 101 \cos(M+\omega-\Omega) - 1430 \cos(M+\omega+\Omega) - 33 \cos(M+2M'+\omega-\Omega) - 88 \cos(2M-2M'+\omega+\Omega) - 133 \cos(2M-2M'+3\omega+3\Omega) - 14 \cos(2M+\omega+\Omega) + 2208 \cos(3M-2M'+3\omega+3\Omega) + 33 \cos(3M+3\omega+\Omega) + 88 \cos(4M-2M'+3\omega+3\Omega)$$

$$\dot{k} = -42 \sin(\omega+\Omega) - 221 \sin(2M'-\omega-\Omega) - 6637 \sin(M-2M'+\omega+\Omega) + 33 \sin(M-2M'+\omega+3\Omega) - 101 \sin(M+\omega-\Omega) + 1430 \sin(M+\omega+\Omega) + 33 \sin(M+2M'+\omega-\Omega) - 88 \sin(2M-2M'+\omega+\Omega) + 133 \sin(2M-2M'+3\omega+3\Omega) + 14 \sin(2M+\omega+\Omega) - 2208 \sin(3M-2M'+3\omega+3\Omega) - 33 \sin(3M+3\omega+\Omega) - 88 \sin(4M-2M'+3\omega+3\Omega)$$

$$\dot{p} = -193\cos(\Omega) + 195\cos(2M'-\Omega) + 11\cos(M-2M'+2\omega+3\Omega) - 11\cos(M+2\omega+\Omega) - 195\cos(2M-2M'+2\omega+3\Omega) + 193\cos(2M+2\omega+\Omega)$$

$$\dot{q} = +193\sin(\Omega) - 195\sin(2M'-\Omega) - 11\sin(M-2M'+2\omega+3\Omega) + 11\sin(M+2\omega+\Omega) + 195\sin(2M-2M'+2\omega+3\Omega) - 193\sin(2M+2\omega+\Omega)$$

$$\dot{\lambda}_0 = -2896 - 101\cos(2M'-2\Omega) + 466\cos(M-2M'+2\omega+2\Omega) + 101\cos(M) - 8882\cos(2M-2M'+2\omega+2\Omega) - 101\cos(2M+2\omega) - 155\cos(3M-2M'+2\omega+2\Omega)$$
$$+t\left\{398\sin(M-2M'+2\omega+2\Omega) + 85\sin(M) - 13269\sin(2M-2M'+2\omega+2\Omega) - 203\sin(2M+2\omega) - 398\sin(3M-2M'+2\omega+2\Omega) - 10\sin(4M-2M'+2\omega+2\Omega)\right.$$

It can be seen in the above series that $\dot{h}$ and $\dot{k}$ (also $\dot{p}$ and $\dot{q}$) are very symmetric. This is due to the special choice of initial conditions for the present example. It can be seen that there are several kinds of terms in these series: constants, short-period terms, solar terms and long-period terms. Besides these terms, we also see that $\dot{\lambda}_0$ has mixed secular periodic terms, with the time t only appearing with the first power.

As was said before, the secular rates are present or can be easily extracted from the previous series. In the case of a planetery satellite perturbed by the sun it is known that the orbit elements a, e, and i have no secular perturbations. The secular rates for M_o, ω, and Ω obtained from the classical element solution as well as the equinoctial solution are as follows, (in radians per canonical time unit):

$$\dot{\bar{M}}_o = -.5008 \times 10^{-7}$$
$$\dot{\bar{\omega}} = +.4327 \times 10^{-7}$$
$$\dot{\bar{\Omega}} = -.2215 \times 10^{-7}.$$

9. Integration of the Derivatives of the Orbit Elements

The integration of the Fourier series given in the previous section presents no essential difficulty. The primitives are all simple sine and cosine functions, because the angles M, M', ω,Ω are treated as ordinary linear functions of time. The rates of these 4 linear functions have to be known. For M' the rate is simply the Keplerian mean motion of the sun's orbit, while for the satellite the rate of change is the perturbed value of the mean motion; i.e., the Keplerian mean motion augmented by the secular perturbation rate of M_o given at the end of the previous section. For the angles ω and Ω, we also use the secular rates given in the previous section.

The most serious problem that arises in the integration of the series is related to the long-period terms. Let us assume that we have the first-order long-period perturbation in the eccentricity:

$$\dot{e} = \varepsilon \cos \omega$$

where ω is equal to $\omega_o + \varepsilon At$. In other words ω is equal to it's constant initial value augmented by a first-order secular perturbation εAt. The symbol ε represents the usual small parameter of the problem and is assumed to be a first-order quantity while εA is the secular rate of ω .

Integrating now the above expression gives (omitting the constant of integration):

$$e = \int \varepsilon \cos \omega \, dt = \sin(\omega_0 + A\varepsilon t)/A = (\sin \omega)/A$$

We see thus that the result is again a long period-term, but with order one. The fact that the order of a long-period term decreases by one unit upon integration is well-known in satellite theory, (reference 8, page 66). This phenomenon is at the origin of a small denominator problem (and thus a loss of precision) in the integration of the series.

If however we would treat $\cos \omega$ as a constant $\cos \omega_o$ during the integration, then the result would be a secular term of first-order

$$e = \int \varepsilon \cos \omega_0 \, dt = \varepsilon t \cos \omega_0$$

The same type of secular term is also obtained if we expand $\cos(\omega_0 + \varepsilon At)$ before integration, (assuming εAt small)

$$\cos(\omega_0 + \varepsilon At) = \cos \omega_o - \varepsilon At \sin \omega_0.$$

Then we have

$$e = \int \varepsilon \cos \omega \, dt = \varepsilon t \cos \omega_0 + \text{second-order term.}$$

A principal conslusion from the above considerations is that $\dot{e}$ must be correct to second-order in ε in order to obtain e correct to first-order in ε.

If we use the equinoctial variables $(a, h, k, \lambda_0, p, q)$ we always have four trivial long-period terms of order zero. These four terms are:

$$h = e \sin(\omega + \Omega),$$

$$k = e \cos(\omega + \Omega),$$

$$p = \tan \frac{i}{2} \sin \Omega,$$

$$q = \tan \frac{i}{2} \cos \Omega.$$

These four trivial terms always appear in the integrated series with large coefficients (e and $\tan \frac{i}{2}$). These terms are similar to the free oscillation terms in the solution of the rigid-body motion problem. For this reason we call these four terms the free oscillation terms.

After the series of the previous section are integrated term by term, we generally obtain the following types of terms that are characteristic for the solar perturbation problem:

1. The semi-major-axis has only short-period periodic terms.
2. The elements e, i, h, k, p, and q have solar periodic terms and short-period terms only.
3. The orbit elements λ_0, M_0, ω, and Ω have solar periodic terms, short-period terms, one secular term and several mixed secular periodic terms.

Besides these terms there are also the four free oscillation terms and a few long-period terms with very small coefficients. We do not know at this moment if these small terms are due entirely or partly to error propagation during the series manipulations, and for this reason, they are not included in the series below.

As was said before, all our integrations are performed in two different ways: with classical or with equinoctial elements. The results in classical elements are then converted to the equinoctial form and this allows us again to perform a verification of the integrations. The conversion formulas that have been used are again obtained by taking derivatives of (3).

We will now give the truncated Poisson series for the orbit elements a, e, i, λ_0, h, k, p, q, and Ω.

$$\begin{aligned} a = & -265\cos(M-2M'+2\omega+2\Omega) \\ & -57\cos(M) \\ & +4423\cos(2M-2M'+2\omega+2\Omega) \\ & +67\cos(2M+2\omega) \\ & +88\cos(3M-2M'+2\omega+2\Omega) \end{aligned}$$

$$\begin{aligned} i = & \quad 1191\cos(2M'-2\omega-2\Omega) \\ & +1128089\cos(2M'-2\Omega) \\ & +23\cos(M-2M'+2\omega+2\Omega) \\ & -23\cos(M+2\omega) \\ & -1193\cos(2M-2M'+2\omega+2\Omega) \\ & +192\cos(2m+2\omega) \end{aligned}$$

$$\begin{aligned} e = & \quad 668\cos(M'-\omega-\Omega) \\ & -32\cos(M'+\omega-\Omega) \\ & +639328\cos(3M-2\omega-2\Omega) \\ & -37\cos(2M'-2\omega-2\Omega) \\ & -33\cos(M-2M'+2\omega+2\Omega) \\ & +6642\cos(M-2M'+2\omega+2\Omega) \\ & -1430\cos(M) \\ & +101\cos(M+2\omega) \\ & -33\cos(M+2M'-2\Omega) \\ & -22\cos(2M-2M'+2\omega+2\Omega) \\ & +736\cos(3M-2M'+2\omega+2\Omega) \\ & +11\cos(3M+2\omega) \\ & +22\cos(4M-2M'+2\omega+2\Omega) \end{aligned}$$

$$\begin{aligned} \Omega = & +639771\sin(2M'-2\Omega) \\ & -65\sin(M''-\omega-\Omega) \\ & +29\sin(M'+\omega-\Omega) \\ & -6444\sin(2M'-2\omega-2\Omega) \\ & +49\sin(2M'+2\omega-2\Omega) \\ & +11\sin(3M'-\omega-3\Omega) \\ & -44\sin(M-2M'+2\Omega) \\ & +133\sin(M-2M'+2\omega+2\Omega) \\ & +88\sin(M) \\ & -132\sin(M+2\omega) \\ & -44\sin(M+2M'-2\Omega) \\ & -1114\sin(2M-2M'+2\omega+2\Omega) \\ & +1105\sin(2M+2\omega) \\ & -14\sin(3M-2M'+2\omega+2\Omega) \\ & +14\sin(3M+2\omega) \\ & -2215\,t\,. \end{aligned}$$

$$
\begin{aligned}
h = {} & +639289 \sin(2M'-\omega-\Omega) \\
& + \quad 37 \sin(2M'+\omega-3\Omega) \\
& + \ 3907 \sin(2M'+\omega-\Omega) \\
& \quad\ \ 32 \sin(M'-2\Omega) \\
& + \ \ 670 \sin(M') \\
& + \ 3907 \sin(2M'-\omega-3\Omega) \\
& - \ 6640 \sin(M-2M'+\omega+\Omega) \\
& - \quad 33 \sin(M-2M'+\omega+3\Omega) \\
& - \ \ 101 \sin(M+\omega-\Omega) \\
& - \ 1430 \sin(M+\omega+\Omega) \\
& - \quad 33 \sin(M+2M'+\omega-\Omega) \\
& - \quad 44 \sin(2M-2M'+\omega+\Omega) \\
& - \quad 66 \sin(2M-2M'+3\omega+3\Omega) \\
& + \ \ 736 \sin(3M-2M'+3\omega+3\Omega) \\
& + \quad 11 \sin(3M+3\omega+\Omega) \\
& + \quad 22 \sin(4M-2M'+3\omega+3\Omega)
\end{aligned}
$$

$$
\begin{aligned}
k = {} & +639289 \cos(2M'-\omega-\Omega) \\
& - \quad 37 \cos(2M'+\omega-3\Omega) \\
& + \ 3907 \cos(2M'+\omega-\Omega) \\
& - \quad 32 \cos(M'-2\Omega) \\
& + \ \ 670 \cos(M') \\
& - \ 3907 \cos(2M'-\omega-3\Omega) \\
& + \ 6640 \cos(M-2M'+\omega+\Omega) \\
& - \quad 33 \cos(M-2M'+\omega+3\Omega) \\
& + \ \ 101 \cos(M+\omega-\Omega) \\
& - \ 1430 \cos(M+\omega+\Omega) \\
& - \quad 33 \cos(M+2M'+\omega-\Omega) \\
& + \quad 44 \cos(2M-2M'+\omega+\Omega) \\
& - \quad 66 \cos(2M-2M'+3\omega+3\Omega) \\
& + \ \ 736 \cos(3M-2M'+3\omega+3\Omega) \\
& + \quad 11 \cos(3M+3\omega+\Omega) \\
& + \quad 22 \cos(4M-2M'+3\omega+3\Omega)
\end{aligned}
$$

$$
\begin{aligned}
p = {} & - \ \ 563 \sin(2M'-2\omega-3\Omega) \\
& - \ 4317 \sin(2M'-3\Omega) \\
& +564080 \sin(2M'-\Omega) \\
& + \quad 11 \sin(M-2M'+2\omega+3\Omega) \\
& - \quad 97 \sin(2M-2M'+2\omega+3\Omega) \\
& + \quad 96 \sin(2M+2\omega+\Omega) \\
& - \quad 11 \sin(M+2\omega+\Omega)
\end{aligned}
$$

$$
\begin{aligned}
q = {} & + \ \ 563 \cos(2M'-2\omega-3\Omega) \\
& + \ 4317 \cos(2M'-3\Omega) \\
& +564080 \cos(2M'-\Omega) \\
& + \quad 11 \cos(M-2M'+2\omega+3\Omega) \\
& - \quad 97 \cos(2M-2M'+2\omega+3\Omega) \\
& + \quad 96 \cos(2M+2\omega+\Omega) \\
& - \quad 11 \cos(M+2\omega+\Omega)
\end{aligned}
$$

$$
\begin{aligned}
\lambda_0 = &- 172\sin(M'-\omega-\Omega) \\
&- 19282\sin(2M'-2\omega-2\Omega) \\
&- 294448\sin(2M'-2\Omega) \\
&+ 865\sin(M-2M'+2\omega+2\Omega) \\
&+ 187\sin(M) \\
&+ 11\sin(M+2\omega) \\
&- 7760\sin(2M-2M'+2\omega+2\Omega) \\
&- 101\sin(2M+2\omega) \\
&- 96\sin(3M-2M'+2\omega+2\Omega)
\end{aligned}
\quad + t\left\{
\begin{aligned}
&-2896 \\
&- 398\cos(M-2M'+2\omega+2\Omega) \\
&- 85\cos(M) \\
&+ 6635\cos(2M-2M'+2\omega+2\Omega) \\
&+ 101\cos(2M+2\omega) \\
&+ 132\cos(3M-2M'+2\omega+2\Omega)
\end{aligned}
\right.
$$

The above series can be used for the calculation of the six orbit elements at any given time. The orbit elements can then be converted to position and velocity by using well-known formulas, (reference 6).

It should be noted that the series for $\lambda_o(=M_o+\omega+\Omega)$ may suffer from a bad error propagation problem due to the presence of mixed secular terms. In order to avoid this problem another version of the program has been prepared. The new program uses the variable $\lambda(=M+\omega+\Omega)$ rather than λ_o. The variables λ and M have no mixed secular terms. However, it is necessary to perform a double integration in order to obtain M or λ (see reference 9, page 285). We found that the double integration results in a rather dangerous error propagation on the long-period terms. Because of this fact we feel that both formulations, with λ_o or with λ, are worthwhile.

Acknowledgements

The author wishes to thank M. Ananda and C. Chao from the Jet Propulsion Laboratory, for several interesting suggestions and some programming assistance.

References

1. R.A. Broucke, "Iterative Perturbations in Rectangular Coordinates," Celestial Mechanics, Vol. 1, no. 1, pages 110-126, 1969.

2. R.A. Broucke and K. Garthwaite, "A Programming System for Analytical Series Expansions on a Computer," Celestial Mechanics, Vol. 2, no. 2, pages 9-20, 1970.

3. R.A. Broucke, "How to Assemble a Keplerian Processor?" Celestial Mechanics, Vol. 2, no. 1, pages 9-20, 1970.

4. R.A. Broucke, "On the Matrizant of the Two-Body Problem," Astronomy and Astrophysics, Vol. 6, pages 173-182, 1970.

5. R.A. Broucke, "Construction of Rational and Negative Powers of a Series," Comm. of Assoc. of Comp. Machines, pages 32-35, January 1971.

6. R.A. Broucke, "Properties of the Equinoctial Orbit Elements," Celestial Mechanics, Vol. 5, no. 1, January 1972.

7. R.A. Broucke and G. Smith, "Expansion of the Planetary Disturbing Function," Celestial Mechanics, Vol. 4, pages 490-499, 1971.

8. J. Kovalevsky, "Introduction to Celestial Mechanics." Springer-Verlag, New York, 1967.

9. D. Brouwer and G. Clemence, "Methods of Celestial Mechanics," Academic Press, New York, 1961.

NUMERICAL DIFFICULTIES WITH THE GRAVITATIONAL N-BODY PROBLEM

by

R.H. Miller

University of Chicago

Chicago, Illinois

Abstract

The gravitational n-body problem is unusually difficult to integrate reliably. In this review, the motion of a set of mass points under their forces of self-gravitation is asserted to be unstable in the sense of a hydrodynamical flow. While this does not imply that an n-body stellar system must collapse or explode, it does mean that we are attempting to treat an unstable system of differential equations by numerical methods. Many of the proofs of convergence or of numerical stability of integration methods are not applicable because they presuppose stability of the differential equations. The history of the problem is reviewed, experimental evidence is given in support of the assertion of instability, and a plea is made for arguments to justify inferences based on numerical experiments even though the differential equations may be unstable.

1. Introduction

The gravitational n-body problem has a long and distinguished history. The case n = 2 is the well-known Kepler problem, while n = 3 is the famous "three-body problem" that has provided the stimulus for the development of several mathematical techniques and has attracted the attention of such celebrated mathematicians as Poincaré. Larger values of n may be expected to lead to even more intractable problems, although the big jump in difficulty appears to be in going from n = 2 to *n* = 3. In asking for the help of mathematicians and of numerical analysts with the numerical integration of *n*-body problems, we are inviting your attention to a difficult, but potentially fruitful, area that has enough to it to challenge any mathematician.

As astronomers, we have some peculiar problems such that computer experiments with *n*-body systems are practically the only recourse we have to study certain kinds of systems. We would like to kick a galaxy to see if it bounces, or to go around to look at the other side. We would like to watch some of these systems change; stars in clusters move, clusters are dynamical objects, but we don't have enough detail to determine the true orbits of the stars. Things just take too long to happen. So we need computer experiments to provide a laboratory in which to study these systems. We need to be able to put a thermometer in the mouth of a stellar system, or to see how it behaves if something is thrown at it. But, as experimenters, we must know that our apparatus works--and this is the problem that we are here to discuss today.

A pretty dismal picture will be painted--at least in part because this is an appeal for help; but the help we need is not necessarily in directions that have been stressed at this meeting. Some of the speakers to follow will paint a rosier picture and will stress that one can get useful results. I, too, use these methods and believe some of the results; but I want to stress that all is *not* well in this business. As the title indicates, there are some serious problems. But we need answers that can only be provided by the technique of computer experiments, so we feel that it is worth considerable effort to understand just what is going on and how the experiments work in practice.

Among the various simulations of star clusters and galaxies that have been run, this discussion will be concentrated on attempts to integrate the exact differential equations with the force computed between each pair of particles, the entire calculation being done as carefully as possible. A lot of interesting approximate methods will be omitted—particle-in-cell methods, lattice methods, spherical shell approximations, and so on.

The calculations considered are initial value problems--the evolution is studied from some (possibly arbitrary but otherwise well-defined) initial state, according to self-consistent equations of motion. This method has produced good results in celestial mechanics and with satellite orbit integrations, as you have heard. But the results, when the method is applied to stellar dynamical problems (simulation of star clusters, galaxies, etc.) have not been as satisfactory.

2. Description of Difficulties

The difficulty observed with gravitational n-body calculations may be described by the trajectory of its representative point in the $6n$-dimensional phase space (Γ - space). If two systems are represented, there will be two points and two trajectories. Suppose these trajectories are arranged to start from points that are very close together; as the systems evolve and the two trajectories will separate as the two points move apart, as in the sketch:

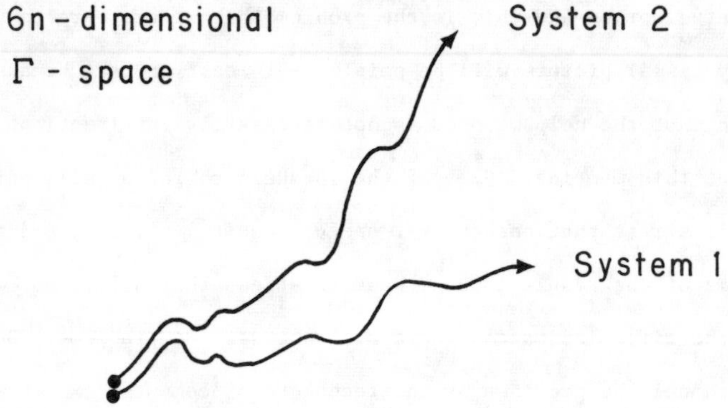

This may be made quantitative by introducing a normed separation,

$$\Delta^2 = \sum_i \{ (x_i^{(1)} - x_i^{(2)})^2 + \frac{T^2}{m^2}(p_i^{(1)} - p_i^{(2)})^2 \} ,$$

where $x_i^{(1)}$, $x_i^{(2)}$; $p_i^{(1)}$, $p_i^{(2)}$ are the coordinates and momenta of the two systems respectively ($i = 1,2, \ldots 3n$ for n particles). The separation, Δ , was computed and served as a measure of the amount by which two systems diverged; it showed a characteristic exponential growth as a function of time, with superposed large "spikes." The spikes were identified with close encounters, but the underlying exponential growth is the feature that indicates trouble with the integration. Typical plots of $\ln \Delta$ as a function of the time-variable, t, are shown in Figure 1 (for an 8-particle system) and Figure 2 (for 12 bodies). The dynamical time ("crossing-time") for the clusters is about 32 for Figure 1 and 90 for Figure 2. The change in $\ln \Delta$ is about 1 per dynamical time unit for the 8-body system of Figure 1, it is about 2 per dynamical time unit for the 12-body system of Figure 2, and increases to about 10 for 32-body systems. Note the large range, and the rapid growth, of Δ. The experiment must be terminated rather soon, or it will no longer be evident that the two systems had a common origin. These results have been confirmed by others working with the gravitational n-body calculation.

What does this mean from a practical standpoint? It certainly indicates troubles in trying to compute some properties of n-body systems. It also implies that the calculation may be dominated by error growth after a fairly short time; all the while maintaining that peculiar deterministic property of computations. Some quantities are well determined--the energy and the other first integrals (constant velocity of centroid, angular momentum = 10 first integrals; the hypersurface in the 6n-dimensional Γ - space on which these 10 integrals are constant will be called the "integral hypersurface"). There is one exact integral--the particle number, n. It would be quite surprising, after an integration step, to find that there were 3 more particles than there were before the step! But other quantities may not be so well determined. The quantity, Δ, is probably much more sensitive to numerical error than most quantities of interest. But Lecar's [1] comparative study showed very dramatically that the effect persists into some quantities that we would like to determine from the calculations. Lecar's comparison was a set of "standard" initial conditions for a 25-body system that each of several different workers ran; the results were intercompared at several standard values of t. The comparison is dramatic because it showed that even simple quantities which one might expect to be reliably computed are surprisingly unreliable. Not only do different numbers of particles escape from the clusters computed by various investigators, but even the identity of the escaping particles is different. Even so simple a parameter as the radius of a sphere containing half the particles varied by as much as a factor of two among the various calculations. Similar difficulties were encountered in the intercomparison of other test quantities.

The test involving Δ is more stringent still; the same computer working on the same program with sets of data that may differ only at the roundoff level soon leads to markedly differing systems. It is not as dramatic as Lecar's study because the tests for differences deal with subtler quantities.

We are dealing with what is apparently one of the most unstable problems that is the object of serious computational effort.

3. The Questions

This leads to the two major questions to be raised: A) Given some property that we want to compute--how can we compute it reliably? Is it possible to compute it reliably? B) As a subsidiary to (A), what is the origin of the growth? An understanding of the origin is necessary to find what remedies may be available. If the problem is strictly numerical, a remedy may be available in better numerical methods. However, if the problem arises from the physical system (or, alternatively, in the structure of the differential equation system), better numerical methods may not be of much help.

We assert that the trouble is of physical origin--that the numerical problem arises from trying to solve an inherently unstable system of differential equations. Unfortunately, this is not an assertion that has been proven mathematically; rather it stands primarily on experimental evidence, obtained through computer experiments. But just as we have been unable to prove instability, so, too, we have been unable to prove stability. And it is noteworthy that all the discussions of stability of numerical methods that we have heard at this meeting presuppose the stability of the differential equations being solved.

The evidence in support of the assertion that the exponential growth of the separation of phase points has its origin in the structure of the differential equations (is of physical origin) is as follows. (1) The perturbation used is substantially larger than roundoff or truncation error in any one step. (2) The result is insensitive to change of integration step size. Runs with half the usual integration step size yielded tracks of $\ln \Delta$ vs t that were identical except for a little more detail when started from identical initial data. (3) Different workers, using different integration methods, have confirmed the result. (4) Standish [2] showed that the basic slope was reduced by the introduction of a near-cutoff in the force-law. The spikes were also affected--they were much less sharp and "spiky." (5) Similar results may be obtained by heuristic arguments of the properties of n-body systems that do not depend on computer integrations of the n-body system[3]. (6) Direct integration of perturbation equations that describe the difference between the two systems also reproduces the effect[4]. More will be said of the direct integration of the perturbation equations later.

It appears that we are dealing with something like a mixing flow. A proof of this conjecture would be very difficult if it is possible at all. (The only proof of mixing properties for a nontrivial system that I know of is that of Sinai for the Boltzmann hard-sphere gas; a full proof has evidently not yet been published.) The proof must be complicated by the fact that the integral hypersurface is not measurable. The "something like mixing" conjecture, coupled with the infinite integral hypersurface implies that almost all points of the phase space lie on escape orbits. Most experimenters would agree with that--escapes are a common feature of the n-body calculations. The conjecture about mixing is not new: Ulam made a similar conjecture at the Thessaloniki conference in 1964 [5], along with some interesting comments on the implications for reliable computation of certain functionals.

To return to question (A), what recourse do we have in the face of a situation that clearly presents substantial difficulties for reliable numerical computation? Abandonment of the entire program of gravitational n-body computation is not regarded as an admissible solution, although it might be by far the most practical.

The first possible recourse is to restrict computations to times so short that numerical errors don't have much effect. This is not practical, because the time scale for numerical errors is on the order of, or less than, the dynamical time scale of the clusters being modelled. Alternatively, it might be admitted that properties that develop on a time scale longer than that of the error growth are not reliably determined. Again, we want to learn about such properties.

A second recourse might be to appeal to ergodicity and say that the calculation allows us to sample the "interesting" part of the integral hypersurface (all particles fairly near each other) in such a way that what we call time averages are actually ensemble averages. (It would be lovely if we could prove ergodicity for this system--a much weaker condition than mixing.) There are several objections to this viewpoint. First, we don't want a true ensemble average, since the infinity in the phase volume on the integral hypersurface occurs for a pair of particles very close together in a tight binary with the others well separated and at infinite distance. Second, the calculation has no proper random element to assure unbiased sampling--it is deterministic. Third, it is not clear that error-driven processes sample the interesting part of the integral hypersurface properly. And, finally, there is a question of how long to wait between measurements to obtain independent samples.

The spirit of this second recourse is almost that of Monte Carlo, and raises the question whether integration brings any real advantage over Monte Carlo techniques. Most of us who do _n_-body integrations do not want to use Monte Carlo methods, because we feel that there are certain built-in prejudices (no particle correlation, for example) in the usual Monte Carlo sampling that might mask or unintentionally avoid the very kinds of things that may be important in stellar dynamics. I've considered trying a calculation in which the system is made to do a random walk around the integral hypersurface; it is fairly straightforward to force the system on the integral hypersurface as closely as desired [6,7]--a technique that might be used in conjunction with a random walk to mimic an _n_-body calculation. But one feels that integrations should have more to recommend them than such peculiar sampling methods. The difficulty is that it is not clear just what the advantage is. Ironically, it turns out that integration is one of the cheapest ways of generating sequences of (phase-space) configurations that are consistent with the same values of all of the first integrals. But that scarcely justifies the attachment of meaning to the integration in an evolutionary sense.

A third recourse is to design the experiments so that there is some measure of the internal consistency of a set of runs. This can be done by running a variety of initial conditions, each consistent with the desired properties of the system being studied, and comparing the results. This is a fairly standard method; the input parameter space is sampled in a Monte Carlo sense. This brings an important advantage in an aspect of the problem that has not been stressed so far in this report; typically, a given initial condition (positions and velocities specified for each particle) specifies far more than is actually known about any stellar system (as distinct from the solar system in celestial mechanics problems, or from input data for satellite orbit integrations), and the parameter space that is consistent with the small amount of available information should somehow be sampled. The results Wielen will report in the next talk rely heavily on this method.

How much effort is it worthwhile to put into improving numerical techniques for this problem? There may be quite a difference of opinion on this matter, so let me label these next few remarks as my opinion, based on the presumption that the underlying reasons for the numerical difficulties is physical. While we want to avoid poor numerical methods, there is probably not much reason to go to a lot of pains to improve the stability of numerical methods.

However, and this is an important point, almost any procedure that would allow us to make valid inferences in spite of the difficulties would be extremely helpful. Some ways of living with the problems are desperately needed. It is sometimes argued that the departure of a computed system from the physical system it is supposed to mimic adequately takes account of real physical effects that are not built into the problem as formulated. In the gravitational *n*-body problem, for example, numerical effects might mimic the effect of the galaxy on the star cluster being studied, or of other irregularities in the background force field. That seems to be a dangerous attitude; it is certainly safer to have all error terms understood, with such effects being intentionally introduced in a controlled and understood manner. It is also frequently argued that departures of the computed system from the physical system occur in such a way that the computed system "heads for more probable regions of the phase space." The same objections hold again--how does one know that a computed system that cannot follow the track of a physical system will populate the phase space with the same or even a similar probability distribution? It would be very helpful to have good, solid answers to these assertions, rather than the mere expressions of discomfort that appear in this paragraph.

Certainly the numerical difficulties make it almost impossible to conduct carefully controlled experiments. For example, it would be difficult to design a numerical experiment to test the effect of surrounding each particle with a hard sphere in addition to its gravitational force field, because differences between systems with and without the hard sphere could easily be no greater than would appear between two systems, neither of which had the hard sphere interaction, but were somehow slightly disturbed to force them down different evolutionary tracks.

While on the topic of numerical methods, it may be pointed out that there are some matters of taste in the methods used--matters for which it may be difficult to demonstrate a practical importance, but which, if incorporated, are an aesthetically pleasing feature of the program. An example of this arises with the Liouville theorem. Consider a set of points in the Γ - space, each point representing a possible state for the $\underline{n}$-body system. The Liouville theorem states that the Lebesgue measure of that set of points is invariant under the mappings of the Γ - space onto itself that are induced by the evolution of the $\underline{n}$-body system according to the equations of motion. As customarily formulated, the theorem presupposes a continuous phase space. But the computer number representation is discrete, so the representable states of the $\underline{n}$-body system lie on a lattice; if magnified suitably, the representable states might be considered to lie on the $6\underline{n}$-dimensional lattice of integers. At each integration step, the point representing a $\underline{n}$-body system jumps from one of these lattice points to another. The properties of computer arithmetic guarantee that any lattice point will map onto some lattice point after the integration step, but they do not guarantee that the mapping is one-to-one. It may happen that several lattice points might map onto one point (several sources have the same target), even in the case where the number representation has not crossed a boundary at which the exponent changes, requiring a rescaling of the lattice. With the interpretation that the measure refers to the number of systems represented by each lattice-point, the number of systems that occupy any one lattice point may change during the calculation, in violation of the Liouville theorem. Certain single-step methods have the property that the mapping is one-to-one (a single step midpoint rule, for example), and thus have a Liouville theorem (and are reversible). They thus have the possibly unimportant, but aesthetically pleasing, feature referred to. Are there other methods that enjoy this property? Can the property be assured for certain multistep, higher-order methods? This would be useful information for the practitioners of the _n_-body calculation art.

4. Recent Experiments Relating to Stability

If the system is unstable numerically as a consequence of a physical instability, that physical instability is in the sense of an unstable flow. It is not a situation in which there is an equilibrium which is being tested to see whether motion in the neighborhood of the equilibrium solution remains in that neighborhood; the usual situation for stability analysis. Rather, stability in the neighborhood of a dynamically changing state is being tested. The system is unstable in the sense that near any phase point (at any given instant) is another phase point with the property that the trajectories passing through the two phase points separate--that the images of the two original phase points at a later time have a much greater separation. At very long times, the separation may be unbounded, but at relatively short times, the separation may grow exponentially. Mixing flows are usually unstable in this sense, and Sinai's proof of mixing flows for the Boltzmann hard sphere gas leads to a similar exponentiation of the separation of phase points. The instability is not one that would lead the star cluster to blow up or to collapse;

but it is an instability in the usual sense of unstable differential equations.

A proof of instability (in this sense) would be useful to remove the matters of opinion from this subject.

The subject can be investigated along the following lines. Let $\underline{p}^{(1)}$, $\underline{q}^{(1)}$, and $\underline{p}^{(2)}$, $\underline{q}^{(2)}$ represent two systems in the Γ - space. These are not to be regarded as representatives of ensembles. Let $\delta p = p^{(2)} - p^{(1)}$ and $\delta q = q^{(2)} - q^{(1)}$ be "small" quantities. The motion of either system is determined by the usual Hamiltonian equations of motion, governed by a Hamiltonian $\underline{H}(\underline{p},\underline{q})$ which is a function of position in the Γ - space, smooth almost everywhere. Equations of motion for $\delta\underline{p}$ and $\delta\underline{q}$ can be constructed by forming a Taylor series expansion for $H(p^{(2)},q^{(2)})$ in terms of $H(p^{(1)},q^{(1)})$ and powers of $\delta\underline{p}$, $\delta\underline{q}$. The linerarized form of these equations is

$$(1) \qquad \delta\dot{q} = \frac{\partial^2 H}{\partial q \partial p}\,\delta q + \frac{\partial^2 H}{\partial p^2}\,\delta p = \frac{\partial \dot{q}}{\partial q}\,\delta q + \frac{\partial \dot{q}}{\partial p}\,\delta p$$

and

$$(2) \qquad \delta\dot{p} = -\frac{\partial^2 H}{\partial q^2}\,\delta q - \frac{\partial^2 H}{\partial p \partial q}\,\delta p = \frac{\partial \dot{p}}{\partial q}\,\delta q + \frac{\partial \dot{p}}{\partial p}\,\delta p$$

where the derivatives are to be evaluated at $p^{(1)}$, $q^{(1)}$. Some care is required to remain in a regime in which the linear terms of the expansion suffice. The actual values to be inserted for the derivatives depend on the details of evolution of the unperturbed system. The second form, obtained by inserting the canonical equations, shows that the equations are just what one would expect them to be.

The perturbation equations can be written in the usual matrix form for a (time-dependent) homogeneous system of equations

$$(3) \qquad \dot{\xi} = M\xi,$$

where ξ is written as a 2-component vector $\binom{\delta q}{\delta p}$ (it actually has 6 $\underline{n}$ components). The elements of the matrix $M = M(t)$ can be read off from Eqs. (1) and (2). The time dependence enters through the unperturbed motion of one of the two systems being considered. In a canonical coordinate system based on cartesian coordinates, with forces that are not velocity dependent (the usual formulation for a computer calculation), the matrix reduces to $3\underline{n} \times 3\underline{n}$ blocks:

$$(4) \qquad M = \left(\begin{array}{c|c} 0 & (\underline{\text{Diag}}\ 1/m) \\ \hline (\underline{\text{Grad}}\ F) & 0 \end{array} \right)$$

The ($\underline{\text{Diag}}$ 1/m) in the upper righthand corner is a $3\underline{n} \times 3\underline{n}$ diagonal matrix with the particle masses on the diagonal, and the element in the lower lefthand corner is the $3\underline{n} \times 3\underline{n}$ gradient of the forces. The matrix, M is the Jacobian matrix of the system, or is the Hessian of the Hamiltonian. This substitution of a linearized problem reduces the problem of an unstable flow to that about an equilibrium: the solution $\xi = 0$ is clearly an equilibrium solution; the problem is to determine whether it is a stable equilibrium.

Eq. (3) does not lend itself to any of the usual methods of studying stability solutions. The matrix M continues to vary with the time forever--it never settles down to a nice constant matrix. It is not periodic with any reasonable period, and has huge variations in its elements, although it is continuous.

Some experimental studies were conducted in which the system (3) was integrated as an $\underline{n}$-body system evolved. The (euclidean)norm of $||\ \xi\ ||$ is just the Δ introduced earlier; the Δ so obtained behaved just as did that obtained from earlier experiments. The numerical calculations do things that the mathematical solution indicates that they should not do. For example, they promptly develop components along the gradients to the first integrals (components not lying in the integral hypersurface). However, the difference-vector ξ, tends to lie principally along the trajectory, swinging from forward to backward (something else that the mathematical solution says it should not do). Its components outside the integral hypersurface are much smaller than the components within the integral hypersurface.

As the difference-vector grows with the passage of time, the relative sizes of the components lying in and orthogonal to the integral hypersurface remain about the same.

A picture emerges in which the numerical difference-vector prefers to go along the trajectory, but it makes only small excursions in directions orthogonal to the integral hypersurface, preferring to make up the remainder of its length in the hypersurface but in directions other than that of the trajectory. The volume in which it can swim about grows with the time, retaining its elongated disk-like shape. This volume is not subject to the Liouville theorem; this is another way in which the calculated system differs from the mathematical system.

Truncation and roundoff error can be introduced as an inhomogeneous part of Eq. (3) in the usual way. Whether the differential equation system Eq. (3) has exponentially growing solutions or not, the numerical systems are experimentally found to have them. These experiments have been reported in detail elsewhere [4].

The system of equation (3) can be integrated explicitly for a rigidly rotating system of particles. This has been done for $\underline{n}$ particles at the vertices of a regular polygon, as a rotating ring. While this configuration is known to be unstable, (for $\underline{n} > 2$), the solution by eigenvalues shows the instability and the unstable modes for $\underline{n} = 3, 4, .., 8$. We will not go into the details here except to note that the most positive eigenvalue of the 8-particle system indicates a blowup by a factor of 2.5 per radian of rotation of the ring--a rather rapid growth of an instability.

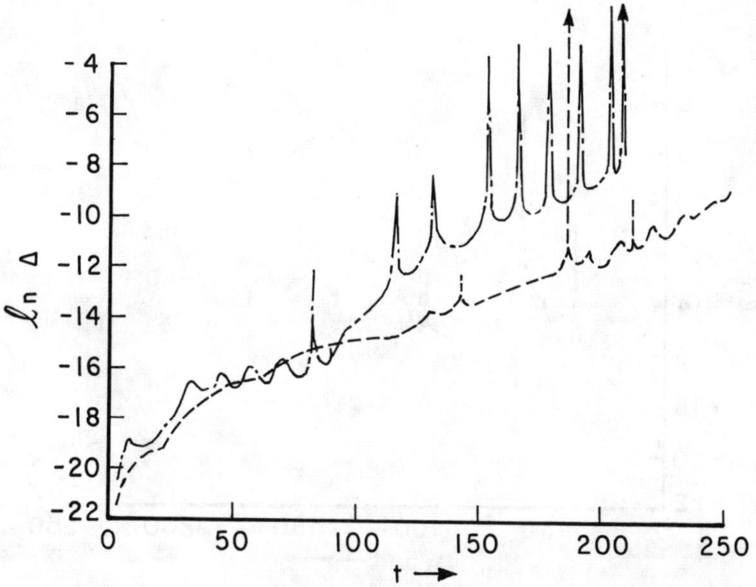

FIG. I

Figure 1. Tracks of the separation of phase points for 8-particle stellar systems. The two tracks represent different initial conditions. The repeated equidistant spikes between $\underline{t} = 150$ and $\underline{t} = 220$ are formed by a single long-lived binary pair, which was destroyed by a close collision of one of its members with another star at $\underline{t} = 225$. A dynamical time unit is about 32 of the units of the abscissa.

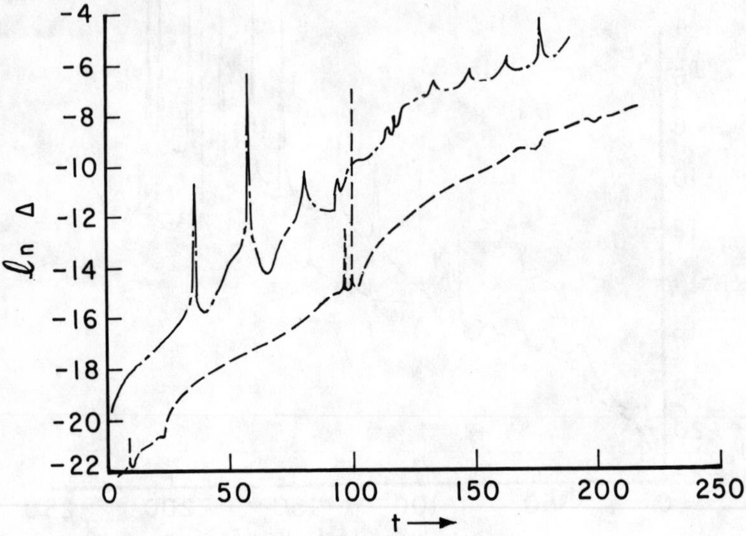

FIG. 2

Figure 2. Twelve particle system; the tracks are labelled as in Figure 1. A dynamical time unit is about 90 of the units of the abscissa.

References

1. M. Lecar, Bull. Astron. (3), (1968), 3, 91

2. E.M. Standish, Thesis, Yale University, 1968, (Unpublished)

3. R.H. Miller, Astrophys. J. (1966), 146, 831

4. R.H. Miller, Journ. Comp. Phys. (1971), 8, 449

5. S.M. Ulam, comment in IAU Symposium No 25, Theory of Orbits in the Solar System and in Stellar Systems, G. Contopoulos, Editor, New York: Academic Press, 1966, p. 140

6. P.E. Nacozy, "The Use of Integrals in Numerical Integrations of the n-Body Problem," in Gravitational N-Body Problem, Proceedings of IAU Colloquium No. 10, M. Lecar, Editor, Dordrecht: D. Reidel, 1972

7. R. H. Miller, Journ Comp. Phys.(1971), 8, 464

ON THE NUMERICAL INTEGRATION
OF THE N-BODY PROBLEM FOR STAR CLUSTERS

by

Roland Wielen

Astronomisches Rechen-Institut

Heidelberg, Germany

Abstract

We describe a numerical procedure of integrating the equations of motion of the gravitational N-body problem, which has been successfully used for integrating systems with up to 500 stars. The method uses polynomials which typically include the fifth time-derivative of the acceleration. The time step varies with time and differs from star to star. A discussion of the observed range of individual time steps indicates a large gain in computational efficiency by using individual time steps instead of integrating all the equations with the same step size.

1. Introduction

We shall describe a numerical procedure for integrating the equations of motion of N mass points under their mutual gravitational attraction. The procedure has been successfully used by Aarseth and by the author for studying the dynamical evolution of star cluster models containing up to 500 stars. In this paper, we shall discuss the basic integration scheme, especially the use of individual step sizes of integration for each particle. Other papers in this volume, especially those by Aarseth, by Ahmad, and by Hayli are devoted to additional refinements, namely to the special handling of close approaches (regularization) and to the introduction of double-individual time steps.

For a detailed discussion of the fundamental problems, the numerical results and the astronomical applications of the N-body experiments with star cluster models, we refer to recent review papers by Aarseth (1973) and by Wielen (1973).

2. Choice of the Method of Integration

The equations of motion for the gravitational N-body problem are given by

$$(1) \qquad \ddot{\underline{r}}_i = \underline{b}_i = \sum_{\substack{j=1 \\ j\neq i}}^{N} \frac{Gm_j}{|\underline{r}_j-\underline{r}_i|^3}(\underline{r}_j-\underline{r}_i) \; ; \quad i = 1,\ldots,N.$$

Here, $\underline{r}_i$ is the position vector of the star no. i, m_i the mass of this star, and G the gravitational constant; $\underline{b}_i$ is the force per unit mass acting on the star. These equations of motion may be extended by adding external gravitational fields or by considering a change of the masses m_i with time. In most cases, these extensions do not introduce any significant additional difficulty for the numerical integration.

For a large cluster, say $N > 100$, the calculations of the forces $\underline{b}_i$ are the most time-consuming part of the computations, mainly for two reasons: (a) each force $\underline{b}_i$ contains N-1 terms, and (b) for each of these terms, a square root operation is necessary, because we derive primarily the square of $|\underline{r}_j-\underline{r}_i|$. The computing time for all other manipulations, beside the force calculations, is negligibly small for high N. Hence, our general rule in constructing an efficient integration scheme for star clusters must be to minimize the number of force evaluations per unit time for a given accuracy.

We get further indications for constructing an efficient scheme by considering the motions of the stars in a cluster. In a typical cluster, we have a strong decrease of the star density from the center outwards. This leads to the following orbital behaviour of the stars: (a) The stars in the dense core of the cluster move very fast (strong mean field) in rather irregular orbits (perturbed by many encounters with other stars); (b) The stars in the outer parts of the cluster, on the contrary, move quite slowly (long orbital periods) on smooth orbits; (c) Many stars move nearly radially in highly excentric orbits, hence they oscillate often between stage (a) and (b); (d) Furthermore, for a few stars (members of binaries, stars involved in close encounters), the time scales of the motions deviate extremely from those of the other stars. These orbital properties of stars in a cluster favor the use of a variable <u>individual</u> step size, differing from star to star and varying with time, if we wish to minimize the total number of force evaluations. If we were adopting the same step size for all the stars, then we would have to choose the smallest of the appropriate individual step sizes. This small collective step size, governed by a few stars in stage (a) or (d), would lead to a large waste of computing time, since the frequency of force evaluations for the other stars, especially those in stage (b), would be unnecessarily high.

The two essential requirements for an efficient integration scheme of the N-body problem of star clusters, namely minimal number of force evaluations and individual step sizes, rule out the use of iterative methods (like those of Runge-Kutta or Bulirsch -Stoer) in which the force has to be calculated at many points within the step interval. Even if these iterative methods could balance their large number of force evaluations per step by a large step size for a given accuracy, they are severely hampered, because they cannot make use of individual step sizes. Iterative methods require a synchronous handling of all the differential equations within each collective step for several reasons. For example, for iterating the motion of a star over a given time interval, we have to know the motions of all the other stars during the whole period. This condition is unrealizable in the case of individual time steps.

It turns out that a finite-difference method seems to be the most efficient scheme for solving numerically the N-body problem for star clusters. First, this method makes efficient use of past information about the forces, thus keeping the number of force evaluations per unit time small. Second, it allows adoption of variable individual time steps. In the following section, we shall outline such a numerical procedure.

3. The Adopted Integration Procedure

The procedure of numerical integration described here is a variation of the general finite-difference predictor-corrector method modified by the introduction of individual and variable step sizes. While the integration scheme has been developed for the gravitational N-body problem, we make no direct use of any property of the forces $\underline{b}_i$, or more generally, of the right-hand side of the system of differential equations. Hence the integration procedure described below can be used for other purposes too. The scheme will work efficiently, if the calculation of the forces $\underline{b}_i$ occupies most of the computing time, and if the natural step sizes for each of the individual differential equations of the system differ drastically.

In his pioneering studies of the gravitational N-body problem, von Hoerner (1960) has used a variable but collective step size for the numerical integration. A. Schlüter (private communications) first proposed the use of individual time steps as well as the semi-iteration described below. While Aarseth (1963, 1971) has presented the formulae for a specific order n of the integration scheme, we shall describe here the general procedure as derived by Wielen (1967) for an arbitrary order n.

Owing to the individual time steps of integration, each star has to be handled almost independently of all the other stars. The coupling between the motions of all the stars is manifest only insofar as we need the instantaneous positions of the other stars when we wish to compute the acceleration $\underline{b}_i$ of the star under consideration from Eq. (1). The needed positions $\underline{r}_j$ have to be derived from some fitting extrapolation polynomials which must be available for all the stars at any time. Each fitting polynomial is valid over a period defined by the current step size h_j of this star. In a table, we list the times t_j at which the validity of the current fitting extrapolation polynomials ends. In order to proceed with the running integration, we search in this table for the smallest time t_j. Let us call this time t_o and the number of the corresponding star i. In the following, we shall suppress the star index i for writing economy.

At the time t_o, we compute first the positions of all the stars from their fitting extrapolation polynomials and then the acceleration $\underline{b}_o$ of star i from Eq.(1). We assume now that we know the acceleration $\underline{b}_\nu$ of star i at n previous instants t_ν (ν=1,2,...,n). We call n the order of our integration procedure. The extrapolating polynomial $\underline{B}(t)$ of degree n through the n+1 pivotal points $\underline{b}_\nu$ (ν=0,1,2,...n) in Newtonian form is given by:

$$\underline{B}(t) = \underline{C}_o + \sum_{k=1}^{n} (\underline{C}_k \prod_{\lambda=0}^{k-1} (t-t_\lambda)) \quad . \tag{2}$$

The coefficients $\underline{C}_k$ are special divided differences:

$$\underline{C}_k = \underline{D}_{o,k} \qquad (k=0,1,\ldots,n) \tag{3}$$

with

$$\underline{D}_{r,o} = \underline{b}_r \qquad (r=0,1,\ldots,n) \ ,$$

$$\underline{D}_{r,s} = (\underline{D}_{r,s-1} - \underline{D}_{r+1,s-1})/(t_r - t_{r+s})$$

$$(r=0,1,\ldots,n-s; s=1,\ldots,n) \quad . \tag{4}$$

For computing efficiency, the Newtonian polynomial (2) must be expressed in terms of powers of $t-t_o$:

$$\underline{B}(t) = \sum_{\lambda=0}^{n} \underline{A}_\lambda (t-t_o)^\lambda \quad . \tag{5}$$

The coefficients $\underline{A}_\lambda$ are derived from the coefficients $\underline{C}_k$ by using generalized Stirling numbers $S_{\mu\nu}$:

$$\underline{A}_\lambda = \sum_{k=\lambda}^{n} \underline{C}_k \, S_{k-\lambda,k} \tag{6}$$

with

$$S_{o,\nu} = 1 \qquad \text{for } \mu=0,$$

$$S_{\mu,\nu} = S_{\mu,\nu-1} - (t_{\nu-1} - t_o) S_{\mu-1,\nu-1} \qquad \text{for } \mu=1,\ldots\ \nu-1$$

$$S_{\nu,\nu} = -(t_{\nu-1}-t_o)S_{\nu-1,\nu-1} \qquad \text{for } \mu=\nu,$$

(7) $$\text{for } \nu = 1,\ldots,n \,.$$

The polynomial $\underline{B}(t)$ is now integrated twice in order to derive the extrapolation polynomial $\underline{R}(t)$ for the position $\underline{r}(t)$ of star i:

$$\underline{R}(t) = \underline{r}(t_o) + \underline{v}(t_o)\,(t-t_o)$$

(8) $$+\sum_{m=2}^{n+2} (\underline{A}_{m-2}/(m(m-1)))\,(t-t_o)^m \,,$$

where $\underline{v} = \dot{\underline{r}}$ is the velocity of star i. Both $\underline{r}(t_o)$ and $\underline{v}(t_o)$ are first provisionally predicted from the former extrapolation polynomial $\underline{R}_{old}(t)$ which was valid up to the time t_o.

In order to improve the values of $\underline{r}(t_o)$ and $\underline{v}(t_o)$, we shall use for a moment the interpolating polynomial $\underline{B}_{int}(t)$ of degree n+1 through the n+2 pivotal points $b_{n+1}, b_n,\ldots,b_1, b_o$. By integrating the difference between $\underline{B}_{int}(t)$ and the former extrapolating polynomial $\underline{B}_{old}$ which was based on the pivotal points $b_{n+1},\ldots,b_1$, we obtain corrections to the provisionally predicted values of $\underline{r}(t_o)$ and $\underline{v}(t_o)$. We get:

$$\underline{r}_{corr}(t_o) = \underline{r}_{pred}(t_o) +$$

$$\underline{C}_{n+1} \sum_{\mu=1}^{n+1} (((t_1-t_o)/(\mu+2))-((t_{n+1}-t_o)/(\mu+1)))S_{n+1-\mu,n+1}(t_1-t_o)^{\mu+1},$$

$$\underline{v}_{corr}(t_o) = \underline{v}_{pred}(t_o) +$$

(9) $$\underline{C}_{n+1} \sum_{\mu=1}^{n+1} (((t_{n+1}-t_o)/\mu)-((t_1-t_o)/(\mu+1)))S_{n+1-\mu,n+1}(t_1-t_o)^{\mu} \,.$$

The coefficient $\underline{C}_{n+1}$ is derived from the highest coefficients of the new and old extrapolating polynomials, $\underline{B}(t)$ and $\underline{B}_{old}(t)$:

(10) $$\underline{C}_{n+1} = (\underline{C}_n-\underline{C}_{n,old})/(t_o-t_{n+1}) \,.$$

The required Stirling numbers follow from an extension of the scheme (7) to $\nu=n+1$. The corrected position $\underline{r}_{corr}(t_o)$ may give rise to a recomputation of the acceleration $\underline{b}_o$, thus starting an iteration. However, experience has shown that the main correction is already applied by the first step using Eqs. (9). This "semi-iteration" yields a considerable increase in accuracy without any remarkable increase in computing time. Hence, the polynomial $\underline{R}(t)$ according to Eq. (8), using the corrected values of $\underline{r}(t_o)$ and $\underline{v}(t_o)$ given by Eqs. (9), represents our numerical solution $\underline{r}_i(t)$ of the gravitational N-body problem for each star.

The step size $h(t_o)$ of the star i limits the validity of the fitting polynomial $\underline{R}(t)$ to the time interval $t_o \leq t \leq t_o+h(t_o)$. We define the time step h by requiring that the difference between the true acceleration $\underline{b}(t)$ and the extrapolating polynomial $\underline{B}(t)$ shall not exceed a given error limit ε in the time interval from t_o to t_o+h. We approximate the true acceleration $\underline{b}(t)$ by the polynomial $\underline{B}_{int}(t)$ discussed earlier, which involves one pivotal point more than $\underline{B}(t)$. Under this assumption, we obtain the following non-linear equation for $h(t_o)$:

$$|\,\underline{b}(t_o+h) - \underline{B}(t_o+h)| \sim |\underline{B}_{int}(t_o+h) - \underline{B}(t_o+h)| =$$

$$|\underline{C}_{n+1}| \prod_{k=0}^{n} (h+t_o-t_k) = \varepsilon. \tag{11}$$

For safety, we limit the relative increase of consecutive time steps: The step size $h(t_o)$ given by Eq. (11) is replaced by $F\cdot h(t_1) = F(t_o-t_1)$ with $F\sim 2$, if $h(t_o)/h(t_1)>F$. For the error bound ε, we use with good success:

$$\varepsilon = \varepsilon_{abs} + \varepsilon_{rel} \cdot |\underline{b}_o| , \tag{12}$$

where ε_{abs} and ε_{rel} are constants. For checking the accuracy of the numerical integration for each individual star, we test at the end of each step whether the difference between $\underline{B}(t_o+h)$ and $\underline{b}(t_o+h)$ derived from Eq. (1) is really of the order of ε as to be anticipated from Eq. (11). Of course, we also use overall error checks by watching the numerically caused change in the values of the integrals of motion, e.g., the total energy E of an isolated cluster.

Starting our integration procedure is no additional task, since the scheme works with arbitrary order n. We use the order n=0 for the first and second step of each star. Then we increase n by one after each step until we reach the prescribed maximum order $n=n_{max}$. Only the initial step sizes h(0) are derived not from Eq. (11) but from Taylor series.

Practical experience and rough theoretical estimates (Wielen 1967) have shown that a suitable order for studying the N-body problem with moderate accuracy is about n=4. Aarseth (1971) has adopted n=3. Using $\varepsilon_{rel}\sim 10^{-3}$, we find a typical relative error of the total energy, $|\Delta E/E|$, per dynamical crossing time, of about 10^{-5}. The computing time on a standard computer (IBM 7090) is of the order of one hour per dynamical crossing time for a cluster of N=100 stars, and is proportional to N^2 for higher N. However, since the actual computing time depends very critically on the detailed structure of each individual cluster, the observed computing times show a very large scatter.

Unfortunately, the gravitational N-body problem is very unsuitable for a convincing test of an integration scheme and for discussing the relative merits of various numerical methods. The reason for that is the basic physical instability of the gravitational N-body problem for star clusters as discussed by Miller elsewhere in this volume. Due to the instability, any small change of the integration scheme can lead after a short time to a very different physical structure of the evolved cluster, and hence to a different error behavior which may fully obscure the intrinsic properties of the numerical method of integration.

4. Individual Time Steps

We shall now discuss in more detail the advantages of using individual step sizes for integrating the gravitational N-body problem. Figures 1 and 2 show the observed distribution of such individual time steps h_i for a few star cluster models. We find that the step sizes h_i are spread over as much as five orders of magnitude;

even the bulk of them is spread over two orders of magnitude. The observed distribution of h_i does not depend strongly on the exact definition of the time step or on the parameters ε and n. Such details affect mainly the absolute values of the step sizes, but not their relative ratios. The relative distribution of the step sizes is usually determined by the physical situation, e.g. the orbital periods of the stars. Furthermore, the distribution of the time steps seems to be quite similar for N=100 and N=500, and it does not depend on the spectrum of stellar masses, comparing clusters which are otherwise very similar to each other. However, even for the same probability distribution of h_i, the ratio between the smallest time step which is actually observed in a cluster and the mean step size decreases with increasing N.

The computing time T_{ind} for a given distribution of individual step sizes h_i is determined by the geometrical average of the time steps:

(13) $$T_{ind} \propto \langle 1/h_i \rangle \quad .$$

If we use a collective step size for all the stars, then the computing time T_{coll} is governed by the smallest individual time step:

(14) $$T_{coll} \propto 1/\min(h_i) \quad .$$

We have

(15) $$(1/\min(h_i))/\langle 1/h_i \rangle \lesssim N.$$

Hence the potential efficiency of the use of individual time steps instead of a collective step size increases proportional to the number of stars in the cluster. Of course, this does not imply that the ratio T_{coll}/T_{ind} will actually reach N because of the following reasons: (a) At least two of the stars will have time steps close to $\min(h_i)$, since the components of a binary or two stars in a close encounter have necessarily almost equal time steps. Hence we have:

(16) $$<1/\min(h_i)>/<1/h_i> \sim N/N_{min} \text{ with } N_{min} \sim 2...5.$$

(b) We lose a factor of two in the computing time, if we use individual step sizes instead of treating all the stars synchronously, because we cannot make use of the fact

(17) $$m_i \underline{b}_{ij} = -m_j \underline{b}_{ji}$$

for the evaluation of the forces.

(c) The absolute accuracy of the integration is higher, if we use $\min(h_i)$ as a collective step size instead of integrating each star with its appropriate individual time step, because the truncation errors for stars with $h_i >> \min(h_i)$ are much smaller in the first case. However, the gained accuracy is more or less useless since, from a physical point of view, it is desirable to compute the orbits of all the stars with the same relative accuracy. This is just done by the individual time steps. Furthermore, the major contribution to the errors of many global quantities like the total energy of a cluster, stems from the stars with the smallest step sizes.

(d) A collective step size allows the use of more sophisticated iterative methods which are perhaps more powerful than the simple method described in Section 3. Since we do not have a reliable estimate how much accuracy we may gain from iterative methods, we shall neglect this effect at present.

From the above considerations, we can estimate the following ratio of the computing times for collective versus individual time steps:

(18) $$T_{coll} / T_{ind} \sim N/2N_{min}$$

Using $N_{min} = 3$, we find a ratio T_{coll} / T_{ind} of about 4 for N=25. Hence for small clusters, the individual time steps do not help very much. For N=500, however, the situation is very different; here we find a ratio T_{coll} / T_{ind} of about 80. For such large clusters, the individual time steps are absolutely necessary for an economic use of computing time. It is also clear that such a large difference in the computing times T_{ind} and T_{coll} cannot be compensated by using a more powerful iterative method in the case of collective time steps.

Figure 1: Distribution of individual time steps h_i in a star cluster model with N = 500 stars. Model FP, described by Wielen (1973), is shown at time $t = 10\ T_{cr}$.

Figure 2: Distribution of individual time steps h_i in star cluster models with N = 100 stars. Unequal masses are represented by model P at time $t = 3\ T_{cr}$, - • - • - , and $t = 10\ T_{cr}$, ---. As an example for equal masses, model E is shown at $t = 18\ T_{cr}$.

FIGURE 1

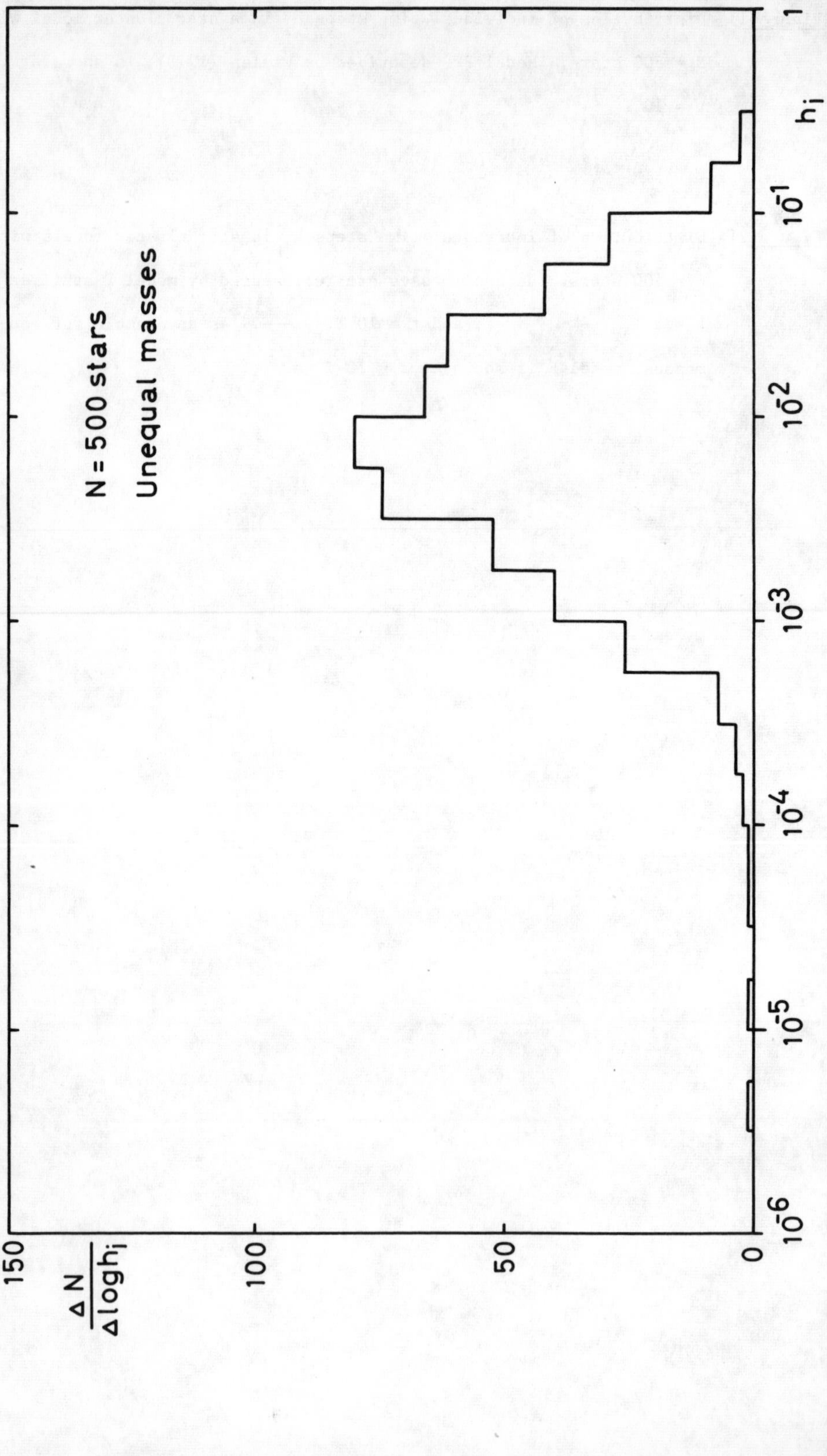
N = 500 stars
Unequal masses
150
100
50
0
ΔN / Δlog h_i
10^-6
10^-5
10^-4
10^-3
10^-2
10^-1
1
h_i

FIGURE 2

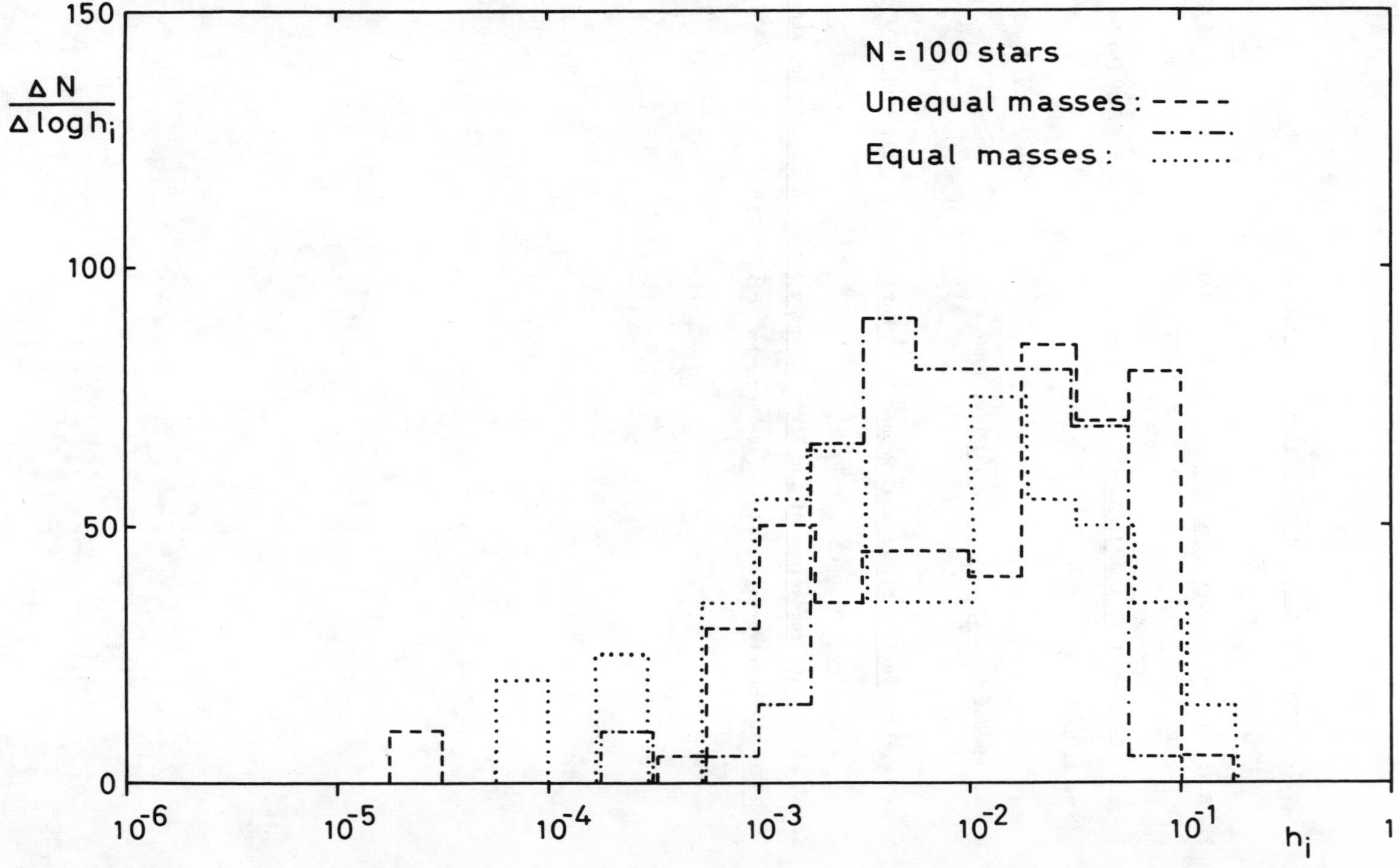
N = 100 stars
Unequal masses :
Equal masses :
ΔN / Δ log h_i
150
100
50
0
10^-6
10^-5
10^-4
10^-3
10^-2
10^-1
1
h_i

References

1. S.J. Aarseth, Monthly Notices Roy. Astron. Soc., 126, 223, 1963

2. S.J. Aarseth, Astrophys. Space Sci. 14, 118, 1971

3. S.J. Aarseth, in Vistas in Astronomy, Pergamon Press, London, 1973 (to be published)

4. S. von Hoerner, Z. Astrophys. 50, 184, 1960

5. R. Wielen, Veröffentl. Astron. Rechen-Inst. Heidelberg No. 19, 1967

6. R. Wielen, in Proceedings of the First European Astronomical Meeting, Athens, 1972, Springer Verlag Berlin- Heidelberg-New York, 1973

A VARIABLE ORDER METHOD FOR THE NUMERICAL INTEGRATION OF THE GRAVITATIONAL N-BODY PROBLEM

by

Guy Janin*

The University of Texas at Austin

Austin, Texas

Abstract

In an efficient computer program for numerical experiments with gravitational N-Body systems, at least one of the two following techniques is used: (1) regularization of the two-body close encounters, (2) variable and individual step size for each particle. Technique (1) involves frequent changes of the equations of motion and suggests the use of a one-step method. Technique (2) implies the solutions to be interpolated and extrapolated on non-step points and suggests the use of a power-series method.

A new tentative technique (2) is proposed, where the inconveniences associated with the individual step size scheme are avoided: the step size is the same for all the particles but the order of integration is variable and individual. A first application of this technique to a N-Body computer program with regularization of close encounters is developed.

The Gravitational N-Body Problem

The N-Body problem consists of studying the motion of a system of N particles interacting in their gravitational field. The nature of the gravitational field is such that the effect of distant particles is

*Present address: European Space Operations Center, Darmstadt, West Germany.

of the same order of importance as the effect of neighboring particles. Thus, one cannot study the behavior of a part of the system and extrapolate the results to the whole system. One has to consider the N particles.

On the other hand, the force between two particles is increasing indefinitely when their distance tends to zero. This is the problem of close approaches.

Gravitational N-body systems are encountered in astronomy, more precisely in celestial mechanics, stellar dynamics and cosmology. There is no direct experimental way to study gravitational N-Body systems. The only way is a simulation by computer experiments consisting of solving numerically the Newtonian equations of motion

$$\frac{d^2\underline{x}_\alpha}{dt^2} = \underline{a}_\alpha(\underline{x}), \quad \alpha = 1, 2, \ldots, N \tag{1}$$

with the initial condition for the positions: $\underline{x}_\alpha(t_o) = \underline{x}_{\alpha o}$,

and for the velocities: $\frac{d\underline{x}_\alpha}{dt}(t_o) = \underline{v}_{\alpha o}$,

where $\underline{a}_\alpha(x)$ is the acceleration of particle α. It depends on the positions of all the other particles through to universal gravitational law:

$$\underline{a}_\alpha = \sum_{\beta\neq\alpha}^{N} \frac{m_\beta}{(r_{\alpha\beta})^3}\,\underline{r}_{\alpha\beta} , \tag{2}$$

where $r_{\alpha\beta}$ is the distance between particle α and β .

If the system of differential equations (1) is reduced to a system of first order equations, the number of differential equations is 6N. The total number of acceleration components (2) appearing on the right-hand side of (1) is 3/2 N(N-1). For certain experiments in astronomy, it is desirable to consider systems of the order of N = 500 bodies. The amount of computing time necessary to investigate the evolution of such systems is prohibitive, unless an efficient algorithm is developed. An efficient algorithm is one that minimizes computer storage and computing time for a prescribed accuracy of the solution.

The accelerations (2), appearing on the right-hand side of the differential equations (1), are characterized by a rapid variation if the particle separations become small. A method used for the solution of the problem must therefore incorporate an effective step-size control.

A mathematical transformation of the equations can be made in order to remove the singularity of the acceleration during close approaches. This transformation, called regularization, increases the number of variables, therefore the number of differential equations. Each time the members of the configuration of the close bodies is changed, the numerical integration has to be reinitiated. As a result, a one-step method has decided advantages for the algorithm since it is self-starting.

A considerable saving of computing time can be made if each body has its individual step size. A body orbiting on a regular path will be integrated with a rather large step size, whereas a body involved in a close encounter will need a smaller step size for computing its orbit to the same accuracy.

An individual step size scheme implies the facility to compute the position of any particle at any time. In other words, one has to interpolate and extrapolate the solution of the differential equations at non-step points. This can easily be done if the solution is given by a series. Extrapolation will still give very unreliable results.

All these requirements for an efficient algorithm for solving the gravitational N-Body problem can hardly be satisfied together. The purpose of this paper is to suggest a method which satisfies all the requirements, provided the idea of the variable and individual step size scheme is replaced by another one which has the same effect and advantages: the variable and individual order scheme.

In this scheme, the step size is still variable but is the same for all the particles (collective step size). The order of integration is supposed to be variable and dependent on the particle. A significant savings of computing time can be realized. Interpolation of the solution on non-step points is still needed, but there is no extrapolation.

A Variable Order One-Step Method

The foremost one-step methods are of the Runge-Kutta type. The higher-order Runge-Kutta methods developed by Fehlberg (1968) with step size control are extremely efficient and have been applied with great success in N-Body programs. Nevertheless, they have two disadvantages: (1) the solution is given with full accuracy only at the step points; (2) the order of integration cannot be changed.

An integration method not having these two handicaps can be easily found amoung the multi-step methods. But if the main objective is to have a one-step method, our choice is restricted to a type of method like the following one.

Given a function x(t) in a certain interval of the independent variable t which we denoted by (t_m, t_{m+1}), and an approximation $x^{(L)}(t)$ of this function such that

$$\text{Max} \mid x(t) - x^{(L)}(t) \mid \leq \varepsilon \ , \ t_m \leq t \leq t_{m+1} \ ,$$

where ε is the accuracy of the approximation, the most suitable form for $x^{(L)}(t)$, from the computational point of view, is a polynomial. To express such a polynomial, it is advisable to use a representation by elementary orthogonal polynomials

$$x^{(L)}(t) = \sum_{i=0}^{L} a_i P_i(t) \, , \tag{3}$$

where $P_i(t)$ are well chosen orthogonal polynomials of degree i, and a_i are coefficients usually defined by the sum

$$a_i = \int_{t_m}^{t_{m+1}} x(t) \; P_i(t) dt \; .$$

We would like to have such a polynomial representation in the time interval (t_m, t_{m+1}) (the step size) for each component of each particle α. The maximum degree L_α of this polynomial approximation should depend on the particle α and should be the lowest degree required to give the accuracy ε for the approximation in (t_m, t_{m+1}).

To build such polynomial approximations, an iterative procedure seems the most suitable for a one-step method.

Let us consider simply one first order differential equation

$$\frac{dx}{dt} = f(x,t)$$

to be integrated between t_m and t_{m+1}, and let us suppose we have, at some stage, an approximation $x^{(L)}(t)$ for the solution in the interval (t_m, t_{m+1}). We can therefore compute the derivative $f(x^{(L)}(t^{(\ell)}), t^{(\ell)})$ on a set of points $t^{(0)}, t^{(1)}, \ldots, t^{(L)}$ situated in (t_m, t_{m+1}) which defines a polynomial approximation $P^{(L)}(t)$ of order L for the derivative. Through integration of $P^{(L)}(t)$, we obtain a polynomial of order $L + 1$, $x^{(L+1)}(t)$, which is an approximation of order $L + 1$ of the solution. The constant of integration is given by the initial value $x(t_m)$. This procedure, so-called Picard iteration (Picard, 1893), is iterated until a convenient accuracy is reached. It is generally convergent when the Lipschitz condition is satisfied.

Lanczos (1938) first used Chebyshev polynomials for solving differential equations. The advantages of Chebyshev polynomials are principally that the error of the approximation is uniformly distributed all along the interval of approximation, and that the convergence of Chebyshev polynomials is faster than for any other class of polynomials (the coefficients a_i in (3) are decreasing fast when i is increasing). We have to interpolate the solution many times for any value of the independent variable and these two properties are therefore extremely useful.

Chebyshev Polynomials (see Fox and Parker, 1968)

Chebyshev polynomials $T(t)$ are defined between -1 and $+1$. We must therefore transform our interval (t_m, t_{m+1}) accordingly. The low order Chebyshev polynomials are given by the expressions:

$$T_0(t) = 1$$

$$T_1(t) = t$$

$$T_2(t) = 2t^2-1$$

$$T_3(t) = 4t^3-t$$

...

This implies the recurrence relation:

$$T_{r+1} = 2t \cdot T_r - T_{r-1} .$$

As usual with such recurrence relations, they are subject to round-off error propagation and it is advisable to compute the necessary values of $T_r(t^{(\ell)})$ in double precision.

A development of Chebyshev polynomials in t can be expressed by a cosine Fourier series in θ by defining the variable $t = \cos\theta$. The $L+1$ zeros of the Chebyshev polynomial of order L are located at the points

$$t^{(\ell)} = \cos\left(\frac{2\ell+1}{L+1}\cdot\frac{\pi}{2}\right), \quad \ell = 0, 1, \ldots, L .$$

The Lth order polynomial approximating a function $x(t)$ between -1 and $+1$ has the representation

$$x^{(L)}(t) = \sum_{r=0}^{L}{}' \; c_r^{(L)} \; T_r(t)$$

where $\sum'$ means that the first term of the sum is taken with a factor 1/2. The coefficients $c_r^{(L)}$ are given by

$$c_r^{(L)} = \frac{2}{L+1} \sum_{\ell=0}^{L} x(t^{(\ell)}) T_r(t^{(\ell)}).$$

The error $| \, x^{(L)}(t) - x(t) \, |$ is bounded by the sum of the absolute value of the neglected coefficients:

$$\sum_{r=L+1}^{\infty} | \, c_r^{(L)} \, | .$$

These coefficients are usually strongly decreasing and the first one is a good approximation of the error bound: $| \, c_{L+1}^{(L)} \, |$.

There are fast algorithms for computing $x^{(L)}(t)$ in terms of the coefficients $c_r^{(L)}$ without evaluating $T_r(t)$. Similar algorithms exist for computing the indefinite integral of $x^{(L)}(t)$.

A Variable Order Method

A zero-order solution in (t_m, t_{m+1})(transformed into interval (-1,+1)) is given by the initial conditions $x(t_m)$ at t_m. A zero-order Chebyshev representation of the derivative $f(x,t)$ is therefore immediately defined and gives, through integration, a first-order Chebyshev representation for the solution $x^{(1)}(t)$. Let us use this representation to compute an approximation of $x(t^{(0)})$ and $x(t^{(1)})$, where $t^{(0)}$ and $t^{(1)}$ are the two zeros of the Chebyshev polynomial of order one. Next we compute $f(t^{(0)}, x^{(1)}(t^{(0)}))$ and $f(t^{(1)}, x^{(1)}(t^{(1)}))$ to define a representation of order one for f. Integration then gives a representation of order two for the solution.

This procedure can be iterated and each time the order of the solution is increased by one; in addition, the number of points for computing the derivatives is also increased by one.

The iterations are terminated when the difference between two successive approximations of the solution at the same point - for instance, the end of the interval (+1) - is less than a certain tolerance. We say in this case that the convergence is satisfied. If the iterations do not converge, the whole procedure has to be resumed with a smaller integration step.

Application to the N-Body Problem

An initial step size $\Delta t = t_1 - t_0$ is first chosen. Successive approximations of the solution of the first order equations of motion are computed. After each iteration, the convergence is checked separately for each particle. If, for instance, the convergence is satisfied for the particle $\alpha = K$, the solution for the equations of motion describing the orbit of this particle (usually 6 solutions $x_1(t)$,

$x_2(t)$, $x_3(t)$, $v_1(t)$, $v_2(t)$, $v_3(t)$) are not improved any more and the subsequent iterations are only applied to the other particles.

The polynomial approximation of the solutions for the motion of particle number K is used each time the force between particle K and the others have to be computed. But if several particles are at the same stage as particle number K (i.e., their convergence is satisfied), the forces should not be computed between them. This procedure leads to a significant saving of computing time and constitutes the main advantage of the method. Usually, the last iterations concern only a few strongly perturbed particles and, therefore, not too many derivative evaluations are required for computing their motion.

Once the convergence is satisfied for all the particles, the integration is terminated and a new step size is computed for the next step.

During each iteration, the coefficients of the Chebyshev polynomials for the derivatives and for the solutions have to be temporarily stored. If n is the total number of differential equations and k is the maximal order, the number of coefficients to be stored is $n(2k+1)$. For n large, the size of the computer memory may limit the maximum value of k. If convergence is not achieved for all particles at order k, the subsequent iterations have to be made with order k. If convergence is not achieved after two or three of these extra iterations, it is recommended to reject the step and to try another one with a smaller step size.

Practically, we limit the order to $k = 9$ and the number of extra iterations to 3.

Step-Size Control

To reject a step leads to a loss of computation time. On the other hand, a too-fast convergence also represents a loss of efficiency because the high-order scheme is not used. Therefore, the step size should be chosen to maintain the maximum order around k. This is a delicate point which requires further discussion.

Let us define an appropriate function $g(\ell)$ having the following properties

$$g(\ell)\begin{cases} < 1 \text{ if } \ell > k \\ = 1 \text{ if } \ell = k \\ > 1 \text{ if } \ell < k \end{cases}$$

where ℓ is the number of iterations. If Δt is the current step size, the recommended step size $\Delta t'$ for the next step is defined by

$$\Delta t' = \Delta t \; g(\ell) \quad .$$

This expression should be used only if the behavior of the functions is similar from one step to the other. In other cases, we still proceed in this way, since the variable order method allows us to make a bad guess of the step size and still obtain a good solution. New steps are seldom rejected.

A multitude of expressions can be suggested for the function $g(\ell)$. One idea which is working very well is

$$g(\ell) = a^{k-\ell} \quad , \quad a > 1$$

with $a = 1.6$. But the choice of the function $g(\ell)$ is not very critical.

A First Application to the N-Body Problem

The variable order scheme has been implemented for a N-Body computer program. All close two-body encounters are regularized by means of the Kustaanheimo-Stiefel transformation, except for the closest encounter where uniformly regular canonical elements are used (Stiefel and Scheifele, 1971). The equations of motion are written in terms of the same independent variable, namely the fictitious time attached to the most critical encounter.

A typical case of a 200-body system, integrated during 1/4 of the mean period of the system, requires 20 minutes of the CDC 6600 computing time if the total energy of the system is conserved to five significant figures throughout the calculation.*

The quoted performance is not especially outstanding because of the two following reasons: (1) The efficiency of the integration method (Picard iteration with Chebyshev polynomials) is rather poor and is in no case comparable to some of the high-order Runge-Kutta schemes. However, the variable and individual order does improve the procedure some what. (2) The variable and individual order scheme is not working in the best conditions. The same fictitious time (new independent variable defined by the regularization transformation) is used for all the equations. This implies that the rapid variation of the distance between the two bodies involved in a close approach is appearing on the right hand side of all the equations (Bettis and Szebehely, 1971), keeping the order of the integration high even for weakly perturbed bodies.

---

* As a comparison, the same case was run with a highly elaborate individual step size scheme (Aarseth, 1971) and cost about three times less computing time for the same accuracy.

Both the disadvantages may be improved by further modifications. In particular, the second disadvantage can be removed by integrating each body with its own physical or fictitious time. However, this procedure leads to practical difficulties which have not yet been completely solved.

Acknowledgments

J'aimerais exprimer ma reconnaissance au Fonds Janggen-Pöhn et au Fonds National Suisse de la Recherche Scientifique pour leur aide financière qui m'a permis de mener cette recherche aux Etats-Unis.

References

Aarseth, S.J., 1971, "Direct Integration Methods of the N-Body Problem," Astrophys. and Space Science, 14, 118.

Bettis, D.G. and Szebehely, V., 1971, "Treatment of Close Approaches in the Numerical Integration of the Gravitational Problem of N Bodies," Astrophys. and Space Science, 14, 133.

Fehlberg, E., 1968, "Classical Fifth-, Sixth-, Seventh-, and Eighth-Order Runge-Kutta Formulas with Stepsize Control," NASA TR R-287.

Fox, L. and Parker, J.B., 1968, "Chebyshev Polynomials in Numerical Analysis," Oxford Univ. Press, London.

Lanczos, C., 1938, "Trigonometric Interpolation of Empirical and Analytical Functions," J. Math. Phys., 17, 123.

Picard, E., 1893, "Sur l'application des méthodes d'approximations successives à l'étude de certaines équations différentielles ordinaires," J. Math., 9, 217.

Stiefel, E.L. and Scheifele, G., 1971, "Linear and Regular Celestial Mechanics," Springer-Verlag, Berlin.

THE METHOD OF THE DOUBLY INDIVIDUAL STEP FOR N-BODY COMPUTATIONS

by

Avram Hayli

Observatoire de Besançon

Besançon, France

Abstract

It is possible to reduce the computation time in the numerical integration of N-body gravitational problem by introducing an individual and variable step, but also by recalculating the force which act on a star only when this force has experienced a significant change. To do so, an individual time scale is introduced for each force.

Forces are classified into categories and each category has its own time scale. Time is noted in a numeration system with base b. Stars are also grouped into categories in such a way that, at a given time one knows, without keeping in the computer's memory and without comparing the $\frac{1}{2}N\ (N - 1)$ forces which act between the stars, what forces must be recomputed.

1. Introduction

The major limitation one has to face when integrating numerically the set of differential equations of the N-body gravitational problem is that of computing time. It is possible to save much time by introducing for each force an individual and variable step. Within this scope, let us consider the force $\underline{F}_{ij}$ acting between stars i and j. It is convenient to introduce a time scale for each $\underline{F}_{ij}$. When this time scale has elapsed the force will be recalculated by means of the equations of motion.

For example, let us consider star number i, with mass m_i. We could keep in the computer's memory each of the forces $\underline{F}_{ij}$ which act on this star as a polynomial of time t, for instance of degree s. The time scale of the force $\underline{F}_{ij}$ once elapsed we would recalculate $\underline{F}_{ij}$, correct the polynomial, and make the corresponding corrections on the total force $\underline{F}_i$ acting on star i. By doing so we would avoid recalculating all the forces $\underline{F}_{ij}$ which have a larger time scale.

But things are not as simple for two reasons. The first one is that ½ N(N-1) forces act between N stars. Each force has three components. If we use polynomials of degree 2, we would have to keep in the memory for each force nine quantities plus the time scale, that is ten quantities. For N = 250 we would need 311,250 fast memories in the computer! The second reason is that we would need to know at each step what forces must be recalculated. This means we would have to compare ½ N(N-1) forces. For large values of N this comparison would take even more time than the calculation of the forces.

2. The Method of the Doubly Variable Step

One can avoid the difficulties mentioned above by introducing a quantization of time. We decide that the time scale associated to a force at some instant can take only the following values:

$$t_o/b \ , \quad t_o/b^2 \ , \quad t_o/b^3 \ , \ \ldots\ldots, \quad t_o/b^m \ , \ \ldots \tag{1}$$

where t_o is a time unit and b an integer greater than one. To be adequate, t_o/b must be of the same order of magnitude as the greatest possible time scale. t_o/b being known, one will have to choose the best values for t_o and b.

All forces will be classified in a number of categories, each category having its own time scale. The category corresponding to time scale t_o/b^m will have the number m. We shall note time in a numbering system with base b. Time will be expressed by a number with at most M significant digits where M is equal to the number of the highest category.

During computation itself, it is convenient to group stars into categories as well. Let us call p the highest category to which one or several of the forces acting on star i belong. Star i will be said to be of category p.

To see how the method operates at time t, let us assume, for example, that b = 10 and that t/t_o = 6,7942. The last significant digit occupies the fourth position. Thus, all the forces acting between stars belonging to categories ≥ 4 will be recalculated exactly, that is by using the equations of motion. We see therefore that one knows what forces are to be recalculated completely at time t. On the other hand, it becomes obvious that it is not necessary to keep a record of the individual forces acting on a star, but only of the sum of these forces.

This classification into categories corresponds roughly to a growing proximity to the central region of the cluster. Let us note, however, that two stars very close to each other will belong to a high category even when they are far from the central region. In general, we have in the cluster a small nucleus with stars belonging to the highest category which appears. The forces acting between these stars are recalculated very often. A larger nucleus contains the stars belonging to the two highest categories. The forces acting between these stars are recalculated $\frac{1}{b}$ times as often, and so on. Therefore, at time t, we integrate the equations of motion for the stars belonging to the corresponding nucleus. The effect due to the stars which do not belong to this nucleus is represented by a polynomial of t. This polynomial is readjusted at the end of the integration step. Last but not least, it is obvious that star i will not remain always in the same category. For each time star i is taken into account in the integration process, it will be permitted to leave its category for a higher or a lower one according to its situation relative to the other stars of the cluster. Of course, new categories of high order will be created during the integration if necessary.

3. The Time Scale and the Determination of Categories

It seems natural to take for the time scale of the force $\underline{F}_{ij}$ which acts between stars i and j, the quantity $\mu r_{ij}/v_{ij}$ where v_{ij} is the relative velocity of the two stars and μ a coefficient to be determined. r_{ij} is the distance between the two stars.

This time scale may be however very inadequate when the relative velocity of stars i and j is almost equal to zero. In that case the time scale could be exceedingly large, because one of the two stars may greatly change its velocity after a close encounter with a third one. The relative velocity of stars i and j will then become larger and the real time scale will become much smaller. In fact, when v_{ij} is very small one can say that the order of magnitude of the real time scale of the force $\underline{F}_{ij}$ is $r_{ij}/\overline{v}$, where $\overline{v}$ is the mean velocity of the stars of the cluster. It is then more adequate to take:

$$\mu[r_{ij}^2/(v_{ij}^2 + \overline{v}^2)]^{\frac{1}{2}} \tag{2}$$

for the time scale of the force $\underline{F}_{ij}$.

The time being quantified the corresponding integration time step t_o/b^m will be such that:

$$t_o/b^m < \mu[r_{ij}^2/(v_{ij}^2 + \overline{v}^2)]^{\frac{1}{2}} < t_o/b^{m-1} \tag{3}$$

that is:

$$m-1 < \log\{t_o/\mu[r_{ij}^2/(v_{ij}^2 + \overline{v}^2)]^{\frac{1}{2}}\} /\log b < m \tag{4}$$

It is clear that m must be greater or equal to 1 and must remain smaller than or equal to a given integer M_{sup} specified at the beginning of the integration. The reason is that the number of fast memories in a computer is limited. Hence, new categories can be created only within this limitation. If formula (4) gives $m \leqslant o$, m is given the value one.

4. Integration Scheme and Procedure

Suppose $t = t_o$. Let C_i be the category to which star i belongs. We call C_{max} the highest category which appears for the moment. Of course $C_{max} \leqslant M_{sup}$. $\underline{F}_i$ is a function of time and we write:

(5) $$\underline{F}_i = \sum_j \underline{F}_{ij} = \sum_j m_i \underline{\Gamma}_{ij} = m_i \sum_j \underline{\Gamma}_{ij} = m_i \underline{\Gamma}_i$$

where the letter $\underline{\Gamma}$ denotes an acceleration. Let us introduce vectors $\underline{T}_{ij}$ and $\underline{T}_i$ defined as

(6) $$\underline{T}_i = \dot{\underline{\Gamma}}_i = \sum_j \dot{\underline{\Gamma}}_{ij} = \sum_j \underline{T}_{ij}$$

where the dot stands for the derivation with respect to time.

We now define ${}^{\ell}\underline{F}_i$ as this part of the force acting on star i which is due to stars belonging to categories smaller than ℓ. We introduce the following notation:

(7) $${}^{\ell}\underline{F}_i = \sum_{j:C_j \leq \ell} \underline{F}_{ij}$$

to indicate that the stars involved in the summation are those for which $C_j < \ell$. In the same manner, we introduce the corresponding ${}^{\ell}\underline{\Gamma}_i$ and ${}^{\ell}\underline{T}_i$.

Now the forces $\underline{F}_{ij}$ which act on star i are divided into two classes which we call the upper and the lower classes. We said previously that at a given time we recalculate completely, i.e. using the positions of the stars, only those forces which act between stars belonging to categories greater or equal to a certian category k. (We shall later see what value is given to the integer k at a given time). If $C_i < k$, star i is not recalculated; neither are the forces $\underline{F}_{ij}$. Only if $C_i \geqslant k$ are the forces $\underline{F}_{ij}$ divided into two classes. The forces due to stars belonging to a category smaller than k are placed in the lower class. Those due to other stars are placed in the upper class. We can write, therefore:

(8) $$\underline{F}_i = \sum_{j:C_j<k} \underline{F}_{ij} + \sum_{j:C_j \geq k} \underline{F}_{ij} = {}^{k}\underline{F}_i + \sum_{j:C_j \geq k} \underline{F}_{ij}$$

k is normally equal to C_{max}, but once every b steps it is not greater than $C_{max}-1$, once every b^2 steps it is not greater than $C_{max}-2$, and so on.

To give more details on the method, let us assume for the sake of simplicity, that stars do not change their category during the first steps of the integration. We shall examine later how the category of a star changes when it becomes necessary. The scheme of integration is represented on fig. 1, where C_{max} is given the value 5 and b the value 2. The coordinates are time and category.

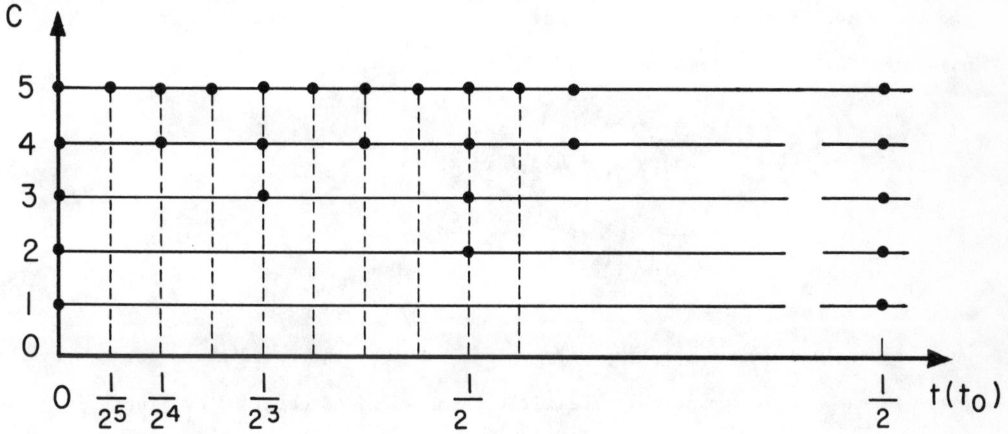

Figure 1

At a given time we know what stars must be integrated at the next step. They belong to the categories which appear with black dots on fig. 1. The positions and the velocities of all the stars are known at $t = t_o$ from the equations of motion.

We calculate the $\underline{F}_{ij}$, $\underline{\Gamma}_{ij}$ and T_{ij} for i, j = 1,2,...,N according to the following formulae:

(9) $$\ddot{x}_{ij} = m_j(x_j - x_i)/r_{ij}^3$$

(10) $$\dddot{x}_{ij} = m_j[(\dot{x}_j - \dot{x}_i) - 3(x_j - x_i)\dot{r}_{ij}/r_{ij}]/r_{ij}^3$$

The same formulae hold for the y and x components. (x_i stands for the x coordinate of star i). These quantities are stored and the numerical integration can start.

Let us thoroughly examine one integration step. The step to be considered is $[n\tau, (n+1)\tau]$ where n is an integer and $\tau = t_o/b^{C_{max}}$. C_{max} is now the highest category which appears at $t = n\tau$. For this step we integrate the equations of motion only for those stars which belong to categories higher or equal to the value of k for $t = (n + 1)\tau$. We need, therefore, to know the forces which act upon these stars. As we have seen before, these forces are distributed into two classes. The position and the velocity of a star i whose category is $C_i \geq k$ are recalculated by first using the following formulae:

(11) $$x_i' = x_i + h\dot{x}_i + \frac{h^2}{2}\ddot{x}_i$$

(12) $$\ddot{x}_i' = f(x_i')$$

where the primes denote the temporary values taken by the coordinates for $t = (n+1)\tau$. $f(x_i')$ is the sum of two terms. The first term is ${}^{k}\ddot{x}_i$ as obtained by linear extrapolation. (Its former value has been kept in memory). The second term is the sum of the $\ddot{x}_{ij}$ $(j{:}C_j \geq k)$ as computed from the temporary positions.

The final values for the x coordinates of the position and the velocity for star i with $C_i \geq k$, are given by

(13) $$\dot{x}_i = x_i + \frac{h}{2}(\ddot{x}_i + \ddot{x}_i')$$

(14) $$x_i = x_i' + \frac{h^2}{6}(\ddot{x}_i' - \ddot{x}_i)$$

At the end of this step and before proceeding to the next integration step, we have to examine what stars must change categories. From what has been said before, it is clear that only stars belonging to categories higher or equal to k are concerned. Such stars can jump only to the next highest or lowest category. Moreover, only a star which belongs to a category higher than k can jump to the next lowest one. Thus, in practice, we must first find the stars which are concerned by this operation. The principle is the same as the one used to determine the categories at the beginning of the integration, that is by using relation (4). When a star has jumped to the next highest category we recalculate the value of $^{\ell}\underline{F}_i$ as well as the values of the $^{\ell}\underline{F}_j$.

Let us suppose that at the end of the step, a new category is created, higher than the already existing ones, that is a category $C_{max} + 1$. In such a case the next time step will be $[(n+1)\tau, (n+1)\tau + \frac{\tau}{b}]$. The value of k will then be $C_{max} + 1$. Let us suppose now that at the end of the step there are no stars belonging to category C_{max}. This can happen only when $(n + 1)$ is a multiple of b. In such a case the next time step will be $[(n+1)\tau, (n+1)\tau + b\tau]$.

This method has been used efficiently to integrate several tens of cluster models. See Hayli [1]. $\mu = 0.05$ and $b = 2$ were used to minimize the computing time. M_{sup} can be chosen at will, but thirty categories are sufficient to handle close encounters as well as the early stages of the evolution of close subsystems. It has been compared with others methods by looking at the computing time and at the errors of integration. See, for example, Lecar [2] where the same cluster was integrated with different methods. Our method is faster but not as precise as the methods which use high order formulae. The time which is necessary to follow a star cluster during a certain physical duration, say years, increases approximately like $N^{2.7}$.

Ahmad and Cohen [3] have recently developed a method which has some similarities with the one described in this paper. One of its refinements is that it handles the effect of a star's remote neighbors as a slowly varying field. This allows the authors to save much computing time when clusters with very large values of N are integrated.

References

1. A. Hayli, Astrophys. and Space Sciences, 13, 1971, 309

2. M. Lecar, Bull. Astron. (3) 3, 91

3. A. Ahmad and L. Cohen, To be published

INTEGRATION OF THE N BODY GRAVITATIONAL PROBLEM BY SEPARATION OF THE FORCE INTO A NEAR AND A FAR COMPONENT

by

Afaq Ahmad and Leon Cohen

Hunter College of the City University of New York

1. Introduction

The main difficulties encountered in numerically integrating N particles acting under their own mutual gravitational attraction stems from the fact that the gravitational force is relatively strong at both close and far distances. The singularity of the inverse square force law is sufficiently strong to cause significant numerical difficulties during a close encounter of two bodies or when a binary is formed. In these instances the force on each particle is changing so rapidly that very small time steps must be taken. This causes a number of difficulties, namely the large number of steps that have to be taken leads to a loss of significant digits and, therefore, a loss in accuracy. Also the frequent recalculation of the force which is necessary because of the small time steps results in an enormous amount of computing time. This aspect of the difficulty in integrating the N body gravitational problem is presently being attacked through a number of techniques. One method[1] involves the use of analytical orbital equations properly modified to take into account the presence of an external field. Another very successful approach is the regularization of the equations of motion[2,3,4,5]. This involves the transformation of the equations of motion into a form such that when two particles get close together physically, the equations are nonetheless regular, that is nonsingular. The other basic difficulty in the N body gravitational probelm is due to the long range of the gravitational force. The inverse square force law decays relatively slowly with distance and hence, the far away stars contribute significantly to the total force. In calculating the force, account of all stars in the system must be taken and therefore the number of force calculations per particle is equal to N and the total number of force calculations is proportional to N^2, severely limiting the number of particles that can be handled. At first sight, this difficulty appears to be basic to the nature of the force of gravity but as we shall see the relatively frequent recalculation of the total force can be avoided to a large degree.

In all methods of integration which use individual time steps the time step determines the amount of time for which the force is accurate enough and hence does not have to be recalculated exactly during that time. At the end of the time step, the exact force on that particle has to be recalculated from

$$\frac{d^2\vec{r}_i}{dt} = - \sum_{j=1}^{n}{}' m_j \frac{\vec{r}_{ij}}{|r_{ij}|^3} \qquad (1.1)$$

$$\vec{r}_{ij} = \vec{r}_i - \vec{r}_j$$

The time step is basically determined by the relative rate of change of the force. The higher the time derivative of the force is, the smaller the time step will be. But it is clear that most often the high rate of change of the force will be caused by only a few nearby particles and hence it should be unnecessary to recalculate the force due to all particles if it is only a few particles which are causing the time step to go bad. We shall describe a method which avoids the recalculation of the total force as frequently as would be demanded if one only looked at the time rate of change of the total force.

To see how this can be done let us briefly examine the nature of the force on a star. It is standard[6] in theoretical work to separate the force into two components. One component (the regular force) due to the smoothed out distribution of stars and another (the irregular force) due to the stars in the vicinity of the star in question. There are two basic differences between the two components which can be exploited for practical purposes. One is that the regular force changes relatively slowly as compared to the rate of change of the irregular force and secondly that only a few particles are responsible for the irregular force. Thus, when the time step goes bad the only revision that is probably necessary is not for the total force force but only for the random part of it. As this is due to relatively few particles the recalculation of it is not costly in terms of computing time. We now proceed by describing the operational definition of the irregular force.

That this natural subdivision of the force can be used in the actual integration of the N body problem has been suggested by a number of authors[7,8]. A scheme which uses double individual time steps using a different approach has been developed by Hayli[9] and is described elsewhere in this volume. The basic flow of our scheme follows that of Aarseth[10].

2. Irregular Force

As a definition of the irregular force, $\vec{S}$, we choose it to be the force due to the particles (neighbors) which fall within a sphere of radius R circumscribed about each particle. As different parts of the cluster will have different densities the radius of the sphere is made variable. We have found that the radius of the sphere should be so chosen as to include at least about 5 particles as neighbors. We have also found it convenient to limit the maximum number of neighbors to about 15. This has been found to be quite sufficient for systems containing up to 1000 particles.

As the idea of subdividing the force into two parts is to pinpoint those particles which are causing a high fluctuation in the force it is clear that we must also include as neighbors particles which are outside the sphere but approaching the particle in question with such a high velocity that they will be in it before the next regular force recalculation. A search for such particles is made within a sphere of radius 2R and a straight line approximation is used to determine the likelihood that it will be inside the sphere before the total force calculation. Thus, if v_o and r_o are the speed and the relative distance respectively, we can find the impact parameter d and the time of the nearest approach t from

$$d = r_o \sin\theta \tag{2.1}$$

$$t = -\frac{r_o}{v_o}\cos\theta \tag{2.2}$$

where θ is the angle between $\vec{v}_o$ and $\vec{r}_o$. The approaching star is included as a neighbor if both

$$t < \tau$$

$$d < R$$

where τ is the time before the next force calculation.

Clearly, the neighbors of any one star will change and a decision has to be made when to check for this. One cannot do it too often for this would be quite expensive. Since the mutual distances between particles has to be calculated during the regular force calculation anyway, we have found that time to be the best time for checking for any change of neighbors. What is actually done is to store the new neighbors and to check against the old to identify those stars which have either moved into the neighborhood of the star or moved out. Of course, the regular and irregular forces have to be corrected for this change, which requires addition or subtraction of the contributions of these stars to the force. Difficulty occurs, however, in the calculation of the derivatives which in our program are calculated by a difference scheme. If the regular or irregular forces are not caused by the same stars at two times we cannot use the difference scheme to find the derivative directly. To circumvent this we do the following. The forces and derivatives are first calculated assuming there has been no change in neighbors. Then they are corrected by using analytic formulas for the derivatives to calculate contributions from old neighbors which are no longer neighbors and the new neighbors. These are then subtracted and added, respectively.

For the irregular force we have followed Gonzalez and Lecar[11] in choosing the time step

$$\delta t_i = \alpha \text{ Min } \{r_{ij}^{3/2}\} \tag{2.3}$$

where α is a constant which controls the accuracy of integration and $\text{Min}(r_{ij})$ is the distance from the i-th particle to its nearest neighbor. This, however, does not take into account a possible fast approaching neighbor. If there does exist such a particle among its neighbors, then the irregular force time step is chosen from

$$\delta t_i = \frac{\eta\alpha}{m_i^{\frac{1}{2}}} \text{ Min } \left\{ \frac{r_{ij}}{v_{ij}} \right\} \tag{2.4}$$

where η is a constant which we have found to be most appropriately taken as 3.3. In practice, both (2.3) and (2.4) are calculated and the smallest is chosen.

3. Regular Force

The regular force, $\vec{K}$, on a particle is of course the force due to the particles which lie outside the sphere which surrounds each particle. The time step for the regular force is found from

$$\Delta T_i = \beta \frac{|\vec{K}_i|}{|\dot{\vec{K}}_i|} \tag{3.1}$$

where β is a constant and $\dot{\vec{K}}$ is the time rate of change of $\vec{K}$. In cases where $\dot{\vec{K}}$ is very small, (3.1) produces too large a time step and in that case, the regular force time step is taken from

$$\Delta T_i = \beta \frac{r_i}{\langle v^2 \rangle^{\frac{1}{2}}} \tag{3.2}$$

where $\langle v^2 \rangle^{\frac{1}{2}}$ is the root mean square velocity of the cluster.

4. Method of Divided Differences

The basic prescription for updating the positions and velocities of the particles is the Taylor series

$$(4.1)\qquad \vec{r}_i(T+\Delta T) = \vec{r}_i(T) + \vec{v}_i(T)\Delta T + \frac{1}{2}\frac{\vec{F}_i(T)}{m_i}(\Delta T)^2 + \frac{1}{m_i}\sum_{k=1}^{\ell}\frac{\vec{F}^{(k)}(\Delta T)^{k+2}}{(k+2)!}$$

$$(4.2)\qquad \vec{v}_i(T+\Delta T) = \vec{v}_i(T) + \frac{\vec{F}_i(T)}{m_i}\Delta T + \frac{1}{m_i}\sum_{k=1}^{\ell}\frac{\vec{F}^{(k)}(\Delta T)^{k+1}}{(k+1)!}$$

where $\vec{F}$ is the total force on the particle and $\vec{F}^{(k)}$ is the k-th derivative. These are evaluated at time T. There are two possible methods by which one can find the derivatives of the forces at a certain time. We could use the analytic formulas which require only the knowledge of the positions and velocities of the particle at that time. This method is used in the scheme described by Lecar elsewhere in this volume. Alternatively, a difference scheme can be used to find these derivatives, although this requires knowledge of the force at a few previous times. We have found the difference method to be quite efficient. As the time steps are taken to be variable the most suitable scheme is found to be the divided difference method which has the additional advantage of semi-iteration, which is not possible with Lagrangian interpolation methods.

Consider a fourth order divided difference polynomial relating force (regular or irregular) at time T to its value at earlier times

$$\begin{aligned} \vec{F}(T) &= \vec{F}(T_0) + \vec{D}[T_0,T_1]\,(T-T_0) + \vec{D}_2[T_0,T_2]\,(T-T_0)\,(T-T_1) \\ (4.3)\qquad &+ \vec{D}_3[T_0,T_3]\,(T-T_0)\,(T-T_1)\,(T-T_2) \\ &+ \vec{D}_4[T,T_4]\,(T-T_0)\,(T-T_1)\,(T-T_2)\,(T-T_3) \end{aligned}$$

where T_0, T_1, T_2, and T_3 are the times, in decreasing order, of the last four exact force calculations. It can easily be shown that (4.3) is satisfied if the divided differences are defined as follows

$$\vec{D}\left[T_0,T_1\right] = \frac{\vec{F}(T_0) - \vec{F}(T_1)}{T_0 - T_1}$$

$$\vec{D}_2\left[T_0,T_2\right] = \frac{\vec{D}\left[T_0,T_1\right] - \vec{D}\left[T_1,T_2\right]}{T_0 - T_2}$$

(4.4)

$$\vec{D}_3\left[T_0,T_3\right] = \frac{\vec{D}_2\left[T_0,T_2\right] - \vec{D}_2\left[T_1,T_3\right]}{T_0 - T_3}$$

$$\vec{D}_4\left[T,T_4\right] = \frac{\vec{D}_3\left[T,T_2\right] - \vec{D}_3\left[T_0,T_3\right]}{T - T_3}$$

We can now compare (4.3) with a Taylor expansion of the force

$$\vec{F}(T) = \vec{F}(T_0) + \sum_{k=1}^{4} \vec{F}^{(k)} \frac{(T - T_0)^k}{k!} \qquad (4.5)$$

to obtain the derivatives of forces in terms of the divided differences

$$\vec{F}^{(1)} = \vec{D} + T_1'\vec{D}_2 + T_1'\ T_2'\ \vec{D}_2 + T_1'\ T_2'\ T_3'\ \vec{D}_4$$

$$\vec{F}^{(2)} = 2!\left\{\vec{D}_2 + (T_1'+T_2')\vec{D}_3 + (T_1'\ T_2' + T_2'T_3' + T_1'\ T_3')\vec{D}_4\right\}$$

$$\vec{F}^{(3)} = 3!\left\{\vec{D}_3 + (T_1' + T_2' + T_3')\vec{D}_4\right\}$$

$$\vec{F}^{(4)} = 4!\ \vec{D}_4$$

where $T_k' = T_k - T_o$ for k=1,2,3 and where we have taken ℓ=4.

The advantage of using (4.3) and (4.4) is the simple manner in which the divided differences at two successive times are related. This results from the definition of the divided difference. From (4.4) it is clear that the successive divided differences require two divided differences of one lower order. One of them is calculated immediately before, as indicated by (4.4), while the other is the divided difference at the end of the previous force calculation.

5. Semi-Iteration

It should be noted that to obtain the last term in (4.3) a semi-iteration procedure is used. This is necessary because the calculation of $\vec{D}_4$ requires a knowledge of the force at time T, which, of course, is unknown at time T_o, the time of the other $\vec{D}_k$ calculations. The force at T can only be found after the position of the star is known. To circumvent this difficulty, we use a semi-iteration procedure. To describe the semi-iteration procedure, it is easier to consider (4.3) first, as it applies to the irregular force polynomial. Similar consideration applies to the regular force polynomial. The position and velocity of the star are updated using (4.1) up to the $F^{(3)}$ term using the values of the derivatives which only have contributions up to D_3. This new position is now used to find the force for that star at time T. Now one can find D_4 and use that to get a better value for the position and the velocity by adding the $F^{(4)}$ term and also adding the D_4 correction for the lower force derivatives.

6. Computer Storage

At the start of the integration or at any other time during it, a certain number of quantities have to be stored for every star in the system. For completeness they are listed here for the i-th star.

t_i, t_i', t_i'', t_i'''	times of the last four irregular force calculations, in decreasing order
T_i, T_i', T_i'', T_i'''	times of the last four regular force calculations, in decreasing order
$\vec{r}_i$, $\vec{v}_i$	position and velocity, respectively, at time t_i
$\vec{F}_i$, $\vec{F}_i^{(1)}$	total force and its first derivative at time t_i

$\vec{K}_i$, $\vec{D}_{1i}{}^k$, $\vec{D}_{2i}{}^k$, $\vec{D}_{3i}{}^k$	regular force and its divided differences computed at time t_i
$\vec{S}_i$, $\vec{D}_{1i}{}^s$, $\vec{D}_{2i}{}^s$, $\vec{D}_{3i}{}^s$	irregular force and its divided differences computed at time t_i
δt_i, ΔT_i	irregular and regular time steps
R_i	radius of sphere defining neighborhood
n_{ij}	matrix of neighbor n_{ij} is the j-th neighbor of the i-th star
n_{bi}	total number of neighbors

The divided differences are stored rather than the force derivatives for purely computational considerations. But the total force and its first derivative are stored because they are frequently needed for synchronization. It should be noted that it requires approximately twice as many storage locations per particle as the scheme used by Aarseth[10] and Wielen[8] but the increase in the speed of computation, even for a moderate number of stars, outweighs this drawback. We will later discuss how these quantities are obtained at the very start of a computation. It is important to remember that at any time after the time of the start, t=0, the stars are not synchronized. This means that t_i, which is the time for the last irregular force calculation and also the time for which the position and the velocity are stored, are all different. Similarly, the T_i's are also not synchronized.

7. Main Integration

Since the particles are not synchronized, we choose the particle whose position and velocity are to be revised first by finding the minimum of $t_i + \delta t_i$. That is, the particle whose irregular force needs recalculation before all others. In a previous stage a check has already been made to assure that the regular force is still good at this stage. The position and velocity are updated to the time

$t_i + \delta t_i$ using the Taylor expansion (4.1) and (4.2) but carrying it out to only the third order in force. The fourth order term is taken into account at a later stage by the semi-iteration procedure described above.

As the positions of the neighbors of the i-th star are not synchronized with it, we update these to the time t_i

$$(7.1) \quad \vec{r}_k = r_k(t_k) + v_k \Delta T + \frac{1}{2m_k} \vec{F}_k (\Delta T)^2 + \frac{1}{6} \frac{1}{m_k} F_k^{(1)} (\Delta T)^3$$

$$\Delta T = t_i + \delta t_i - t_k$$

for each neighbor k, and t_k is the time for which the position $\vec{r}_k$ is known. We use a low order extrapolation for $\vec{r}_k$ here because it is found that it is more efficient to control the accuracy by α than to go to the higher order for synchronization. We can now use these $\vec{r}_k$s to calculate the irregular force and its derivatives. The positions and velocities of i are now corrected by semi-iteration which involves adding the contribution of $\vec{F}_i^{(4)}$ and also the additional contribution from the lower force derivatives due to the $\vec{D}^4$ term as shown in (4.6). The semi-iteration correction to the position and velocity, r_s and v_s, respectively, are

$$(7.2) \quad \begin{aligned} \Delta\vec{r}_s &= \frac{\delta t_i^{\,3}}{m_i} \left[\frac{F_i^{(4)}}{6!} \delta t_i^{\,3} + \frac{\Delta F_i^{(3)}}{5!} \delta t_i^{\,2} + \frac{\Delta F_i^{(2)}}{4!} \delta t_i + \frac{\Delta F^{(1)}}{3!} \right] \\ \Delta\vec{v}_s &= \frac{\delta t_i^{\,2}}{m_i} \left[\frac{F_i^{(4)}}{5!} \delta t_i^{\,3} + \frac{\Delta F_i^{(3)}}{4!} \delta t_i^{\,2} + \frac{\Delta F_i^{(2)}}{3!} \delta t_i + \frac{\Delta F^{(1)}}{2!} \right] \end{aligned}$$

where the ΔF_is are semi-iteration corrections to the lower force derivatives given by

$$(7.3) \quad \begin{aligned} \vec{\Delta F}_i^{(1)} &= t_1' t_2' t_3' \vec{D}_4' \\ \vec{\Delta F}_i^{(2)} &= 2!(t_1' t_2' + t_2' t_3' + t_1' t_3') \vec{D}_4 \\ \vec{\Delta F}_i^{(3)} &= 3!(t_1' + t_2' + t_3') \vec{D}_4 \end{aligned}$$

and

$$\vec{F}_i^{(4)} = 4!\vec{D}_4 \tag{7.4}$$

where t_1', t_2', and t_3' are the times of the last three irregular force calculations measured from t_i, and $\vec{D}^4$ is the fourth divided difference.

We now check whether the next time around the regular force would need revision. This is done by checking $t_i + 2\delta t_i$ against $T_i + \Delta T_i$. If the latter is greater, then the regular force polynomial and its derivative are extrapolated to $t_i + \delta t_i$. The new δt_i for that particle is determined and the loop is begun again by finding the new particle with the minimum $t_i + \delta t_i$.

If the regular force polynomial has to be revised, then all the particles are synchronized to $t_i + \delta t_i$ using (4.1) only to the term containing the first derivative of the force. The neighbors of i are determined and any change of the previous neighbors noted. If there are no new neighbors and all of the old neighbors are still neighbors, one straightforwardly calculates the forces and the derivatives. As the local distribution of stars about a certain star changes in time it is probable that there will be a change in the neighbors. For the success of the scheme it is crucial that this change be properly taken into account. It is an easy matter to check the new neighbors against the old and identify those stars which have either moved into the neighborhood of the star or moved out of it. Of course, the regular and the irregular forces have to be corrected for this change, which requires addition or subtraction of the contributions of these stars to the forces. If there has been a change in neighbors, the forces and the derivatives are first calculated assuming there has been no change. Then these are corrected by using analytic formulas to calculate contributions from old neighbors which are no longer neighbors and/or the new neighbors. These are then subtracted and added, respectively. New ΔT_i and δt_i are calculated for the particle and the cycle is repeated.

As the above method is not self-starting, a separate initial routine is used to start.

8. Starting Mechanism

The initial conditions can be given to the integration routine to be evolved in time. This routine, however, needs the force and its three derivatives to start and continue the integration. The simplest way to find these is to use analytic formulas. The force and its first derivative are found first and they are used to find the second and the third derivatives of the force. During these calculations the neighbor matrix n is also found, where n_{ij} is the j-th neighbor of the i-th particle. Similarly, the regular and irregualr time steps ΔT_i and δt_i, respectively, are also found. All this information is stored on a disc or a tape to be used as input for the main integration scheme. This method eliminates the need for a specialized starting procedure.

One possible starting procedure, used by Aarseth[10] requires the positions and velocities of all stars in the system to be updated by a few small, equal time steps. The forces at each of these times are calculated analytically, and these are then used to find the derivatives by numerical differentiation. We have found the analytical calculation of all the derivatives in the beginning to be more convenient.

9. Units and Scaling

A variety of units have been used by different investigators. We have found a particularly convenient set is to take unity for the total mass, M, and the gravitational constant, G. Also, the "radius," R, of the model cluster defined by

(9.1) $$E = -\frac{1}{2}\frac{GM^2}{R}$$

is taken to be 4. The relationship between the model units and the astronomical units is then as follows

$$yr = 1.897\ 10^{-8}\left(\frac{M}{R_0{}^3}\right)^{\frac{1}{2}}$$ model unit of time

$$pc = 2/R_0$$ model unit of distance

$M_{\odot} = 1/M$ model unit of mass

where R_o is the radius in parsecs and M is in solar mass. In our units the total energy is

(9.2) $$E = -.125$$

and assuming the virial theorem for an equilibrium cluster the potential energy V and the kinetic energy T are

(9.3) $$V = 2E = -.25$$ $$T = -E = .125$$

The crossing time for the cluster T_c is given by

(9.4) $$T_c = \frac{R}{\overline{V}} = GM^{5/2}(-2E)^{-3/2} = 8$$

where $\overline{V}$ is the root mean square velocity for the cluster.

10. Generation of Initial Conditions

Computer generation of initial velocities and positions to conform to a given distribution function f(r,v) requires considerable programming and therefore it is best done in a separate program. The methods described below are undoubtedly fairly standard procedures but for the sake of completeness we shall nonetheless give specific details.

The mass density $\rho(r)$ can be found by

(10.1) $$\rho(\vec{r}) = \int f(\vec{r},\vec{v})d\vec{v}$$

For spherically symmetric systems in equilibrium for which the distribution function is a function of energy E only, (10.1) reduces to

$$\rho(\vec{r}) = 4\pi \int f(E)\, v^2\, dv \tag{10.2}$$

where

$$E = \frac{1}{2} v^2 + \psi(r) \tag{10.3}$$

where ψ is the gravitational potential at r. Combining 10.2 with Poisson's equation we get

$$\frac{1}{r^2} \frac{d^2}{dr^2} \left(r^2 \frac{d\psi}{dr}\right) = (4\pi)^2 G \int_0^\infty f(E)\, v^2\, dv \tag{10.4}$$

To generate a cluster with density $\rho(r)$ we need m(r), the mass interior to the radius. This can be done by numerically solving (10.4) by a method such as Taylor series expansion, which gives ψ and its derivatives. The mass interior to r, m(r), can then be found by

$$m(r) = -\frac{r^2}{G} \frac{d\psi}{dr} \tag{10.5}$$

This m(r) is inverted to yield r(m), or radius containing a certain mass. This can now be used to generate the radius r_i for the i-th star

$$r_i = r\left(\frac{i}{N} M\right) \tag{10.6}$$

where M is the total cluster mass and N is the total number of stars. Once a radius r_i has been assigned to each star, it is only necessary to give it a random direction in space. This is most easily done by picking three random numbers, a,b, and c between -1 and +1 such that

$$r_o = \sqrt{a^2 + b^2 + c^2} < 1$$

(10.7)
$$x_i = \frac{ar_i}{r_o}$$

$$y_i = \frac{br_i}{r_o}$$

$$z_i = \frac{cr_i}{r_o}$$

To avoid very close pairs which may lead to numerical instabilities in the beginning, it is advisable to check for such, since they are not dynamically balanced with proper angular momentum and may lead to very close collisions, causing large deviations from the required energy. An additional advantage in removing these close pairs is that the final total potential energy is very close to the theoretical value. Since the radii generated by (10.6) are in ascending order, it is necessary to check r_i against the positions of only a few preceding particles. If r_i is very close to one of these, then its direction is again chosen according to (10.7), and a new check for close pairs is made. This process is repeated until no such close pair is found. This minimum separation distance between two stars is taken to be a small fraction (about a tenth) of the average nearest neighbor distance.

Once the stars have been distributed according to the above procedure, the coordinates of the center of mass R_c are found by

(10.8)
$$\vec{R}_c = \frac{1}{N} \sum_{i=1}^{N} \vec{r}_i$$

It is advantageous to have a system which has its center of mass at the origin and hence, the system is translated by the distance

(10.9)
$$\vec{r}_i' = \vec{r}_i - \vec{R}_c$$

It is important to have a small R_c to prevent the transformation (10.9) from altering the original density profile. This can be done by generating many systems by the above method and selecting the one with the smallest value of $|\vec{R}_c|$. As a further insurance against distortion of the density by the center of mass correction we could use the following transformation

$$\vec{r}_i = \vec{r}_i - \rho_i \vec{R}_c \tag{10.10}$$

where ρ_i is a random number between 1 and 2. Unlike (10.9), this system does not yield a system with the center of mass at the origin. This correction, however, can be applied repeatedly and one can make the center of mass as close to the origin as desired by this fast converging process.

Once the stars have been distributed inside the sphere of radius R as described above, we must assign velocities to them according to

$$\rho(\vec{v}) = f(\vec{r},\vec{v}) \tag{10.11}$$

The velocity of the center of mass V_c given by

$$\vec{V}_c = \frac{1}{N} \sum_{i=1}^{N} \vec{v}_i \tag{10.12}$$

is usually non-zero and must be corrected to zero by a method analogous to the center of mass correction.

Although spherically symmatric distribution functions have no angular momentum, the cluster generated on a computer has an angular momentum

$$\vec{L} = \sum m_i \vec{r}_i \times \vec{v}_i \tag{10.13}$$

which is finite. If it is desirable to have a cluster without angular momentum, then the following transformation can be applied

(10.14) $$\vec{v}_i = \vec{v}_i - \vec{r}_i \times \vec{w}$$

where $\vec{w}$ is the rotational velocity found by solving

(10.15) $$\vec{L} = \Pi \vec{w}$$

and Π is the usual moment of inertia tensor. It is clear that (10.14) leaves the center of mass at rest if it was initially coincident with the origin.

This completes the simulation of the initial conditions for a given distribution function. It must be emphasized that although some restrictions are placed on completely random generation, the restrictions and the corrections are rather slight and do not cause the system to depart considerably from the theoretical model. The potential and kinetic energies and the virial ratio of the simulated cluster are found to be within a few percent of their theoretically predicted values, even for $N = 100$.

11. Polytropic Spheres

Generation of a model cluster from an arbitrary distribution function of energy is usually quite difficult for one must first solve for the distribution of mass and velocity. One particular set of distribution functions for which this can be done are the Emden polytropes. They are particularly useful in testing integration programs for we have found them to be relatively stable. Special cases of the polytropes have been used as initial conditions[8,12]. We shall give here the relevant equations for the polytropes which may be helpful to investigators wishing to program specific cases.

Following Camm[13] we consider

(11.1) $$f(r,v) = A(E_o - E')^n \quad E' = \frac{1}{2} v^2 + \psi < E_o$$

where E' and ψ are the binding and potential energies per unit mass, respectively. A and E_o are constants which are related to the mass M and the radius R of a given cluster of stars. The mass density $\rho(r)$ is given by

$$\rho(r) = \int f(r,v)\, d\vec{v}$$

(11.2)

$$= 4\Pi A \int_0^{v_{max}} \left(\frac{v_{max}^2}{2} - \frac{v^2}{2} \right)^n v^2\, dv$$

with

$$v_{max} = \sqrt{2(E_o - \psi)}$$

Setting

$$y = \frac{v}{v_{max}}$$

$$\rho = A(E_o - \psi)^{n + 3/2}\, I_n$$

where

$$I_n = 4\Pi\, 2^{3/2} \int_0^1 (1 - y^2)^n\, y^2\, dy \qquad (11.3)$$

the Poisson equation gives

$$\frac{1}{r^2} \frac{d}{dr}\, r^2 \frac{d\psi}{dr} = 4\Pi G A (E_o - \psi)^{n + 3/2}\, I_n \qquad (11.4)$$

Let us define a dimensionless variable

$$\theta_m = \frac{E_o - \psi}{E_o - \psi_o} \qquad (11.5)$$

where ψ_o is the central potential.

Thus

$$\theta_m(o) = 1$$

$$\theta'_m(o) = 0$$

Also setting

$$x = \alpha r \tag{11.6}$$

where

$$\alpha^2 = 4\Pi GA\ (E_o - \psi_o)^{n + \frac{1}{2}} \tag{11.7}$$

we get

$$\frac{1}{x^2} \frac{d}{dx}\ x^2 \frac{d}{dx}\ \theta_m = \theta_m^{\ n + 3/2} \tag{11.8}$$

This is Emden's Equation with index

$$m = n + \frac{3}{2} \tag{11.9}$$

It is well known that for $m < 5$, the sphere given by (11.8) has both finite radius and finite mass. For $m = 5$, this represents the Plummer model which has finite mass though infinite radius. For $m > 5$, both the mass and the radius are infinite. The index n must also be greater than -1 for finite density. The density is given by

$$\rho(r) = \rho_c\ \theta_m^{n + 3/2} \tag{11.10}$$

where ρ_c, the central density, is

$$\rho_c = AI_n(E_o - \psi_o)^{n + 3/2} \tag{11.11}$$

We can identify the first zero x_o of the Emden function as the radius of the cluster. We will also need

$$y_o = -\left(x^2\ \frac{d\theta_m}{dx}\right)_{x = 0} \tag{11.12}$$

Obviously, the scaling factor is

(11.13) $$\alpha = \frac{x_o}{R}$$

(This is not strictly true for the Plummer model where x_o, R are infinite.) Since

(11.14) $$\rho(R) = 0$$

combining (11.5) and (11.10) we get

(11.15) $$E_o = \psi(R) = -\frac{GM}{R}$$

Also

(11.16) $$y_o = -x^2 \left.\frac{d\Theta_m}{dx}\right|_{x=0} = \frac{\alpha r^2}{(E_o - \psi_o)} \left.\frac{d\psi}{dr}\right|_{r=R}$$

Using Gauss' Law

(11.17) $$y_o = \frac{\alpha MG}{(E_o - \psi_o)}$$

But (11.3) and (11.6) give

(11.18) $$\psi_o = -\frac{MG}{R}\left(1 + \frac{x_o}{y_o}\right)$$

Now A can easily be related to M and R by combining (11.18), (11.15) and (11.7)

(11.19) $$A = \frac{1}{4\Pi I_n G}\left(\frac{R}{x_o}\right)^{n-3/2}\left(\frac{y_o}{MG}\right)^{n+1/2}$$

Thus, for a given n we can find a constant A for a cluster of mass M and radius R, since x_o and y_o are tabulated in the literature for many polytropes[14].

It is well known that a polytropic sphere of index m has the gravitational potential energy V given by

(11.20) $$V = \frac{3}{5-m} \frac{GM^2}{R}$$

The virial theorem then gives the total energy

(11.21) $$E = \frac{V}{2} = -\frac{3}{2}\,\frac{1}{5-m}\,\frac{GM^2}{R}$$

The velocity distribution $\rho(v)$ for a given $f(\vec{r},\vec{v})$ can be found by

(11.22) $$\rho(\vec{v}) = \int f(\vec{r},\vec{v})\, d\vec{r}$$

For the polytropes, we have

(11.23) $$\rho(\vec{v}) = 4\Pi A \int_0^{R_m} r^2 dr \left[E_o - \frac{1}{2}v^2 - \psi\right]^n$$

where R_m is given by

(11.24) $$E_o - \frac{1}{2}v^2 = \psi(R_m)$$

Although the system of equations cannot, in general, be solved analytically, one could get a $\rho(v)$ for any value of v by numerical methods.

The spectrum of binding energy for different polytropes can be found by a general formula which gives the probability density P(F) for a function F of r and v.

(11.25) $$P(F) = \int f(\vec{r},\vec{v})\, \delta\left[F - F(\vec{r},\vec{v})\right] d\vec{r}\, d\vec{v}$$

Thus, taking (11.1) for f(r,v) and binding energy E' for F

(11.26)
$$P(E') = 16A\Pi^2 \int r^2 dr \int_o^{v_m} dv\, v^2\, \delta\left(E' - \frac{1}{2}v^2 - \psi\right)\left(\frac{v_m^2}{2} - \frac{v^2}{2}\right)^n$$

$$v_m = \sqrt{2(E_o - \psi)}$$

integrating over the velocities we get

(11.27) $$P(E') = 16A\Pi^2\sqrt{2}\,(E_o - E')^n \int_o^{R_{max}} v_m\, r^2 dr$$

with the restrictions

$$E' < E_o$$

and

$$\psi < E'$$

The first inequality is an obvious consequence of (11.1); the second inequality gives the upper limit of the r integration by

(11.28) $$\psi(R_{max}) = E'$$

As in the case of the velocity distribution, the system of equations (11.27) and (11.28) has, in general, no analytic solution and must be numerically solved. This is quite straightforward. One must first invert equation (11.28) for a particular E' to find R_{max} and then integrate (11.27) to this upper limit.

A particularly simple case is the case of n = 0. This gives an Emden polytrope of index 3/2.

$$f(E) = \begin{cases} A & E < E_o \\ 0 & E > E_o \end{cases}$$

where

(11.29) $$E_o = -\frac{GM}{R}$$

In our units it can be shown using (11.19) and (11.20) that

(11.30) $$R = 24/7$$
$$A = .01217$$

12. Conclusion

It has become standard to use the total energy of the system as a measure of the accuracy of integration. To maintain a given accuracy during the integration what we do is to check the energy a certain number of times per crossing times. At each of these times the whole memory is temporarily stored on disc and the total energy is calculated. If the total energy meets the criterion set for it the integration is continued. If not then we pick up the memory from the previous energy check and reintegrate that part of the evolution with more stringent control on the time steps. This is done via α and β . In actual practice we have found that when the energy criterion is not met it is always better to just reduce α and keep the old value of β . Only as a last resort is β reduced.

In the units described previously a good range for α is from .5 to 1.5 and for β from .1 to .3. These figures are for systems of 100 bodies and gives an error in the energy of the order of .1 per cent per crossing time. For larger numbers of particles α should be increased by the square root of the relative number of particles.

We have made a number of tests to ascertain the amount of computing as a function of the number of particles. We have found a good fit to be a power law

$$\gamma \left(\frac{N}{100}\right)^{n}$$

with the exponent equal to approximately 1.6 and γ is about 40 seconds on the CDC 6600 and a relative energy error of about 10^{-4}, per crossing time.

Acknowledgement:

This work was supported by a grant from the National Science Foundation.

References

1. S. Aarseth, Astronomy and Astrophysics, 9, 64, 1970

2. S. Aarseth, Gravitational N Body Problem, ed. M. Lecar, D. Reidel (1972) p. 373

3. C.F. Peters, Bull. Astr., 3, 1967, 1968

4. V. Szebehely and D.G. Bettis, Gravitational N Body Problem, ed. M. Lecar, D. Reidel (1972) p.388

5. P.C. Heggie, Gravitational N Body Problem, ed. M. Lecar, D. Reidel (1972) p. 148

6. S. Chandrasekhar, Rev. Mod. Phys., 15, 1, 1943

7. S. Aarseth and F. Hayli, Astr. Norvegica, 9, 313, 1964

8. R. Wielen, Veroffentlichungen Astr. Rechen-Institut Heidelberg, Nr. 19, 1947

9. A. Hayli, Bull. Astron., 2, 67, 1967

10. S. Aarseth, Bull. Astron,, 3, 209, 1968

11. C. Gonzalez and M. Lecar, Bull. Astron., 3, 209, 1968

12. M. Henon, Gravitational N Body Problem ed. M. Lecar, D. Reidel (1972)

13. G. Camm, M.N.R.A.S., 6, 289, 1950

14. S. Chandrasekhar, An Introduction to the Study of Stellar Structure, Dover, New York, 1967

NUMERICAL EXPERIMENTS ON THE STATISTICS OF THE GRAVITATIONAL FIELD

by

Leon Cohen and Afaq Ahmad

Hunter College

of the City University of New York

Introduction

There are currently two approaches in the computer simulation of the N body gravitational problem. The first to have been historically developed is to directly integrate the exact equations of motion. That is, starting with initial positions and velocities of the N stars one attempts by numerical integration to obtain the positions and velocities at a later time. Having the positions and velocities as a function of time allows one to investigate particular physical phenomena. The direct approach is free from any simplifying assumptions and therefore has the advantage that, in principle, it offers a complete specification of the system.

Unfortunately, the direct approach is severely limited by the fact that even with the largest of computers one cannot simulate systems with large numbers of particles for significant dynamical times. Nonetheless, a number of physical phenomena can be studied by this method and the improvement in both mathematical technique and computer speed has allowed the integration of systems containing a fairly large number of stars (250 - 1000).

The other approach is to use a Monte Carlo scheme. The general approach of these methods is to use some simplifying assumptions derived from theoretical considerations to avoid the direct microscopic integration of the system. One probabilistically averages over some of the microscopic effects by theoretical means and uses these results in the computer simulation. The possible disadvantages of the Monte Carlo methods are that they may be too gross and may miss out on phenomena which may be important (e.g., binaries) and also that the results obtained are, of course, dependent on the statistical assumptions made.

For the above reason and for the inherent theoretical interest we have been investigating some of the stochastic aspects of the gravitational force. The most deeply developed theory of the stochastic properties of the gravitational field is that of Chandrasekhar and von Neumann[1,2,3,4]. Although the theory is not derivable from first principles (i.e., from Liouville's Equation) it has the advantage that its predictions are quite sharp and can thus be compared with results obtained from numerical experiments.

The numerical experiments we have performed concentrate on the crucial aspects of their theory, particularly those which bear on the basic assumptions. We shall present results relating to the probability distribution of the random force, the distribution of the time rate of change of the random force, the two time autocorrelation function of the force, and the existence of dynamical friction.

The starting point of the theory is to divide the force on a star into two parts: one part due to the smoothed out distribution of matter and another part due to the stars in the immediate vicinity of the star in question. The force due to the smoothed out distribution of stars, the mean field force, changes deterministically and can be derived from a time dependent potential. On the other hand, the force due to the nearby stars oscillates very rapidly due to chance encounters and the relatively rapid fluctuations in the distribution of the nearby stars. The mean field force can, of course, be studied by Newton's equations or their fluid dynamic counterpart, but the random force must be studied stochastically.

The theory of Chandrasekhar and von Neumann deals both with the probability distribution of the random force and the joint distribution of the random force with other dynamical variables such as the rate of change of the force or the random force at a different time. Since the gravitational force, $\vec{F}$, depends only on the relative positions of the other stars, obtaining the probability distribution for it depends only on the probability distribution of the relative distances of the other stars. Chandrasekhar and von Neumann assume that the stars surrounding the star on which the force is being calculated are distributed from a uniform probability density. This is the most crucial assumption of the theory. It implies no correlation among the stars, either in positions or between positions and velocities. They also take the number of stars to be infinite but keep the density constant. With these assumptions it is straightforward to derive the probability distribution of force which is known as the Holtsmark distribution.

It is clear that a number of physical quantities depend not only on the force but on the time rate of change of the random force

$$\dot{\vec{F}} = \vec{f} = -G \sum_{j=1}^{N} m_j \left\{ \frac{\vec{v}_j}{r_j^3} - 3 \frac{\vec{v}_j \cdot \vec{r}_j}{r_j^5} \vec{r}_j \right\} \tag{1.1}$$

Since the velocity of the field stars appears in (1.1), a further assumption has to be made about their distribution to derive the joint distribution of $\vec{F}$ and $\vec{f}$. Chandrasekhar and von Neumann assume that the distribution of velocities is independent of the position and is spherical in velocities. Mathematically, the problem now becomes the following. Given the probability distribution, $P(\vec{r},\vec{v})$, of position and velocity for an arbitrary star i derive the probability distributions of the relevant physical quantities which are functions of $\vec{r}$ and $\vec{v}$.

Besides deriving the joint probability distribution of $\vec{F}$ and $\vec{f}$, Chandrasekhar[3] also derived the joint distribution of $\vec{F}$ and $\vec{F}_t$ where $\vec{F}_t$ is the force on the star at time t later. To do this, he further assumed the field particles move in straight line orbits. We should point out that due to the mathematical complexity of the problem, explicit expressions for the joint distribution have never been obtained; but expectation values of relevant physical quantities can be obtained since they very often involve integrations over the distributions which can in many cases be done explicitly.

We now proceed to describe a number of results they have obtained and the experiments we have done to verify these aspects of the theory.

Random Force

As we have published[5] a detailed paper on the probability distribution of the random force we shall here, for the sake of continuity, only briefly describe some of the results. We shall also describe some further experiments we have performed on the probability distribution of the force due to the nearest neighbor of the particle on which the force is being calculated. Using the assumptions mentioned in the introduction yields the Holtsmark distribution, which is the probability of obtaining a force of strength F.

$$W(\beta) = \frac{2}{\pi}\beta \int_0^\infty y \sin \beta y \, e^{-y^{3/2}} \, dy \qquad (2.1)$$

where

$$\beta = \frac{|F|}{Q}$$

$$Q = \left(\frac{4}{15}\right)^{2/3} 2\,\pi G\, m\, n^{2/3} \qquad (2.2)$$

and m and n are the mass of each star and the number density respectively.

To experimentally study the distribution of the random force in a system of gravitational bodies, the integration scheme described elsewhere was used. We evolved systems containing up to 1000 particles for a few crossing times. Since the crossing time is the time in which a star with an average velocity traverses the system, it insures that particles are thoroughly mixed during this interval. Since the calculation of β for a star involves a knowledge of the space density in its neighborhood, it is necessary to have initial conditions which are in or close to equilibrium. This insures that there are no violent changes in the density of the cluster. For most of the systems evolved we have used the Emden polytrope of index 1.5 for initial conditions. During the evolution of the system on the computer, such quantities as the positions, the velocities, and the total forces were sampled every 1/64 of a crossing time and stored on a tape. The random force is then obtained by subtracting the mean field force from the total. Since our systems are spherically symmetrical the mean field force is calculated from

$$\vec{F}_{mean} = - \frac{GM(r)}{r^3}\, \vec{r} \qquad (2.3)$$

where r is the position of the stars and M(r) is the mass interior to it. We tried two different methods of obtaining M(r). One, by merely counting the particles interior to the one in question, and the other, by integrating the density distribution which was averaged for a particular fraction of the crossing time. Both methods give substantially the same results. To find β, however, we should also have the density of stars in the system at different distances from the origin. A time average density is calculated for fractions of the crossing time (usually a quarter). This density is then used to find Q for each star for the samples that are contained in that interval.

Special care must be taken in the comparison of the experimental results with the theory. Since the number of particles is finite here, we cannot compare them directly with the Holtsmark distribution (2.1). One[5] can derive the finite N Holtsmark distribution but even then a direct comparison is not possible since different parts of the cluster have different densities. A means of circumventing this is to generate clusters randomly from the same space density distribution as the evolved case. This is a "static" cluster and has no effect of dynamic evolution and no correlations between nearby stars. The model cluster so simulated satisfies the assumptions of the theory. The simplest way to generate a cluster which has the same density distribution as the evolved cluster is to keep the distances of the stars from the origin the same but choose a new direction. We have tried other methods of simulating the theoretical curve resulting in the same results.

Figure 1 shows a typical result. The solid curve shows the "theoretical" curve obtained by removing the effects of the dynamics as described above. The points represent the experimental values. We should point out that the simulated theoretical curve is indeed very close to the Holtsmark distribution particularly for systems containing 1000 particles. As can be seen, the agreement with theory is excellent. This comparison of theory with experiment is typical of most of the results obtained.

For small particle systems (systems containing 100 particles or less) we have sometimes found a discrepancy. This is illustrated in Figure 2 for a 100 body system. We thought that a possible cause for the discrepancy may be the existence of binaries as they would violate the assumption of no correlations. But when we removed the binaries from the statistics no significant change occurred in the experimental points. This was expected since the number of binaries was relatively small.

Another possibility for the discrepancy is that since we are calculating the random force on a star the star may have modified the field stars so that the assumption of the field stars' being uniformly distributed does not hold. A simple way to check for this is to calculate the force at randomly distributed points rather than on stars. We randomly selected points in the cluster and the force due to all the stars was calculated and from which the random force was found by the method described earlier. When this was done the experimental distribution always agreed with the theoretical curve. Figure 3 is a duplication of Figure 2 with the distribution of force on the random points superimposed. The discrepancy between the distribution of force at points where there are particles and the points where there are none indicates that there might be a modification of the field due to the presence of a star at that point.

We tested for the possibility that when the discrepancy between the theoretical curve and the experimental points appears, the effect is due to the possible enhancement of the force due to the nearest neighbor. This would be true, for example, if the particles in the system were undergoing numerous close collisions. The experiment we performed was to compare the distribution of the force for nearest neighbors for the real system with the "theoretical" distribution obtained by randomizing as described before. Figure 4 illustrates the results for one of the 100 body cases. Not surprisingly, the discrepancy still exists. We should point out that the simulated curve agrees almost identically with the theoretically derived nearest neighbor distribution[6].

But although the nearest neighbor accounts for some of the discrepancy, it does not account for all of it. What we did was to look at the distribution of the total random force minus the nearest neighbor force also. This is illustrated by Figure 5. As can be seen, there is a discrepancy there also. The nearest neighbor is thus not solely responsible for the discrepancy. A possible cause may be due to clumping or the formation of subsystems.

Distribution of the Time Rate Change of the Random Force

Chandrasekhar and von Neumann also derived the joint probability distribution of the random force and its time rate of change. Numerically testing for the joint distribution would be quite difficult and we have therefore checked the distribution of f which is obtained from the joint distribution by averaging over the force. The result obtained by Chandrasekhar and von Neumann for the probability distribution of f is

$$P(\vec{f}) = \frac{1}{\pi^2} \frac{c^2}{(c^2 + f^2)^2} \qquad (3.1)$$

where

$$C = \frac{2}{3} \pi^{3/2} G Q_0 m n \langle v \rangle$$

and where

$$Q_0 = \frac{1}{2\sqrt{3}} \log (2 + \sqrt{3})$$

and <v> is the root mean square velocity.
If f is expressed in units of c and if we let $\alpha = f/c$ then the probability distribution for α is

$$P(\alpha) = \frac{4}{\pi} \frac{\alpha^2}{(1+\alpha^2)^2} \tag{3.2}$$

To experimentally test for the above distribution we divided our system into spherical shells and for each shell the root mean square velocity of the particles was obtained. This allows the determination of c for each shell. The time derivative of the force was obtained by numerical differentiation of the random force and then depending on which shell the particle was in, the time derivative was divided by the appropriate c. The statistics were then combined for various fractions of a crossing time. We also performed the experiment by surrounding each particle by a sphere which on the average contained about 20 particles and calculated the force derivative directly from the analytic formulas. We obtained substantially the same results. Figure 6 represents the results of this experiment for the 1000 body case and as can be seen, the agreement with theory is excellent.

Dynamical Friction

One of the main results of the theory of Chandrasekhar and von Neumann is the prediction of the existence of dynamical friction in gravitational systems. They obtained the following result

$$<\frac{d|\vec{F}|}{dt}> = \frac{4}{3} \pi G\ m\ n\ B(\beta) \frac{\vec{v} \cdot \vec{F}}{|F|} \tag{4.1}$$

where v is the velocity of the particle and B(β) is a positive function related to the Holtsmark distribution H(β) by

$$B(\beta) = 3 \frac{\int_0^\beta H(\beta)\ d\beta}{\beta H(\beta)} - 1 \tag{4.2}$$

Clearly,

$$<\frac{d|\vec{F}|}{dt}> >0 \qquad v \cdot F > 0$$

$$<\frac{d|\vec{F}|}{dt}> <0 \qquad v \cdot f < 0 \tag{4.3}$$

That is, if F has a positive component in the direction of the velocity then the force will tend to increase; and conversely, if F has a negative component it will tend to decrease.

The experiments performed were to test directly for equation (4.1). As in the previous section the time derivative was calculated both by numerical differentiation and by surrounding the particle with a sphere containing from 10 to 20 particles and calculating the force derivatives analytically and again the results were substantially the same. $<\frac{d|\vec{F}|}{dt}>$ is a function of three variables: the magnitudes of $\vec{v}$ and $\vec{F}$ and the included angle. The equation was tested by integrating out two of the variables and examining $<\frac{d|\vec{F}|}{dt}>$ against the remaining one. Now, in integrating out the angle, one cannot average over the whole range from 0 to π since that would obviously give zero. What must be done is to average the cosine from 0 to π/2 and in the experiment use only those particles which have a cosine within that range. In practice, of course, not to waste the statistics for half the particles when the cosine is between $\pi/2$ and π the sign of $\frac{d|\vec{F}|}{dt}$ is changed and is counted in the same statistics. This is under the assumption that the cosine between v and F is uniformly distributed, and this is indeed the case experimentally. Since

$$<\cos\theta>_{\theta\to\pi/2} = \frac{\int_0^{\pi/2} \cos\theta \sin\theta \, d\theta}{\int_0^{\pi/2} \sin\theta \, d\theta} = \frac{1}{2} \tag{4.4}$$

we have

$$<\frac{d|F|}{dt}>_{\beta,v} = \frac{2}{3}\pi G \; m \; n \; v \; B(\beta) \tag{4.5}$$

To express $<\frac{d|\vec{F}|}{dt}>$ as a function of v only one should integrate B(β) with the Holtsmark distribution, but unfortunately this yields an infinite value. What we have thus chosen to do is to look at particles only up to a certain value of the force. For practical reasons we use a maximum value of $\beta = 10.7$. This involves 97% of the particles. The expected value of B(β) in that range is then

$$<B(\beta)> = \frac{\int_0^{10.7} B(\beta)\, H(\beta) d\beta}{\int_0^{10.7} H(\beta)\, d\beta} = 3.89 \tag{4.6}$$

We therefore have

$$<\frac{d|\overline{F}|}{dt}> = 3.89 \cdot \frac{2}{3}\pi G \; m \; n \; v \tag{4.7}$$

As discussed previously in Chandrasekhar and von Neumann's theory the number of field particles is allowed to go to infinity while the density is kept constant. Since we are dealing with finite number of particle systems we should take that into account. We have calculated the equivalent of the above equations when the random force on the test particle is produced by only one field particle. It turns out that for the one particle case the equivalent of equation (4.7) is

$$\left\langle \frac{d|F|}{dt} \right\rangle = 1.88 \cdot 3.89 \cdot \frac{2}{3} \pi G\, m\, n\, v \tag{4.8}$$

Figure 7 shows a typical result of a system containing 1000 stars. The normalization is such that the slope for the infinite particle case is 1 and the slope then for the one body case is 1.88. The fact that the experimental results are slightly above the theoretical curve for the infinite particle case must obviously be attributed to the finite N nature of our system. But as the figure clearly shows there is no question about the existence of dynamical friction in gravitational systems.

A comparison of $\left\langle \frac{d|F|}{dt} \right\rangle$ as a function of the force yields similar results.

The Two Time Autocorrelation Function

The autocorrelation function plays a central role in the theory of stochastic processes. In particular, for example, the root mean square velocity deviation can be expressed as an integral of the correlation function with respect to time. Chandrasekhar has derived the two time correlation function for the random force with the surprising result that for large times it only falls off inversely with the time. The customary definition of the autocorrelation function is

$$A(t_1, t_2) = \langle \overline{F}(t_1) \cdot \overline{F}(t_2) \rangle \tag{5.1}$$

where $F(t_1)$ and $F(t_2)$ are the forces at two different times. In deriving the autocorrelation function it is assumed that the field particles move in straight lines. $A(t_1, t_2)$ is then a function of $t_2 - t_1$ and no generality is lost if we take $t_1 = 0$. We hence have

$$A(t) = \langle \vec{F}(0) \cdot \vec{F}(t) \rangle \tag{5.2}$$

with

$$F(0) = \sum_{i=1}^{N} \frac{G\, m_i\, \vec{r}_i}{r_i^3}$$

$$F(t) = \sum_{i=1}^{N} G\, m_i \frac{\vec{r}_i + \vec{v}_i\, t}{|\vec{r}_i + \vec{v}_i t|^3} \tag{5.3}$$

A(0) is then the expected value of the square of the force and this is well known to diverge if the Holtsmark distribution for the force is used. To avoid this, Chandrasekhar defined the autocorrelation coefficient as

$$R(t) = \left\langle \frac{F(0) \cdot F(t)}{F^2(0)} \right\rangle \tag{5.4}$$

But we shall compare here our experimental results with A(t) rather than R(t) for the explicit expression for R(t) is quite involved. We have though made a direct comparison with R(t) and this is published elsewhere[7].

Using the same assumptions Chandrasekhar used in deriving R(t) one[8] can readily derive an explicit expression for A(t) as defined by (5.2)

$$A(t) = 4\pi G^2\, m^2\, n \int \frac{P(\vec{v})\, d\vec{v}}{v\, t} \tag{5.5}$$

where P(v) is the probability distribution of the velocities. If a Gaussian distribution for v is taken then

$$A(t) = \frac{8j\ \sqrt{\pi}\ G^2\ m^2\ n}{t} \tag{5.6}$$

where $j^2 = \frac{3/2}{\langle v^2 \rangle}$. The velocities are measured with respect to a local standard of rest.

For the purpose of comparison with experiment it will be more convenient to use dimensionless units defined by

$$t = t_0 \tau$$

$$t_0 = \ell j$$

$$\ell = \left(\frac{15}{4}\right)^{1/3} \frac{n^{-1/3}}{(2\pi)^{1/2}} \tag{5.7}$$

and in that case A(τ) becomes

$$A(\tau) = \frac{15}{2\pi}\ \frac{1}{\tau} \tag{5.8}$$

Figure 8 shows the results obtained for a 1000 body case. The fact that for A(0) the experimental points do not match is understandable since of course in reality the expected value of $\langle F^2 \rangle$ is not infinity and as Chandrasekhar has pointed out this is due to the invalidity of the assumption of complete randomness of the field stars for all regions of space. This assumption cannot be valid when the distance between field and test particles is very small. As for the experimental results for large we have always found that the experimental points seem to drop somewhat faster than the 1/t prediction. We are currently investigating whether this may be due to the finite size of the system.

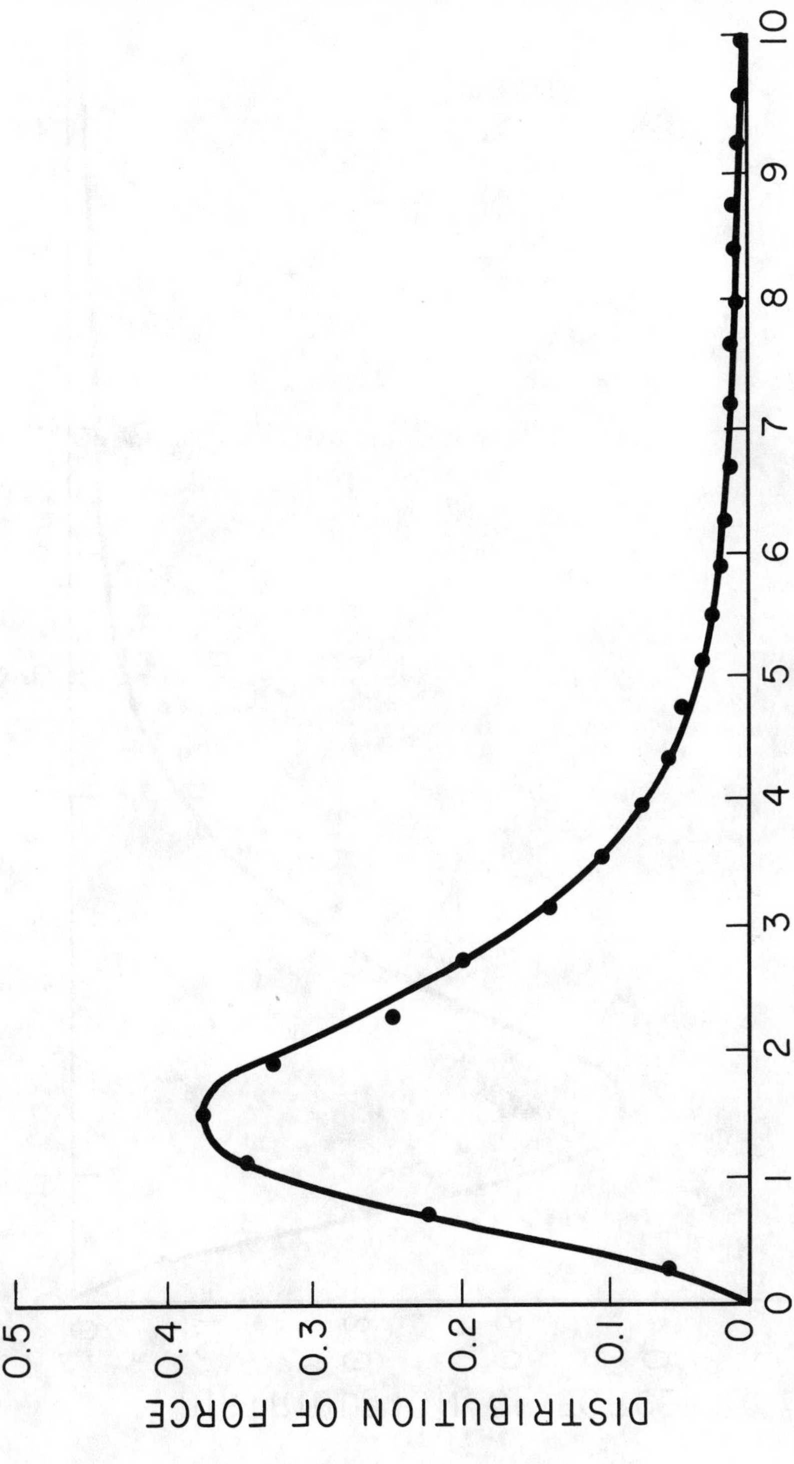

Figure 1.

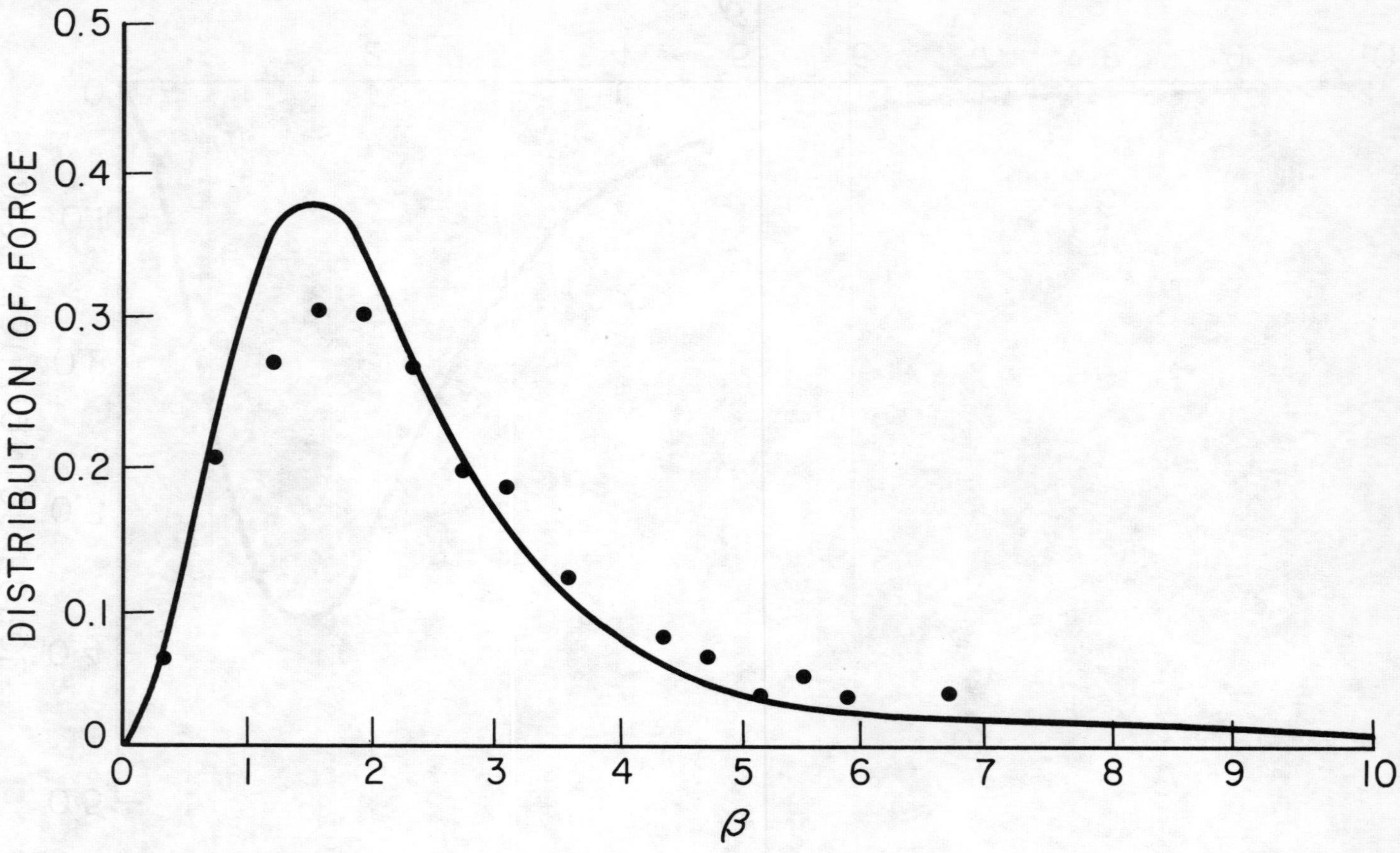

Figure 2.

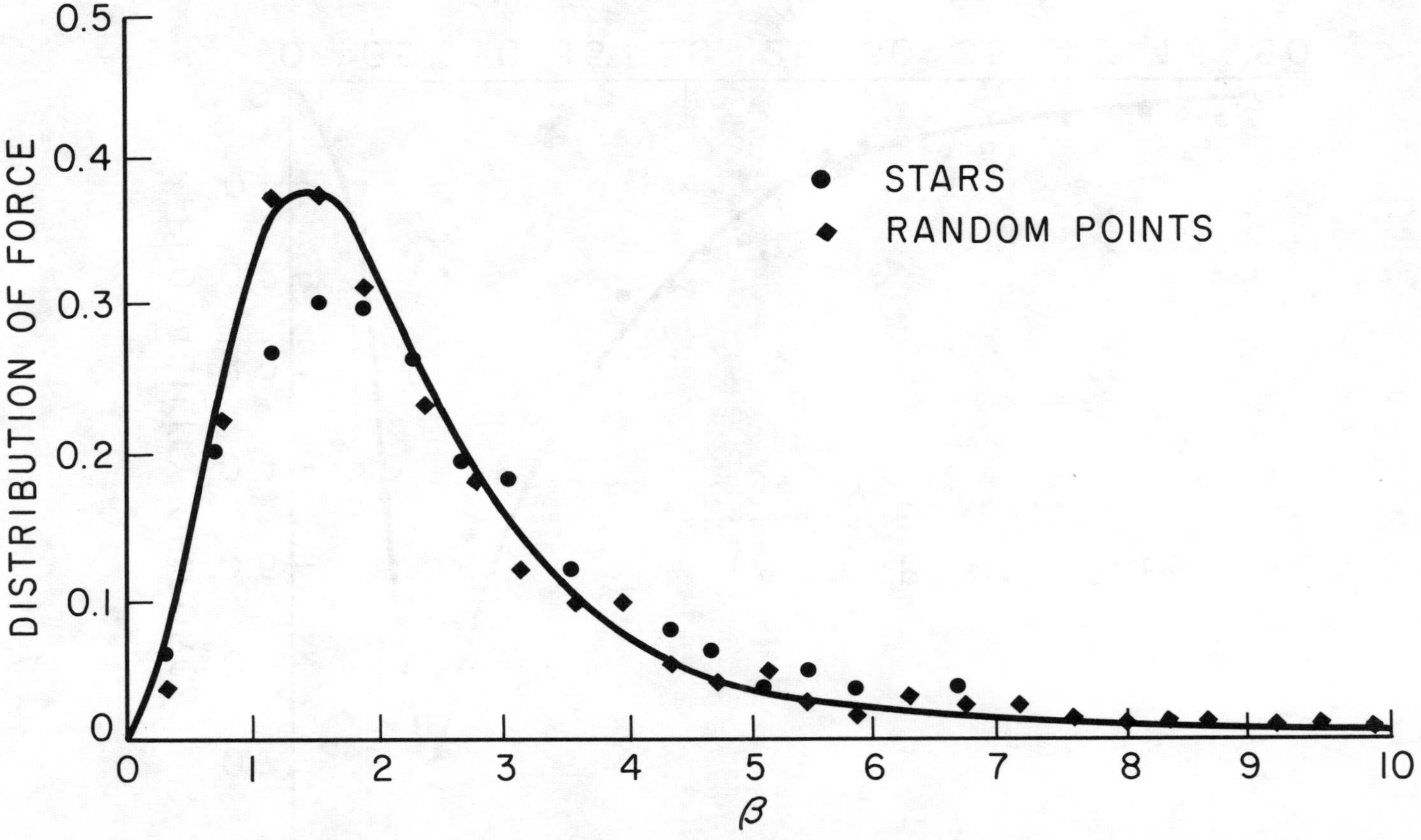
STARS
RANDOM POINTS
DISTRIBUTION OF FORCE
0.5
0.4
0.3
0.2
0.1
0
0
1
2
3
4
5
6
7
8
9
10
β

Figure 3.

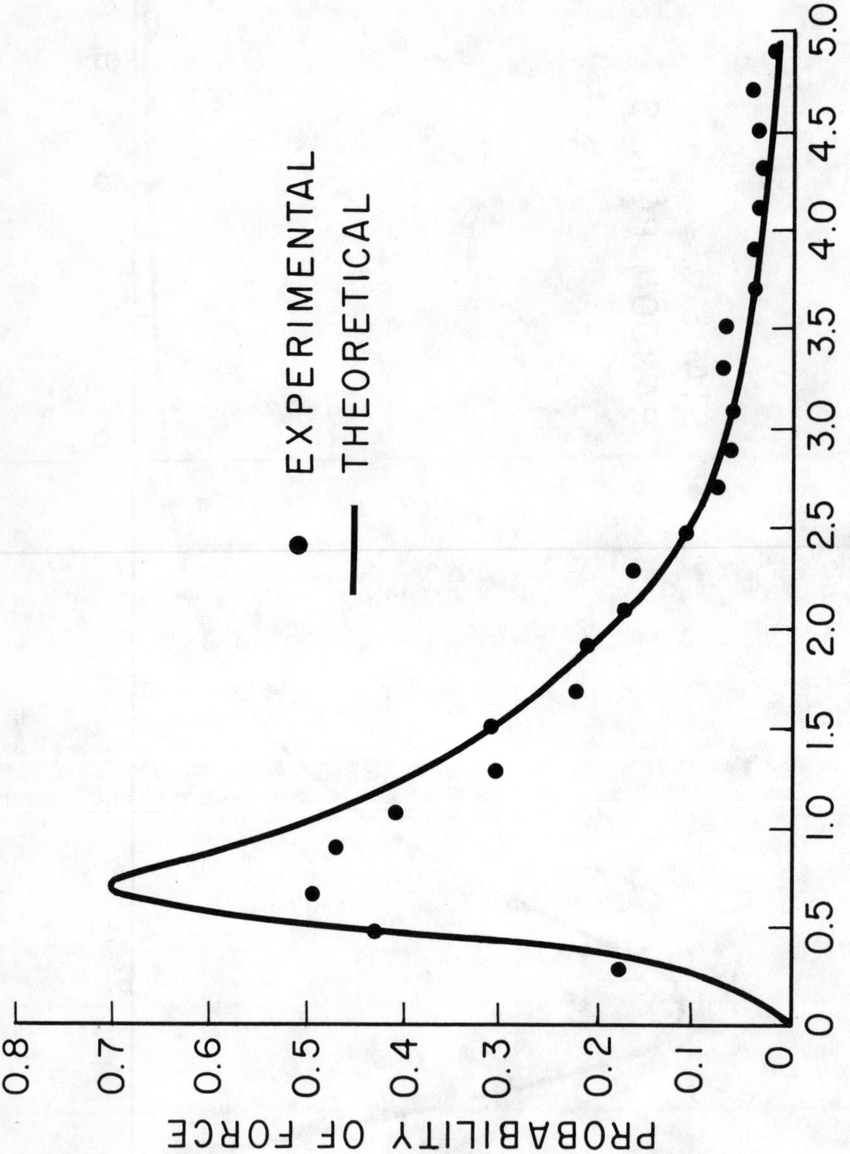

PROBABILITY OF FORCE
0.8
0.7
0.6
0.5
0.4
0.3
0.2
0.1
0
0
0.5
1.0
1.5
2.0
2.5
3.0
3.5
4.0
4.5
5.0
β
EXPERIMENTAL
THEORETICAL

Figure 4.

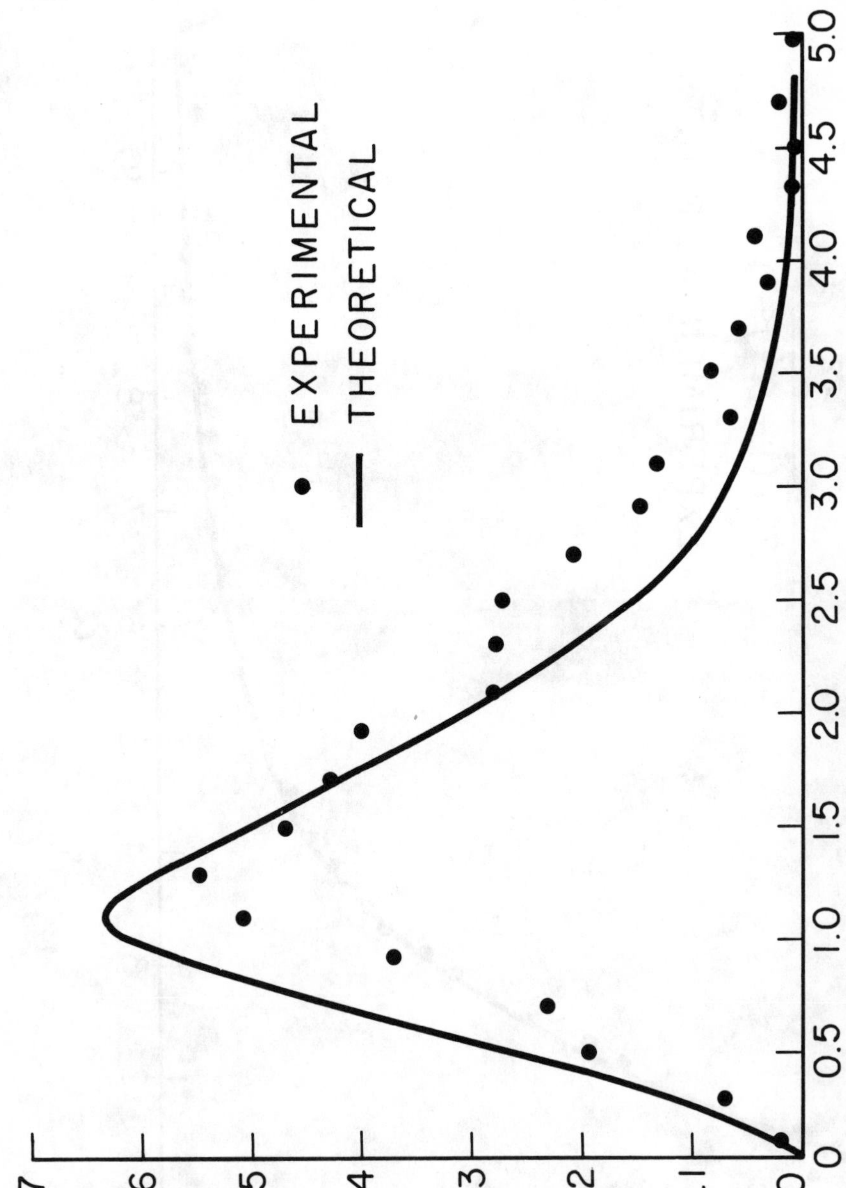
PROBABILITY OF FORCE
0
0.1
0.2
0.3
0.4
0.5
0.6
0.7
0
0.5
1.0
1.5
2.0
2.5
3.0
3.5
4.0
4.5
5.0
β
EXPERIMENTAL
THEORETICAL

Figure 5.

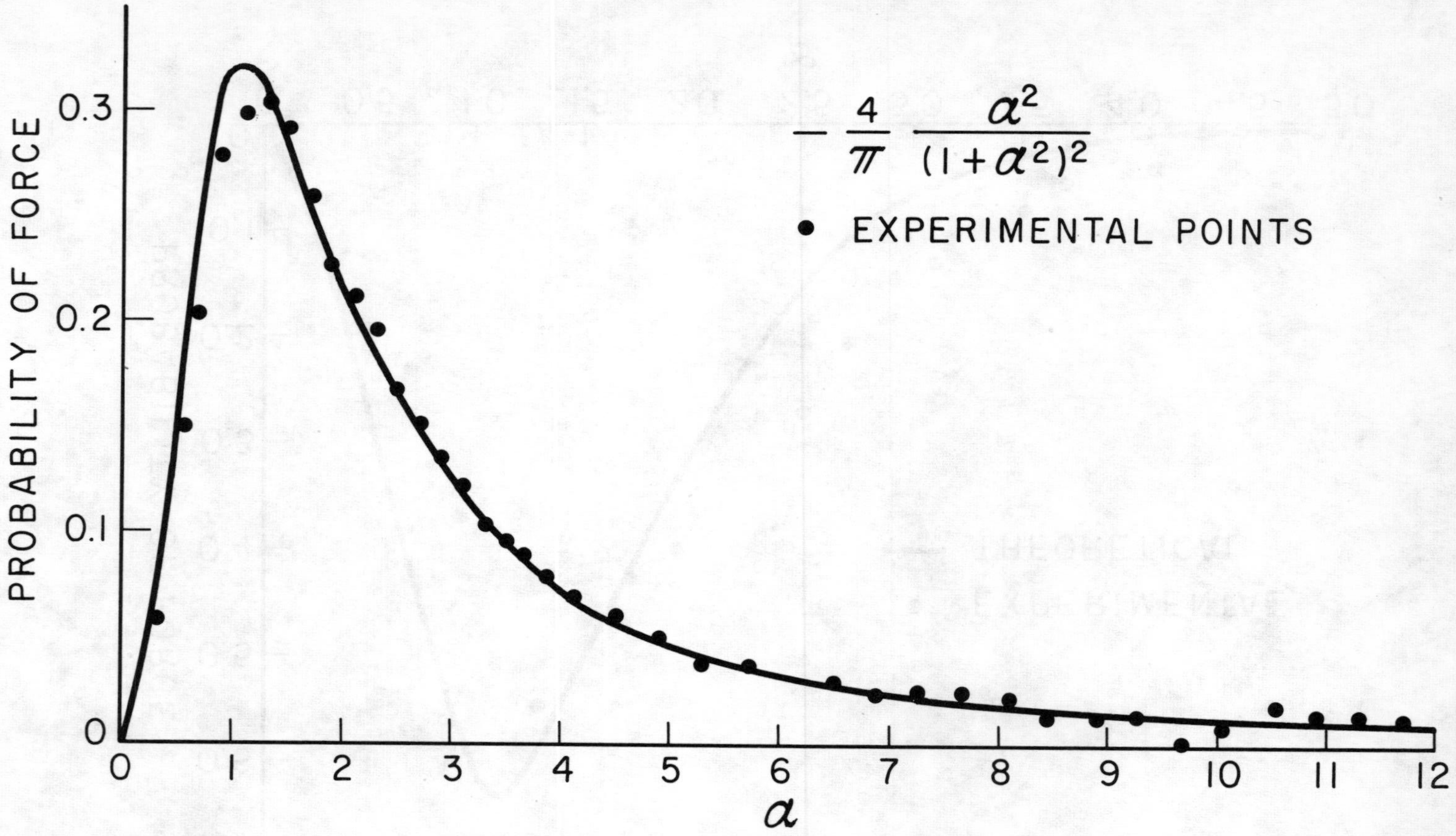

PROBABILITY OF FORCE
0.3
0.2
0.1
0
0 1 2 3 4 5 6 7 8 9 10 11 12
α
— 4/π α²/(1+α²)²
● EXPERIMENTAL POINTS

Figure 6.

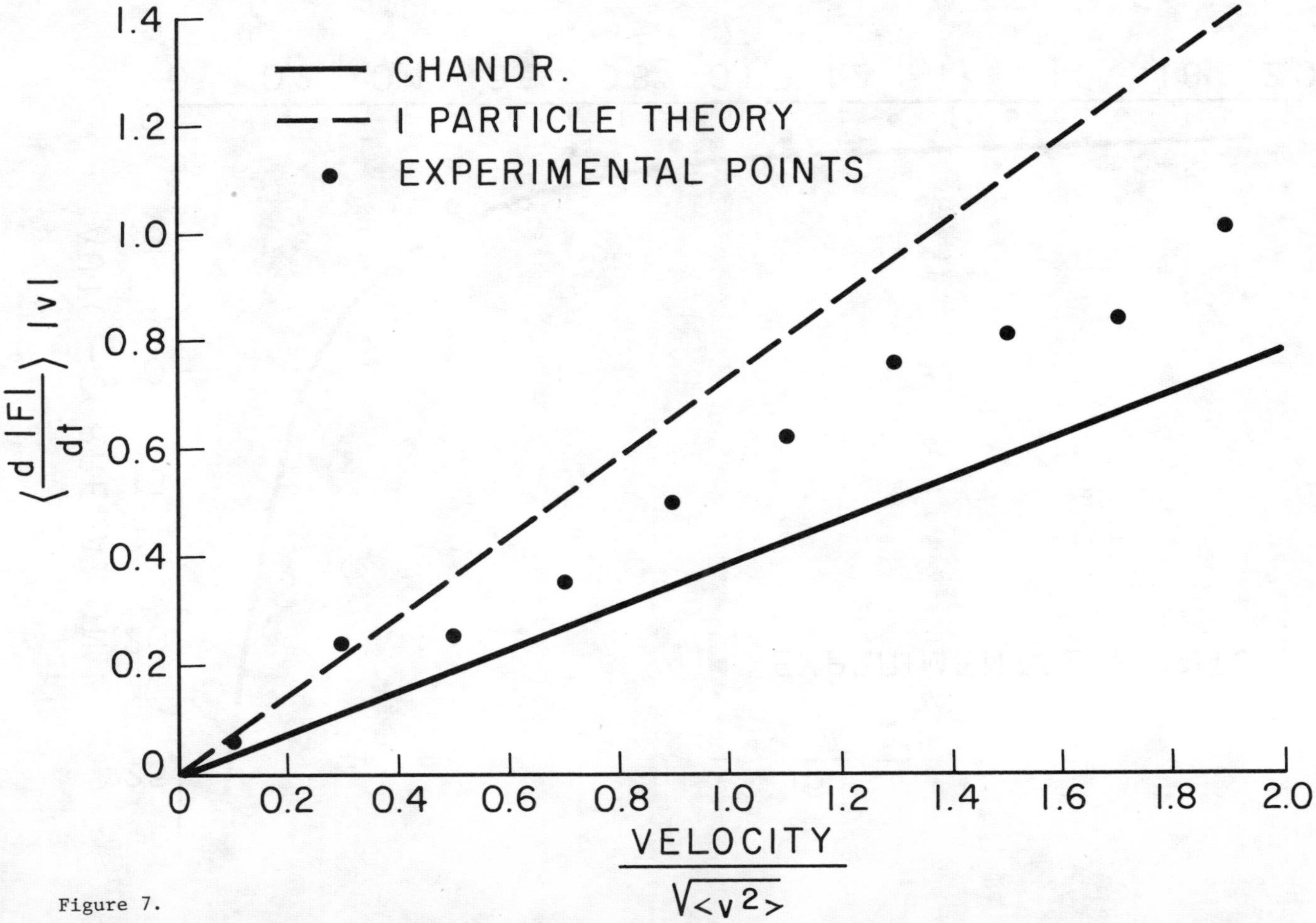
CHANDR.
I PARTICLE THEORY
EXPERIMENTAL POINTS
1.4
1.2
1.0
0.8
0.6
0.4
0.2
0
0.2
0.4
0.6
0.8
1.0
1.2
1.4
1.6
1.8
2.0
$\left\langle \frac{d\,|F|}{dt} \right\rangle |v|$
$\frac{\text{VELOCITY}}{\sqrt{<v^2>}}$

Figure 7.

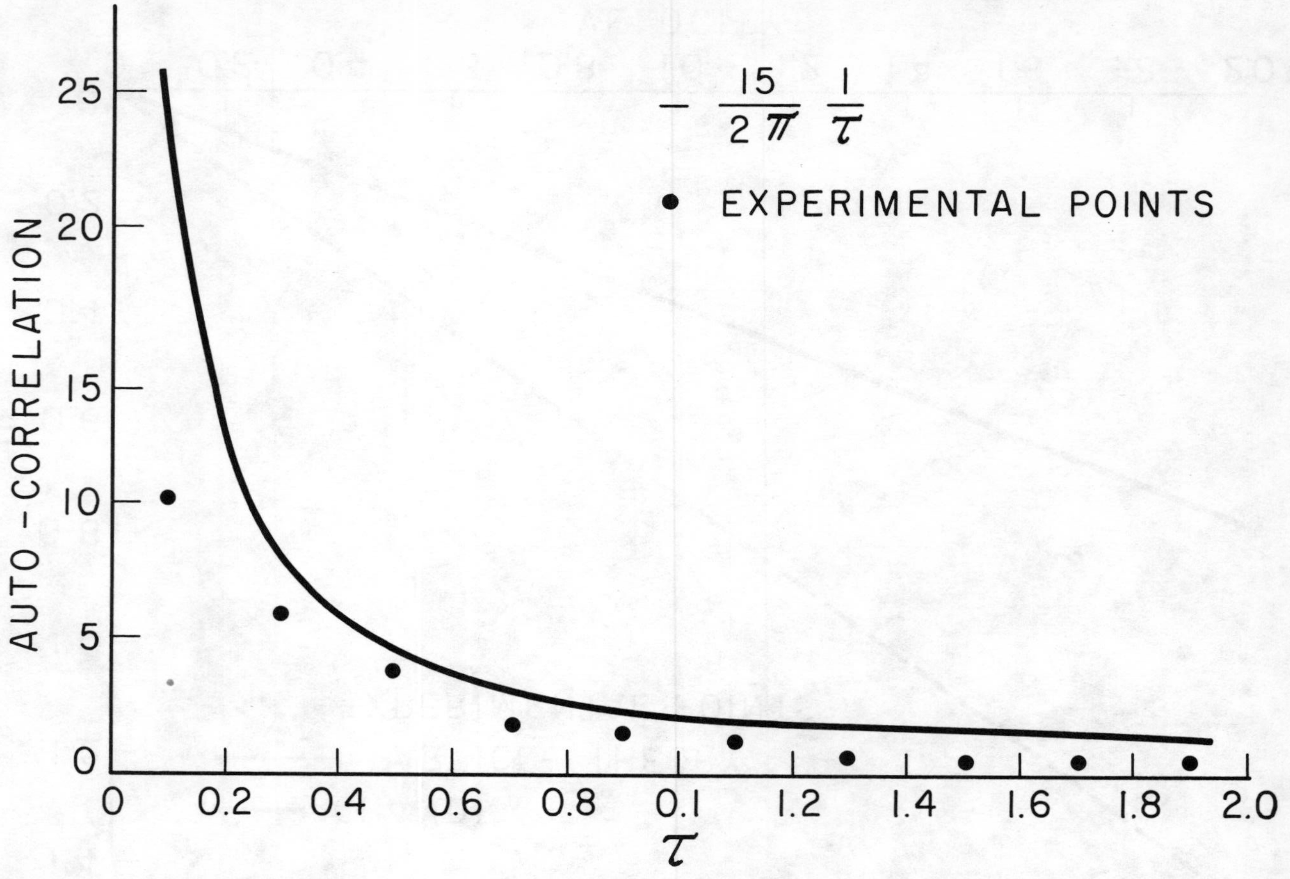

$-\frac{15}{2\pi}\frac{1}{\tau}$
EXPERIMENTAL POINTS
AUTO - CORRELATION
25
20
15
10
5
0
0.2
0.4
0.6
0.8
0.1
1.2
1.4
1.6
1.8
2.0
τ

Figure 8.

References

[1] Chandrasekhar, S., and von Neumann, J., Ap. J. 97, 1, 1943.

[2] Chandrasekhar, S., and von Neumann, J., Ap. J. 95, 489, 1942.

[3] Chandrasekhar, S., Ap. J., 99, 25, 1943.

[4] Chandrasekhar, S., Ap. J., 99, 47, 1943.

[5] Ahmad, A., and Cohen, L., Ap. J., Feb. 1, 1973.

[6] Chandrasekhar, S., Rev. Mod. Phys., 15, 1, 1943.

[7] Ahmad, A., and Cohen, L., Physics Letters (to appear).

[8] Lee, F., Ap. J., 151, 687, 1968.

Acknowledgement

This work was supported by a grant from the National Science Foundation.

INTEGRATION ERRORS AND THEIR EFFECTS ON MACROSCOPIC PROPERTIES OF CALCULATED N-BODY SYSTEMS

by

Haywood Smith, Jr.

University of South Florida

Tampa, Florida

The calculated N-body systems referred to in the title are more specifically systems of N bodies accelerated by their mutual (Newtonian) gravitational attraction. It has been well known since the work of Miller (1964) that the calculated trajectories of individual particles deviate more and more markedly from the trajectories of the particles in the real system corresponding to the same set of initial positions. However, the question of whether these individual deviations lead to a significant distortion of the overall properties of the calculated system has not been studied hitherto in any quantitative way.

In order to examine this question, a set of five 16-body examples was calculated with varying degrees of accuracy. All examples were started in virial equilibrium (in contrast to the I.A.U. "standard problem" (Lecar 1968)) and with uniform density inside a sphere of unit radius; the only difference from one example to the next was in the random numbers specifying each particle's initial position and velocity. The calculating technique was a very simple second-order predictor-corrector with individual timesteps for each particle. The timesteps are essentially given by

$$\Delta t_i = \frac{1}{\sqrt{\frac{Gm_j}{r_{ij}^3}}} \cdot \frac{1}{\mu} \tag{1}$$

where particle "j" is the nearest particle to "i" and r_{ij} is the separation of the two. The masses of all particles were equal, so m_j is of no importance. The parameter μ determines the fineness of the timestep-mesh (details are given by Smith (1972).)

With this integration method there are the customary two types of errors: 1) truncation error, caused by omission of the higher-order terms in the series expansion for the acceleration of each particle, and 2) roundoff error, caused by the limited number of significant digits carried by the computer. The first terms neglected in the series expansion give errors proportional to $(\Delta t)^4$. If we denote the relative energy error $(E-E_o)/E_o$ by ε, where E is the system total energy (kinetic + potential) at some instant and E_o is its initial value, and if by analogy with the two-body problem we take it to be a linear combination of the energy errors of the individual particles, then to first order

$$\varepsilon \cong \sum_{i=1}^{N} \left(a_i \frac{\Delta r_i}{r_i} + b_i \frac{\Delta v_i}{v_i} \right) \qquad (2)$$

which gives $|\varepsilon| \sim \mu^{-4}$. (Here Δr_i and Δv_i are the errors in position and velocity coordinates, respectively). Thus we see that increasing μ (using a finer timestep-mesh) reduces the truncation error. However, if we increase μ to a very large value, so many steps will be taken to cover a given time interval that the roundoff error will accumulate. We expect, then, a sharp decrease in $|\varepsilon|$ initially for μ increasing, followed by a gradual increase in $|\varepsilon|$ at larger μ. This is what is observed, as shown for two 32-body experiments after 1 relaxation time (figure 1). These systems were calculated at single precision ($d \approx 7$, where d is the number of significant figures), so the roundoff error at each step is substantial. In general, for roundoff error we expect $|\varepsilon| \sim 10^{-d}$. This is observed, as shown for 8-body systems after 3 relaxation times in figure 2. The slanted portion of the curve represents roundoff error, while the horizontal portion represents a dominant contribution from truncation error (at μ=6). For a given computer and with any method, the general dependence of $|\varepsilon|$ on the fineness of the mesh and on d should be qualitatively the same, with a maximum attainable accuracy at a given order of integration.

It is not to be expected that truncation and roundoff will have the same effect. For this reason the calculations have been made with μ = 3, 5, 8, 12, and 20 at double precision (d=17) and with d = 5, 6, 7 at μ = 20; at small μ truncation will dominate, and at small d (and large μ) roundoff will dominate. The calculations covered 3 initial mean relaxation times for each example, which is sufficient time for the errors to become appreciable, as shown in table 1.

The microscopic deviations are measured by a quantity ΔP which is the same as Miller's (1964) phase-space separation except for the normalization:

$$\Delta P = \frac{1}{2}\ln\left[\frac{1}{2N}\sum_{i=1}^{N}\left\{\left(\frac{\delta r_i}{r_o}\right)^2 + \left(\frac{\delta v_i}{v_o}\right)^2\right\}\right] \tag{3}$$

with r_o the mean interparticle separation initially and v_o the initial r.m.s. velocity. For $\Delta P \simeq 0$ the deviations δr_i and δv_i, which are respectively the difference in position and in velocity of the ith particle in a given calculation from the position and velocity of the same particle in the most accurate calculation (viz., that with μ=20 and d=17), have become approximately equal to the mean values r_o and v_o, and the less accurate calculation may be said to be unacceptable. Actually, the value of ΔP found by taking the differences in position and velocity of two different examples is $\Delta P \simeq +0.3$, so that when ΔP has reached that value the less accurate calculation has become totally uncorrelated with the more accurate one.

The macroscopic properties to be measured describe the density and energy distributions. The following "statistical quantities" S_i were arbitrarily defined:

$$S_1 \equiv \bar{\rho}_{1/4} \tag{4a}$$

$$S_2 \equiv r_{3/4} \tag{4b}$$

$$S_3 \equiv r_{1/2} \tag{4c}$$

$$S_4 \equiv E_{1/4} \tag{4d}$$

$$S_5 \equiv E_{3/4} \tag{4e}$$

The subscripts refer to the respective quartiles. All radial distances are referred to the "density center" defined by von Hoerner (1963). The energy E is the sum of kinetic and potential energies for a given particle. Relative deviations of these quantities from their values with the "standard calculations" (μ=20, d=17) were computed and compared with the relative energy error ε for each set of calculations (figure 3). The horizontal lines marked "σ" in the figure indicate the value of the r.m.s. deviation σ_i of each of the S_i for the "standard calculations" divided by the respective S_i, as $\log(\sigma_i/S_i)$. In this way we see that, because of the intrinsic scatter of the S_i, in most cases there is no advantage in calculating to $|\varepsilon| \approx 10^{-4}$ - the integration errors will be at most a minor perturbation. A notable exception is S_1; but if the logarithm of $\bar{\rho}_{1/4}$ is used instead (which is more in keeping with actual practice in these experiments), the situation is improved, and an error $|\varepsilon| \approx 10^{-2}$ is tolerable. Table 2 shows the result of replacing $\bar{\rho}$ by $\log \bar{\rho}$.

Over the range of energy errors considered ($|\varepsilon| = 10^{-4}$ to 10^{-2}) the statistical results do not appreciably differ. But are these results correct? There are two arguments in support of their correctness. First, if one calculates with very small μ (μ=1.5, for example), the energy error jumps to a very large value ($\varepsilon \approx 10^{1}-10^{3}$), and the system "explodes". (The same is true for small d). This corresponds to the limiting case of domination by the errors. Second, and more indirectly, we see that in the limit where the error of one particle dominates, $\log \varepsilon$ (equation 2) and ΔP (equation 3) look very similar. The principal difference between the two is that ε includes the errors Δr_i and Δv_i of the i th particle, while ΔP contains instead δr_i and δv_i (i.e., the differences between a system calculated at lower accuracy and the most accurate calculation). If the errors Δr_i and Δv_i for the most accurate calculated system are much smaller than the deviations δr_i and δv_i, we expect $\log \varepsilon$ and ΔP to be well-correlated. Indeed, as shown in figure 4, this is the case for the truncation error. The correlation thus at least suggests convergence of the results to those for the physical system.

These results have implications for the planning of N-body experiments. It is obvious that there is no advantage <u>as far as the statistical quantities are concerned</u> in calculating to an accuracy of $|\varepsilon| \approx 10^{-8}$. Since higher accuracy usually involves a greater expenditure of computer time, less stringent accuracy requirements make possible more and/or longer integrations (von Hoerner 1969). Higher accuracy would be desirable only if individual trajectories were of interest, which they usually are not if $N>5$. In terms of the present method, this means calculating with moderate μ ($5<\mu<10$). A second point is that, unlike truncation error, roundoff error seems to introduce a significant bias in the statistical results, as can be seen merely from the distribution of signs (table 3). For this reason the largest computer word-length practicable should be used; with the present method at $\mu=10$ a word length $d \geq 12$ is sufficient unless the integration is unusually lengthy (e.g., covering 50 or more initial mean relaxation times). Certainly the customary accuracy of $|\varepsilon| \simeq 10^{-3}$ is an acceptable value for meaningful estimates of macroscopic properties of N-body systems.

The author gratefully acknowledges the support of the National Radio Astronomy Observatory*, which generously supplied computer time for this project. Also, the author is deeply indebted to Dr. Sebastian von Hoerner of NRAO for encouragement and guidance of the work.

*Operated by Associated Universities, Inc., under contract with the National Science Foundation.

Table 1

Times* at Which Relative Deviations

Exceed 1% Level

Parameters	Example No.	QUANTITY						
		r	v	S_1	S_3	S_4	S_5	ε
$\mu = 3$,	6	0.6	0.2	0.7	0.7	0.4	0.7	1.3
$d = 17$	7	0.6	0.4	0.5	0.7	1.0	1.0	1.5
	8	1.2	0.7	1.0	1.4	1.5	1.2	2.1
	9	1.1	0.1	0.9	1.0	1.2	1.5	1.0
	10	1.0	0.4	0.7	1.3	1.3	1.7	0.9
$d = 6$,	6	0.7	0.3	0.7	0.7	0.5	0.7	0.9
$\mu = 20$	7	0.7	0.5	0.6	0.9	0.8	1.2	0.7
	8	1.5	0.7	1.1	1.8	1.5	2.2	1.7
	9	1.1	0.9	1.0	1.1	1.5	2.1	3.0
	10	1.5	0.9	1.2	1.5	1.5	1.8	1.6

*Units are initial mean relaxation times.

Table 2

Results for Alternative Definition of S_1

μ	$\mathrm{Log}\lvert\bar{\delta}S_1/S_1\rvert$	$\mathrm{Log}\lvert\varepsilon\rvert$
12	$-1.23^{+.15}_{-.23}$	-3.75
8	$-1.05^{+.18}_{-.33}$	-3.28
5	$-0.54^{+.13}_{-.19}$	-2.87
3	$-0.33^{+.10}_{-.14}$	-2.05
σ	-0.31	——

Table 3

Relative Occurrence of Deviations δS_i of Positive Sign

	QUANTITY			
Parameters	S_1	S_3	S_4	S_5
μ=12	0.48±.08	0.47±.10	0.38±.07	0.45±.09
μ=3	0.50+.08	0.47±.06	0.47±.04	0.45±.08
d=7	0.33±.03	0.73±.08	0.21±.06	0.28±.10
d=6	0.39±.07	0.67±.07	0.28±.07	0.23±.07

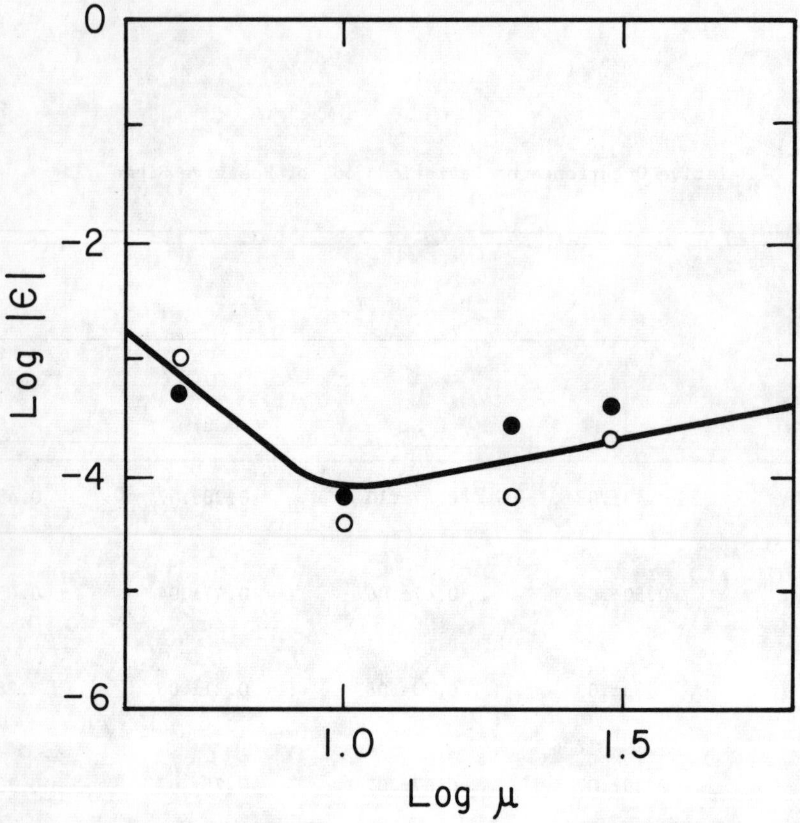

Figure 1: Relative energy error ε as a function of μ for the two 32-body examples after 1 initial mean relaxation time. The different symbols refer to the t two examples. The solid curve is a predicted dependence, $\varepsilon \sim \mu^{-4}$ for truncation and $\varepsilon \sim \mu$ for roundoff (Smith 1972).

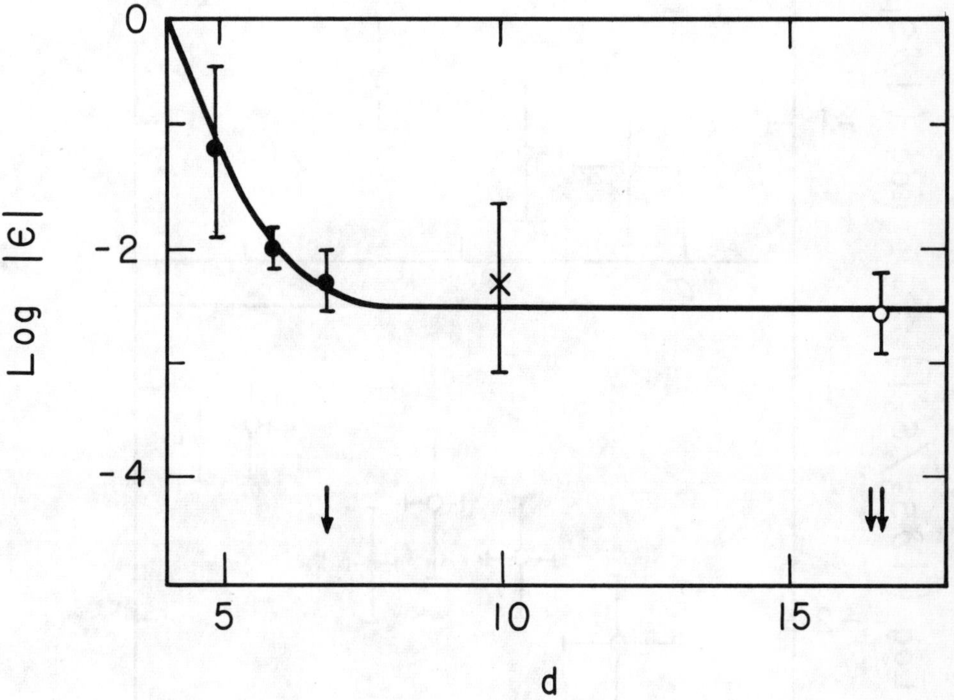

Figure 2: Relative energy error ε as a function of d for some 8-body examples. Filled circles refer to one set of examples, the open circle to another, and the cross to an example calculated by von Hoerner (1960). The solid curve is a predicted dependence, $\varepsilon \sim 10^{-d}$ for roundoff and ε = const. for truncation (Smith 1972).

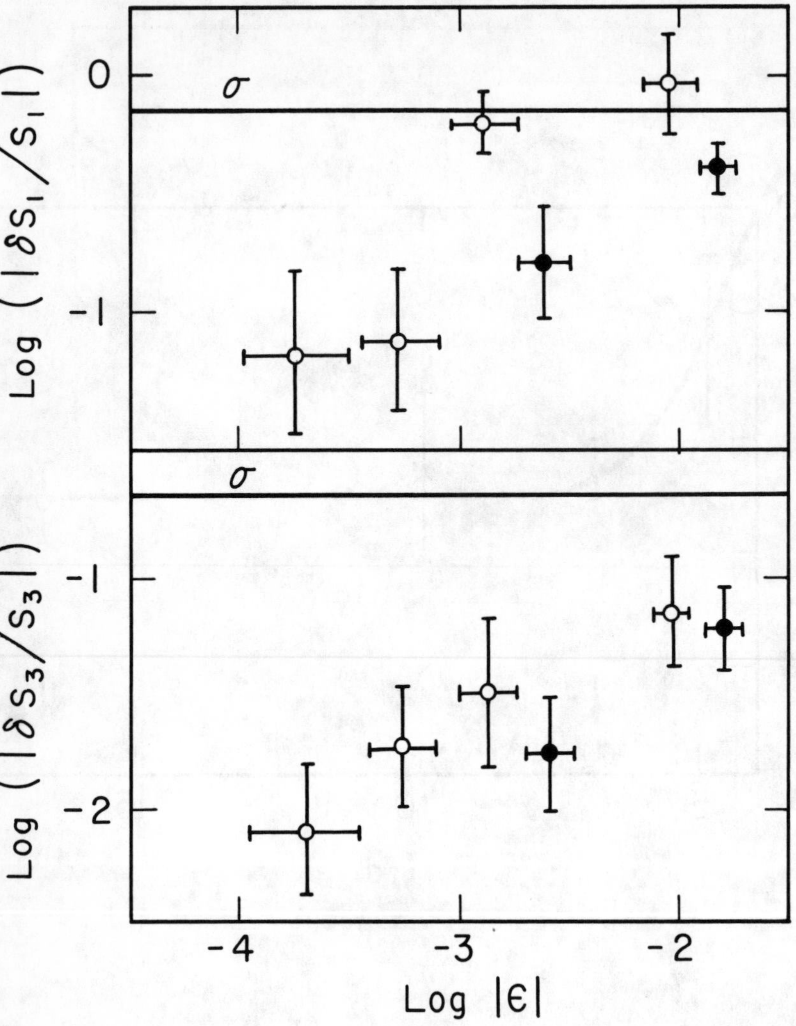

Figure 3: Ensemble averages of log $|\delta S_i/S_i|$ (quantities defined in equation 4 (a-e)) versus averages of log $|\varepsilon|$. Open circles are for domination by truncation error, filled circles for roundoff. The horizontal lines marked "σ" refer to the intrinsic scatter of the "standard" values (see text).

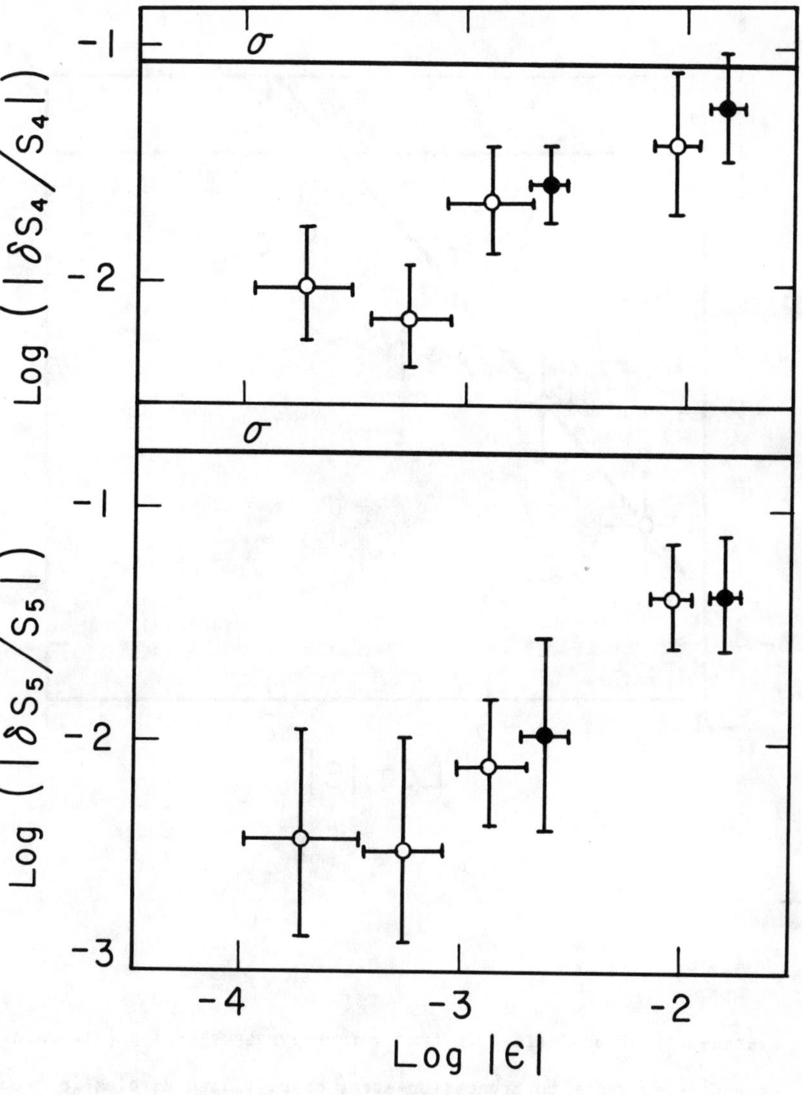

Figure 3 - Second Part.

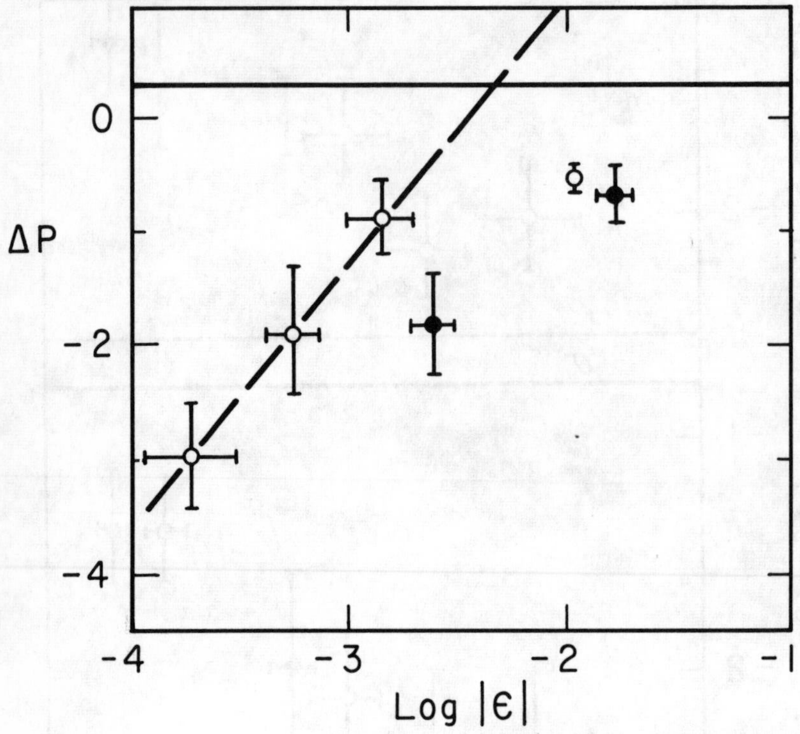

Figure 4: Averages of ΔP as defined in text versus averages of log $|\varepsilon|$. Again open circles refer to truncation-error case, filled circles for roundoff. The solid horizontal line represents the value appropriate to totally uncorrelated examples. The dashed line is actually a line of unit slope.

References

1. M. Lecar, Bull. Astr., Series 3, 3, 91, 1968
2. R.H. Miller, Ap. J., 140, 250, 1964
3. H. Smith, unpublished Ph.D. dissertation, University of Virginia, 1972
4. S. von Hoerner, Zs. f. Ap., 50, 184, 1960
5. S. von Hoerner, ibid, 57, 47, 1963
6. S. von Hoerner, personal communication, 1969

USE OF GREEN'S FUNCTIONS IN THE NUMERICAL SOLUTION OF TWO-POINT BOUNDARY VALUE PROBLEMS *

by

L. J. Gallaher and I. E. Perlin

Georgia Institute of Technology
Rich Electronic Computer Center

Atlanta, Georgia

Abstract

This study investigates the use of Green's functions in the numerical solution of the two-point boundary value problem. The first part deals with the role of the Green's function in solving both linear and nonlinear second order ordinary differential equations with boundary conditions and systems of such equations. The second part describes procedures for numerical construction of Green's functions and considers briefly the conditions for their existence. Finally, there is a description of some numerical experiments using nonlinear problems for which the known existence, uniqueness or convergence theorems do not apply. Examples here include some problems in finding rendezvous orbits of the restricted three body system.

I. Introduction

This report is devoted to the investigation of the use of Green's functions for the numerical solution of second order ordinary differential equations with boundary conditions.

If L_2 represents a second order differential operator, say

$$L_2 \equiv \frac{d}{dx}p(x)\ \frac{d}{dx} + r(x),$$

then the solution to the linear equation

$$L_2 u = f$$

* Supported by the National Aeronautics and Space Administration - Washingtion, D.C.

can be represented by

$$u = L_2^{-1} f = -Gf,$$

where this last equation represents

$$u(x) = -\int_a^b G(x,y)f(y)dy.$$

The function G is called the Green's function and plays the role of the inverse of the operator $-L_2$. The function G is not unique, but depends on the boundary conditions. The existence of G needs also to be investigated in each individual case and depends on the character of p, r and the boundary conditions.

Historically, the first and still most common method of solving second order differential equations with boundary conditions (the two point boundary value problem) is the shooting method. That is, one starts at one end and assumes enough information about the function and its derivative at that end to be able to integrate the differential equation as an initial value problem. By repeatedly integrating to the other end and correcting the assumed initial conditions, one hunts for a solution that satisfies the boundary conditions at the other end. While this method works well for some problems, for others it does not. It has a tendency to fail badly when the solution is extremely sensitive to the assumed initial conditions.

The Green's function method can be useful either in place of the shooting method or as an adjunct or preliminary step to the use of the shooting method. That is, the Green's function method might be used to obtain an initial estimate for the shooting method.

Another useful technique is to introduce a discrete approximation for the operator L_2. The function G then is the inverse of the matrix approximating $-L_2$. Low order approximations to L_2 cause no special trouble except that large matrices result if high accuracy is desired. High order approximations for L_2 tend to introduce extraneous solutions and must be treated with care. (Varga and co-workers [7-9] have given new high order methods for the nonlinear two point boundary value problem).

The Green's function method, on the other hand, can be of an arbitrarily high order, depending only on the order and quality of the methods available for the integration of differential equations (the initial value problem).

The first part (Chapter II) of this paper is concerned with the definition and use of the Green's function. Both single equations and systems of equations are discussed. Applications to nonlinear problems where iterative techniques are needed, are discussed. Here, problems of existence, uniqueness and convergence or stability of the solutions arise.

Chapter III describes procedures for generating Green's functions numerically. Some discussion concerning the question of existence of G is included here.

Chapter IV deals with specific numerical experiments in the use of Green's functions for solving second order nonlinear ordinary differential equations, and systems of such equations, with various boundary conditions. Several types of problems were chosen. The first type was single simple nonlinear equations where multiple solutions were known to exist and problems in stability occur.

The other class of problems investigated is associated with the search for orbits in the restricted three body system. Here one deals with a pair of rather complicated nonlinear equations. Rendezvous type orbits were sought, that is, point to point in a fixed time. Multiple solutions can exist in this case and some solutions are more stable than others.

II. Use of Green's Functions

A. Single Equations

Consider a single second order linear ordinary differential equation of the form

(1) $$\left(\frac{d}{dx}p(x)\frac{d}{dx} + r(x)\right) u(x) = f(x),$$

with boundary conditions

$$u(a) = u(b) = 0.$$

This will also be written as

(1') $$L_2 u = f.$$

Provided that sufficient restrictions are placed on p, r, and f, this equation can be solved for u as

(2) $$u = L_2^{-1} f = -Gf,$$

where the above is a shorthand notation for

(2') $$u(x) = -\int_a^b G(x,y)f(y)dy.$$

The function G(x,y) is known as the Green's function for the operator $\frac{d}{dx}p(x)\frac{d}{dx} + r(x)$ and boundary conditions $u(a) = u(b) = 0$. The restrictions on p and r sufficient for the existence of G will be discussed later. The function -G plays the role of the inverse of the operator L_2. The function G must satisfy the condition

$$L_2 G = -\delta,$$

meaning

(3) $$\left(\frac{d}{dx}p(x)\frac{d}{dx} + r(x)\right) G(x,y) = -\delta(x-y),$$

where $\delta(x-y)$ is the Dirac delta. The function G also satisfies the boundary conditions

(4) $$G(a,y) = G(b,y) = 0$$

for all y, $a \le y \le b$. We note that while G itself is a function of two variables, $L_2 G$ is not a function, but is a distribution.

There are straightforward methods for constructing G that will be discussed later (Chapter III). We point out here that the inverse of L_2 is not unique, but depends on the boundary conditions: different boundary conditions give different Green's functions.

Consider now that the function f is also a (nonlinear) function of u and its derivatives; i.e., let

(5) $$L_2 u = f(u,u'),$$

or

(5') $$\left(\frac{d}{dx}p(x)\frac{d}{dx} + r(x)\right) u(x) = f(x,u(x),u'(x)),$$

where $u'(x) = \frac{d}{dx}u(x)$. In this case, one can still construct an inverse or Green's function for L_2, but the function u(x) is exhibited as the solution of a (nonlinear) integral equation:

$$u(x) = -\int_a^b G(x,y)f(y,u(y),u'(y))dy, \quad (6)$$

or

$$u = -Gf(u,u'). \quad (6')$$

Still, in numerical work this integral equation can be useful in solving for u by iteration. Some theory exists for the existence, uniqueness and convergence of integral equations such as (6). Even when existence, uniqueness or convergence theorems are not available, solutions to equation (6) can sometimes by obtained by iteration of

$$u^{n+1}(x) = -\int_a^b G(x,y)f(y,u^n(y),u'^n(y))dy, \quad (7)$$

also written as

$$u^{n+1} = -Gf(u^n,u'^n). \quad (7')$$

There are two kinds of convergence theorems of interest for equation (7), that will be referred to as local and global convergence.

A global convergence theorem states the circumstances under which, for any initial u^o, iteration of equation (7) converges to a solution of equation (6).

A local convergence theorem states the circumstances under which there exists a neighborhood of the solution, such that if u^o is chosen to lie in this neighborhood, iteration of (7) converges to u, a solution of (6).

Global convergence is much stronger than local convergence. Local convergence requires that the initial trial solution u^o lie close to the true solution before convergence can be guaranteed.

Bailey, Shampine and Waltman [3] examine in considerable detail the convergence properties of the iterative scheme:

$$u^{n+1}(x) = -\int_a^b G(x,y)f(y,u^n(y),u'^n(y))dy,$$

where G is the Green's function for the operator $\frac{d^2}{dx^2}$ and zero boundary conditions. Contraction mapping techniques are used to obtain convergence conditions on f. The same techniques can be used to establish similar results for other Green's functions provided that the quantities

$$\max_{a \le x \le b} \int_a^b |G(x,y)|\,dy,$$

and

$$\max_{a \le x \le b} \int_a^b \left| \frac{d}{dx} G(x,y) \right| dy,$$

can be established or bounded. The theorems of Bailey, Shampine and Waltman are of the global type; that is, they give the conditions on f such that convergence is obtained for any initial u^o.

There are also some local stability conditions of interest. Local stability implies the existence of a finite neighborhood about a solution u such that if u^o lies in this neighborhood the iteration scheme converges to the solution.

With local convergence theorems, uniqueness of the solution is not required. For example, it can happen that there are many solutions and that in the neighborhood of some solution there is a convergence region for that particular solution, but other solutions have no regions of convergence.

For example, consider the differential equation

$$\frac{d^2}{dx^2} u(x) = - \sin u(x)$$

with boundary conditions

$$u(a) = u(b) = 0.$$

The equivalent integral equation is

$$u(x) = \int_a^b G(x,y) \sin u(y)\,dy,$$

where G is the Green's function for $\frac{d^2}{dx^2}$ with the zero boundary conditions at a and b. Iteration gives

$$u^{n+1}(x) = \int_a^b G(x,y) \sin u^n(y)\,dy.$$

The methods of Bailey, Shampine and Waltman can be used to show that this iternation scheme can be guaranteed to converge if

$$\max_x \int_a^b |G(x,y)|\,dy = \left(\frac{b-a}{\pi}\right)^2 < 1.$$

Under these circumstances the solution is unique ($u(x) = 0$), and global convergence to this solution is guaranteed.

However, if $\left(\frac{b-a}{\pi}\right)^2 > 1$ then there are multiple solutions. Furthermore, the iteration scheme is then unstable in the neighborhood of the solution $u(x) = 0$ but may be stable in the neighborhood of some of the other solutions.

B. Multiple Equations

For systems of second order linear differential equations much the same sort of results can be obtained. Consider the system

(8) $$\sum_{1 \le k \le n} \left(\frac{d}{dx} p_{jk}(x) \frac{d}{dx} + r_{jk}(x)\right) u_k(x) = f_j(x), \; j = 1,2..n,$$

with boundary conditions

$$u_k(a) = u_k(b) = 0, \; k = 1,2..n.$$

This will also be written as

(8') $$L_2 u = f.$$

Again provided that sufficient restrictions are placed on p and r, an inverse for L_2 will exist so that

(9) $$u = L_2^{-1} f = -Gf,$$

where this stands for

(9') $$u_k(x) = -\int_a^b \sum_{i \le j \le n} G_{kj}(x,y) f_j(y)\,dy, \; k = 1,2..n.$$

The Green's function, G, is now an n-by-n matrix of functions, and satisfies the system of equations

(10) $$\sum_{1 \le k \le n} \left(\frac{d}{dx} p_{jk}(x) \frac{d}{dx} + r_{jk}(x)\right. \; G_{km}(x,y) = -\delta_{jm} \delta(x-y),$$

and the boundary conditions

$$G_{km}(a,y) = G_{km}(b,y) = 0, \; k,m = 1,2..n.$$

Again there are straightforward but more involved methods for constructing G. These methods will be described in Chapter III.

If the vector f is also a function of vector u and its derivatives, that is

$$(11) \qquad L_2 u = f(u,u'),$$

or

$$(11') \qquad \sum_{1 \le k \le n} \left(\frac{d}{dx} p_{jk}(x) \frac{d}{dx} + r_{jk}(x) \right) u_k(x) = f_j(x,u(x),u'(x)), j = 1,2..n,$$

(here u represents the set $u_1, u_2 \ldots u_n$ and u' represents $\frac{du_1}{dx}, \frac{du_2}{dx} \ldots\ldots$), then the Green's function for L_2 does not give a solution for u but gives the integral equation

$$(12) \qquad u = -Gf(u,u'),$$

or

$$(12') \qquad u_k(x) = -\int_a^b \sum_{1 \le j \le n} G_{kj}(x,y) f_j(y,u(y),u'(y))dy, k = 1,2..n.$$

Again this equation can be used in numerical work by iterating with u^n on the right side and obtaining u^{n+1} on the left. Most of the existence, uniqueness, and convergence theorems can be extended to cover the vector case.

One interesting case that occurs in connection with the vectors case is when p and r of equation (11') are both diagonal. Then G is also diagonal and is much easier to find than if there is coupling between the equations in L_2. Then

$$(13) \qquad u_k(x) = -\int_a^b G_{kk}(x,y) f_k(y,u(y),u'(y))dy, k = 1,2..n,$$

and all of the coupling between the equations occur in the functions f only. Because of the simplicity of (13) compared to (12') and the ease in finding G when it is diagonal as opposed to when it is not diagonal, it is often desirable even when dealing with linear equations to remove the coupling from L_2 and put it in f, and iternate equation (13) with a diagonal Green's function. That is, suppose one has an equation of the form

$$(14) \qquad L_2 u = f,$$

where u and f are vectors, (f does not depend on u or u') and L_2 is a non-diagonal matrix operator. Split L_2 so that $L_2 = L^{Diag} + L^{Off}$, where L^{Diag} contains the diagonal terms of L_2 and L^{Off} contains the off diagonal terms. Then

$$L^{Diag}u = f - L^{Off}u \tag{15}$$

and

$$u = -G^{Diag}(f - L^{Off}u), \tag{16}$$

where G^{Diag} now is the diagonal Green's function for the operator for L^{Diag}. Iteration of (16) may be easier and faster than finding the non-diagonal Green's function of the original operator L_2.

III. Construction of Green's Functions (Numerical)

A. Single Equations

Here we will consider how to construct, or calculate numerically, the Green's function for the differential operator

$$L_2 = \left(\frac{d}{dx}p(x)\frac{d}{dx} + r(x)\right), \tag{1}$$

with boundary conditions

$$G(a,y) = G(b,y) = 0, \quad a \leq y \leq b.$$

The Green's function can be constructed from the solutions of the equation

$$\left(\frac{d}{dx}p(x)\frac{d}{dx} + r(x)\right) g(x) = 0, \tag{2}$$

or

$$L_2 g = 0, \tag{2'}$$

together with the appropriate boundary conditions on g.

Introducing the definition

$$J(x) \equiv p(x)\frac{d}{dx}g(x),$$

equation (2) can be written as the pair of coupled ordinary differential equations,

$$\begin{aligned} g'(x) &= J(x)/p(x), \\ J'(x) &= -r(x)g(x), \end{aligned} \tag{3}$$

(prime indicating derivative with respect to x). There are two families of solutions of interest for this system of equations; one family satifies the boundary condition $g(a) = 0$; the other satisfies the condition $g(b) = 0$. The solution satisfying $g(a) = 0$ will be designated $g_a(x)$, and to make this solution unique, the condition $J_a(a) \equiv p(a)g_a'(a) = 1$ will be appended to this solution. Likewise the solution satisfying $g(b) = 0$ will be designated $g_b(x)$, and the condition $J_b(b) \equiv p(b)g_b'(b) = 1$ will be appended to the solution. This gives the two sets of simultaneous equations and boundary conditions:

$$
\begin{aligned}
g'_a(x) &= J_a(x)/p(x), \\
J_a'(x) &= -r(x)g_a(x), \qquad (4) \\
g_a(a) &= 0,\ J_a(a) = 1,
\end{aligned}
$$

and

$$
\begin{aligned}
g_b'(x) &= J_b(x)/p(x), \\
J_b'(x) &= -r(x)g_b(x), \qquad (5) \\
g_b(b) &= 0,\ J_b(b) = 1.
\end{aligned}
$$

Now provided that sufficient restrictions are placed on p and r, each set of equations and boundary conditions has a solution. From the theory of ordinary differential equations, both existence and uniqueness of these solutions can be guaranteed. Furthermore, there are straightforward numerical methods, such as Euler's method or Adams', or Runge-Kutta methods [13] that can be used to generate these solutions in tabular form on the interval [a,b].

The solutions $g_a(x)$ and $g_b(x)$ are of course also solutions to equation (2), and it is straightforward to show that any pair of solutions of (2) satisfies the Wronskian condition,

$$J_a(x)g_b(x) - J_b(x)g_a(x) = A,$$

where A is a constant. From the boundary conditions chosen here for equations (4) and (5)

(6) $$A = g_b(a) = -g_a(b).$$

It can now be shown that the Green's function can be constructed from the solutions $g_a(x)$ and $g_b(x)$ as follows:

(7) $$G(x,y) = A^{-1}\begin{cases} g_a(x)g_b(y), & x \leq y, \\ g_a(y)g_b(x), & y \leq x. \end{cases}$$

To show this we note that the conditions on the G(x,y) are:

(8) $$\left(\frac{d}{dx}p(x)\frac{d}{dx} + r(x)\right) G(x,y) = -\delta(x-y),$$
$$G(a,y) = G(b,y) = 0, \quad (a \leq y \leq b).$$

First, note that the boundary conditions on G are satified by virtue of the boundary conditions $g_a(a) = g_b(b) = 0$, and that G(x,y) is continuous at x = y.

Next, note that since both $g_a(x)$ and $g_b(x)$ satisfy equation (2), then

$$\left(\frac{d}{dx}p(x)\frac{d}{dx} + r(x)\right) G(x,y) = 0, \quad \text{if } x \neq y\ ;$$

that is, G satisfies equation (8) everywhere except (possibly) at the points x = y, where L_2G is not defined except in the sense of a distribution.

Finally, in the vicinity of the points x = y, G must satisfy the condition

$$\lim_{\varepsilon\to 0}\int_{y-\varepsilon}^{y+\varepsilon}\left(\frac{d}{dx}p(x)\frac{d}{dx} + r(x)\right) G(x,y)dx = -1.$$

This reduces to

$$\lim_{\varepsilon\to 0}\ p(x)\frac{d}{dx}G(x,y)\Bigg|_{x = y-\varepsilon}^{x = y+\varepsilon} = -1,$$

or

$$A^{-1}p(y)\left(g_b'(y)g_a(y) - g_a'(y)g_b(y)\right) = -1.$$

By the Wronskian condition, this is satisfied.

From the above account it can be seen that the conditions for the existence of the Green's function then are essentially the same as the conditions for the existence of solutions to the differential equations (4) and (5). Sufficient conditions for the existence of g_a and g_b are known from the theory of ordinary differential equations and are that r(x) and 1/p(x) be piecewise continuous and finite on the internal $a \leq x \leq b$.

One additional condition is required for the existence G and that is that the Wronskian constant, A, have an inverse, i.e., $A \neq 0$. This is equivalent to requiring that the two solutions g_a and g_b be linearly independent. If g_a and g_b are not linearly independent, then $g_b(a) = 0$, $g_a(b) = 0$, $A = 0$, and there will be no Green's function for these boundary conditions. If g_a and g_b are linearly independent, then $g_a(b) \neq 0$, $g_b(a) \neq 0$, $A \neq 0$, and there will be a Green's function for these boundary conditions.

Green's functions for other boundary conditions can be calculated in a similar way. All that changes are the boundary conditions associated with the equations determining $g_a(x)$ and $g_b(x)$, although it should be pointed out that only those boundary conditions of the form $g_a(a) = 0$, $g_b(b) = 0$, $g'_a(a) = 0$, $g'_b(b) = 0$, or some linear combination of these, will necessarily have an associated G.

However, it should be noted that with boundary conditions of the form $u(a)=u_a$, $u(b) = u_b$, solutions to the equation $L_2u = f$ can be obtained in the form

$$u = -Gf + v,$$

that is,

$$u(x) = -\int_a^b G(x,y)f(y)dy + v(y).$$

Here G is the same Green's function as for the zero boundary conditions. The function v, called the particular solution, satisfies $L_2v = 0$ with $v(a) = u_a$ and $v(b)=u_b$, One can see that

$$v(x) = \frac{g_a(x)u_b}{g_a(b)} + \frac{g_b(x)u_a}{g_b(a)} \quad ,$$

$$= A^{-1} \left(g_b(x)u_a - g_a(x)u_b\right) \quad ,$$

where the g's are the same as those defined for the zero boundary condition.

B. Systems of Equations

The Green's function for the matrix operator L_2 can be obtained in a fashion similar to that described above for the scalar case.

The function G satisfies the system of equations and boundary conditions

$$\sum_{1 \le k \le n} \left(\frac{d}{dx} p_{jk}(x) \frac{d}{dx} + r_{jk}(x) \right) G_{km}(x,y) = -\delta_{jm} \delta(x-y), \tag{9}$$

$$G_{jm}(a,y) = G_{jm}(b,y) = 0, \; j,m = 1,2..n,$$

which will be shortened to

$$(p(x)G'(x,y))' + r(x)G(x,y) = I\delta(x-y),$$

where p, r and G are n-by-n matrices and the primes indicate differentiation with respect to x.

Introduce now the square n-by-n matrix functions g_a, g_b, J_a, and J_b that satisfy the matrix equations and boundary conditions

$$\begin{aligned} g_a'(x) &= p^{-1}(x) J_a(x), \\ J_a'(x) &= -r(x) g_a(x), \\ g_a(a) &= 0, \; J_a(a) = I, \end{aligned} \tag{10}$$

and

$$\begin{aligned} g_b'(x) &= p^{-1}(x) J_b(x), \\ J_b'(x) &= -r(x) g_b(x), \\ g_b(b) &= 0, \; J_b(b) = I. \end{aligned} \tag{11}$$

It is clear that g_a and g_b both satisfy the equation $L_2 g = 0$ or

$$(p(x)g'(x))' + r(x)g(x) = 0. \tag{12}$$

These equations can be solved as systems $2n^2$ coupled ordinary differential equations, integrating from a on the first set, and from b on the second set. The standard theorems for systems of ordinary differential equations apply and can be used to guarantee existence and uniqueness of the solutions provided p and r meet the appropriate conditions discussed later. Numerical methods can be used to tabulate g_a and g_b.

Also needed are the adjoint equations for the same boundary conditions. Let

$$(h'(x)p(x))' + h(x)r(x) = 0, \tag{13}$$

where h is an n-by-n matrix. Introduce h_a, h_b, K_a and K_b that satisfy the matrix equations and boundary conditions

$$\begin{aligned} h_a'(x) &= K_a(x)p^{-1}(x), \\ K_a'(x) &= -h_a(x)r(x), \\ h_a(a) &= 0,\ K_a(a) = I, \end{aligned} \tag{14}$$

and

$$\begin{aligned} h_b'(x) &= K_b(x)p^{-1}(x), \\ K_b'(x) &= -h_b(x)r(x), \\ h_b(b) &= 0,\ K_b(b) = I, \end{aligned} \tag{15}$$

so that h_a and h_b satisfy the adjoint equation (13) and indicated boundary conditions.

Solutions of (12) and (13) satisfy Wronskian conditions. For g, J, h, and K these are

$$\begin{aligned} h_b(x)J_a(x) - K_b(x)g_a(x) &= A_1, \\ K_a(x)g_b(x) - h_a(x)J_b(x) &= A_2, \\ h_a(x)J_a(x) - K_a(x)g_a(x) &= A_a, \\ h_b(x)J_b(x) - K_b(x)g_b(x) &= A_b, \end{aligned} \tag{16}$$

where the A's are constant matrices and from the given boundary conditions

$$\begin{aligned} A_1 &= h_b(a) = -g_a(b), \\ A_2 &= g_b(a) = -h_a(b), \\ A_a &= A_b = 0. \end{aligned} \tag{17}$$

It is clear that equations (14) and (15) can be integrated numerically as systems of ordinary differential equations with initial conditions, and tables of the h's constructed on the interval [a,b]. The existence and uniqueness conditions are straightforward and are the same as those for the g's.

The Green's function for the matrix operator L_2 and these boundary conditions can now be constructed as

$$G(x,y) = \begin{cases} g_a(x)A_1^{-1}h_b(y), & x \leq y \\ g_b(x)A_2^{-1}h_a(y), & x \geq y. \end{cases} \tag{18}$$

It is easily seen that:

1) G satisfies the boundary conditions,

2) $(p(x)G'(x,y))' + r(x)G(x,y) = 0$, if $x \neq y$.

It is a bit more complicated but only an exercise in algebraic manipulation to show that the third and fourth conditions required are satisfied, i.e.,

3) $$\lim_{\varepsilon\to 0} p(x) \frac{d}{dx}G(x,y) \Bigg|_{x = y-\varepsilon}^{x = y+\varepsilon} = -I,$$

and

4) $g_a(y)A_1^{-1}h_b(y) = g_b(y)A_2^{-1}h_a(y)$, (the continuity condition for G(x,y) at $x = y$).

Consider now the questions of existence of the matrix G. One needs restrictions on the matrices p and r to guarantee the existence and uniqueness of the g's and h's. A sufficient condition for the existence and uniqueness of the g's and h's is that $r(x)$ and $p^{-1}(x)$ exist and be piecewise continuous on the interval [a,b].

One further condition is necessary for the existence of G and this is that the Wronskian constants, the matrices A_1 and A_2 have inverses. It is not clear how these conditions reflect back on p and r since $A_1 = h_b(a) = -g_a(b)$ and $A_2 = g_b(a) = -h_a(b)$, and the existence of the inverses of the g's and h's at the end points is difficult to determine except possibly numerically.

Thus, one can state sufficient conditions for the existence and uniqueness of of G as being

a) existence and piecewise continuity of r and p^{-1},

b) existence of $A_1^{-1}$and A_2^{-1},

but note that these may be difficult to establish in any particular case-- especially the existence of A_1^{-1} and A_2^{-1}.

It was noted before that other boundary conditions give different Green's functions for the same L_2 operator. While a particular set of boundary conditions were considered here, the same techniques apply to finding the G matrices for the various other boundary conditions.

We conclude this section by noting that the numerical calculation of the matrix G can be a monumental task. One integrates four sets of $2n^2$ simultaneous differential equations (or 4n sets of 2n equations) to obtain the g's and h's, which then must be stored or tabulated at each point appropriate to the interval [a,b].

In practice it is usually convenient to deal with the matrix Green's function only when G is diagonal. This will occur if it can be arranged that p and r are diagonal. The g's, h's, J's. K's and A's then are also diagonal and all commute. Thus if L_2 is diagonal, the problem decomposes into an uncoupled collection of n, one-dimensional problems. That is, one can look for the $G_{kk}(x,y)$ independently by integrating for n separate scalar g's. Also the corresponding h and g functions are equal and the adjoint equations need not be solved separately.

For L_2 diagonal, G is of the form

$$G_k(x,y) = A_k^{-1}\begin{cases} g_{ka}(x)g_{kb}(y) & x \leq y, \\ g_{kb}(x)g_{ka}(y) & x \geq y, \end{cases}$$

$$k = 1,2,\ldots n,$$

where the g's are solutions of the equations

$$g'_{ka}(x) = J_{ka}(x)/p_k(x) \quad ,$$

$$J'_{ka}(x) = -r_k(x)g_{ka}(x) \quad ,$$

$$g_{ka}(a) = 0, \; J_{ka}(a) = 1 \; ,$$

$$k = 1,2.,,,n,$$

and

$$g'_{kb}(x) = J_{kb}(x)/p_k(x),$$

$$J'_{kb}(x) = -r_k(x) g_{kb}(x)$$

$$g_{kb}(b) = 0, \; J_{kb}(b) = 1,$$

$$k = 1,2,..n,$$

and the A's are given by the Wronskian condition

$$J_{ka}(x) g_{kb}(x) - J_{kb}(x) g_{ka}(x) = A_k, \; k = 1,2..n.$$

Here the k index designates the diagonal element of the corresponding diagonal matrix.

Unless the number of dimensions is small, only the diagonal Green's functions are of much practical use in numerical work.

IV. Computer Experiments

A. Introduction

A series of computer experiments were carried out to investigate convergence of various iteration schemes for solution of the two-point boundary value problem. Particular attention was paid to cases where the standard convergence theorems did not apply; that is, the cases where there were multiple solutions, or where existence of even one solution was an open question.

Three groups of problems were examined. The first of these is typified by the equation

$$\frac{d^2}{dx^2} u(x) = -2\pi^2 \sin u(x), \tag{1}$$
$$u(0) = u(1) = 0.$$

This problem is characterized by having one solution at $u(x) = 0$, and at least two other solutions. The iterative scheme

$$u^{n+1}(x) = 2\pi^2 \int_0^1 G(x,y) \sin u^n(y)\, dy \tag{2}$$

is unstable in the neighborhood of $u(x) = 0$ but could be expected to be locally stable in the neighborhood of each of the other two solutions.

The second group of problems is associated with finding orbits of the restricted three body system characterized by a fixed time between two fixed end points. Here one has a pair of differential equations (that can be written as a single second order differential equation of a complex variable) with two-point boundary conditions.

The equations represent the motion of a very light body in the gravitational field of two massive bodies. The two massive bodies are at a constant separation (circular orbit) and the lighter body is restricted to move in the plane of their rotation. In the rotating coordinate system in which the two massive bodies appear to be at rest, the equations of motion for the restricted three-body problem are

$$\text{(3a)} \qquad \ddot{x} = x + 2\dot{y} - \mu' \frac{(x+\mu)}{((x+\mu)^2+y^2)^{3/2}} - \mu \frac{(x-\mu')}{((x-\mu')^2+y^2)^{3/2}} ,$$

$$\text{(3b)} \qquad \ddot{y} = y - 2\dot{x} - \mu' \frac{y}{((x+\mu)^2+y^2)^{3/2}} - \mu \frac{y}{((x-\mu')^2+y^2)^{3/2}}$$

Here the two massive bodies are located on the x axis with the center of mass of the system at the origin, μ is the ratio of the mass of the body located on the positive x axis to the mass of the entire system, and μ' is the ratio of the mass of the body located on the negative x axis to the mass of the entire system ($\mu + \mu' = 1$). The units of distance here are chosen so that the distance between the two massive bodies is unity, and the unit of time is chosen so that the angular velocity of the rotating reference frame is unity (period = 2π).

The third group of problems is that of finding <u>periodic</u> orbits of the restricted three body system. Here one has the same equations as in the second group but with different boundary conditions.

Arenstorf [2] has shown the existence of periodic orbits for this system but there are practical problems in actually finding such orbits. This problem was considered in the hope that the Green's function method would be useful in finding these Arenstorf orbits. But one of the characteristics of this problem is that not only are there multiple solutions, but these solutions are densely packed. That is, some solutions have the property that in every neighborhood of the solution there are other solutions. Thus, instead of converging to a particular solution, the iterative scheme has a tendency to wander or drift through a family of solutions. This wandering continues until a solution is encountered that is more stable than any of its neighbors. The more stable solutions seem to be ones with the largest radii of curvature or the ones with the least number of axis crossings.

B. Some Simple Nonlinear Problems

The first group of problems run are some simple examples of nonlinear two-point boundary value problems where the usual existence and uniqueness theorems are not valid but for which local stability might be expected. These are of the type

$$\frac{d^2}{dx^2}u(x) = f(x,u(x)), \qquad u(0) = u(1) = 0, \tag{4}$$

with various forms of f. The Green's function is elementary.

A computer program was written (in Algol for the B 5500) to solve equation 4 by iterating on the Green's function integral

$$u^{n+1}(x) = -\int_0^1 G(x,y)f(y,u^n(y))dy. \tag{5}$$

A variation of this with a relaxation parameter was also used; that is,

$$u^{n+1}(x) = (1-\omega)u^n(x) - \omega\int_0^1 G(x,y)f(y,u^n(y))dy. \tag{5'}$$

ω is called the relaxation parameter ($\omega>1$ is called over-relaxation, $\omega<1$ is called under-relaxation) and can be used to control the speed of convergence.

Starting with an initial trial solution $u^o(x)$ the program iterates to find successive $u^n(x)$ stopping when the maximum difference in two consecutive iterations drops below a given threshold. The function $u^n(x)$ is constructed as a table of values on the interval [o,1] and, through interpolation, values at points between tabulated values are obtained.

The results for various f functions are given in Figure 1.

A technique used to reduce computation time is the progressive refinement of the mesh size, interpolation, and quadrature procedures. One starts with a coarse mesh and crude interpolation and quadrature procedures and, as convergence progresses, proceeds to a finer mesh, higher order interpolation and more accurate quadrature.

Figure 1a shows the sequence of approximations for the case

$$f(x,u) = -2\pi^2 \sin u.$$

This case is known to have multiple solutions. The solution $u(x) = 0$, $0 \le x \le 1$, is unstable, but at least two others are locally stable, and the initial trial solution, $u^0(x) = \sin\pi x$, is shown here converging to one of these stable solutions.

Figure 1b shows the sequence of approximations for the case

$$f(x,u) = -\pi^2(2-\cos\pi x) \sin u.$$

Again there is an unstable solution, $u(x) = 0$. The initial trial solution $u^0(x) = \sin\pi x$ converges to a locally stable solution.

Figure 1c shows the case

$$f(x,u) = -\pi^2((u-3u^{-1/3})/4 + 2\sin^{3/2}\pi x).$$

Here an analytic solution is known, $u(x) = \sin^{3/2}\pi x$. The initial trial solution of $u^0(x) = \sin\pi x$ converges to this solution even though f is singular at the boundaries where $u(0) = u(1) = 0$.

C. Three Body Orbits

The second group of problems is associated with finding solutions of the restricted three-body problem that pass through two given points in a fixed time (rendezvous problem).

The differential equation [2] is

$$\ddot{z}(t) + 2i\dot{z}(t) - \gamma^2 z(t) = -f(z(t)), \tag{6}$$

where

$$f(z) = \frac{\mu'(z+\mu)}{|z+\mu|^3} + \frac{\mu(z-\mu')}{|z-\mu'|^3} - (1-\gamma^2)z, \tag{7}$$

with the boundary conditions

$$z(a) = z_a, z(b) = z_b.$$

Here γ, μ, and μ' ($\mu+\mu' = 1$) are constants and $i = \sqrt{-1}$. $z = x + iy$ is a complex variable so that this equation represents a pair of coupled real second order differential equations with two-point boundary conditions.

1) The Green's function for the restricted three body orbits

The Green's function for the operator

$$L_2 \equiv \frac{d^2}{dt^2} + 2i\frac{d}{dt} - \gamma^2, \tag{8}$$

with boundary conditions

$$G(a,s) = G(b,s) = 0, \tag{9}$$

can be written as

$$G(t,s) = \frac{1}{W(s)}\begin{cases} v(t-a)\; v(s-b), & t \le s, \\ v(t-b)\; v(s-a), & t \ge s, \end{cases} \tag{10}$$

where

$$W(s) = \dot{v}(s-a)\; v(s-b) - \dot{v}(s-b)\; v(s-a). \tag{11}$$

(The Wronskian is not constant if L_2 is not self-adjoint).

The function v is a solution of $L_2 v = 0$; $v(0) = 0$; i.e., $v(t) = e^{\alpha t} - e^{\beta t}$; $\alpha,\beta = -i(1 \pm \sqrt{1-\gamma^2})$. The parameter γ is artificially introduced. Its value can be adjusted to control the rate of convergence of the iteration.

Since the boundary conditions are not zero, the integral equation form of (6) is

$$z(t) = V(t) + \int_a^b G(t,s)f(z(s))ds. \tag{12}$$

Here G is the Green's function given by (10) above, V(t) is the particular solution satisfying $L_2V = 0$, and the boundary conditions $V(a) = z_a$, $V(b) = z_b$, i.e.,

$$V(t) = \frac{v(t-a)}{v(b-a)}z_b + \frac{v(t-b)}{v(a-b)}z_a. \tag{13}$$

2) <u>The computer program</u>

A computer program was written in Fortran for the UNIVAC 1108 to integrate the equation

$$z^{n+1}(t) = V(t) + \int_a^b G(t,s)f(z^n(s))ds, \tag{14}$$

where V, G and f are given in (13), (10), and (7) respectively. Starting with an initial guess of $z^0(t)$, the program iterates to find successive $z^n(t)$, stopping when the maximum difference in two consecitive iterations drops below a prescribed threshold.

The function $z^n(t)$ is approximated by constructing a table of its values on the interval [a,b] and using cubic splines [1, 25-27] to interpolate for the in-between points. These splines are also used to do the quadrature.

A relaxation parameter was also introduced to help convergence; that is, instead of (14), one uses

$$z^{n+1}(t) = (1-\omega)z^n(t) + \omega\left(V(t) + \int_a^b G(t,s)f^n(z(s))ds\right). \tag{15}$$

where ω is the relaxation parameter. This gives in effect two parameters, γ and ω, to be adjusted to speed and control convergence.

3) <u>Results</u>

Figure 2 shows a typical example of the sequence of approximations from this iteration. The boundary conditions are $z(0) = 1.2$, $z(3.06) = -1.5$. The system constants μ and μ' ($\mu+\mu' = 1$) characterize the earth moon system, $\mu = 0.012277471$; here $\gamma = 0.95$ and $\omega = 0.5$. The initial orbit is marked with an I, and the final orbit with an F.

One notes here the dramatic and rather violent departure from the initial approximation (the first iteration leaves the page for most of the orbit). However, succeeding iterates come back on the page and quickly settle down to an almost circular uniform speed orbit.

There is an orbit satisfying the boundary conditions in the neighborhood of the initial trial solution, but it is either highly unstable for this iterative scheme or else the initial guess was not close enough to have been in the stable region.

This behavior is typical of this particular problem, that is, orbits having a fixed time between two fixed points in the neighborhood of the earth-moon system. Only those orbits that were very smooth appeared to be stable. The more complex orbits between the same two points appeared to be unstable for this iterative scheme.

V. Results and Conclusions

A. Summary of Results

The primary result of this study is to show that the Green's function method of solving the two-point boundary value problem can be an effective tool in numerical work.

There is a straightforward prescription for producing the Green's function for both the single equation and for the system of equations case. It can be given in terms of solutions to sets of initial value problems which in turn can be generated to arbitrarily high order and accuracy by standard techniques such as Runge-Kutta, Adams, or other methods.

For linear differential equations the solution can be given directly in terms of the integral over the Green's function. For the nonlinear case, the Green's function provides an iterative scheme only. Convergence must be investigated in each individual case. There exists a literature on convergence theorems for a variety of classes of problems, but even when convergence cannot be guaranteed a priori, the method can often be used when combined with a relaxation method or other devices.

For single second order ordinary differential equations the work involved in finding the Green's function can be considered nominal. For systems of such equations the work involved goes up as the square of the number of equations and may be considered excessive if the number of equations is large. In this case, a technique of splitting the original differential operator into a diagonal and off-diagonal component can be used. The Green's function for the diagonal component is then just the set of Green's functions for the individual (uncoupled) diagonal elements and the work involved is only linear in the number of equations. The solution now involves a sequence of iterations, even in the linear case, and raises additional questions of convergence, but this could still turn out to be less work than finding the entire Green's function for the original matrix operator, especially if the problem is nonlinear, and iteration will be required anyway.

The particular numerical experiments carried out involved problems for which there were multiple solutions or for which the standard existence, uniqueness, and convergence theorems were not applicable. A search for orbits of the restricted three body system was investigated for the rendezvous type orbits. The orbits showing most stability with respect to the iteration scheme were those having the largest radii of curvature or that stayed farthest from the singular points. The problem of singling out a special one by this method needs further investigation.

B. Recommendations for Further Study

Continued work is needed in the theoretical area of convergence and stability of iterative methods for the two point boundary value problem. In the cases where multiple solutions exist, methods need to be developed to determine which solutions are locally stable, which are more stable than others and what are the regions of convergence or stability. These investigations could also examine the role of the relaxation parameter in the convergence process.

In the area of numerical experiments, more work should be done in comparing the Green's function method directly to the shooting method of solving the two-point boundary problem to see which takes less computer time, storage space, etc.

There is another way in which the Green's function method can be used, but there has been little or no numerical experimentation undertaken. The theory is reasonably straightforward and goes as follows:

Let L_2 be a second order linear differential operator and the function u satisfy the equation

$$L_2 u = f(u) \tag{1}$$

with zero end point boundary conditions, $u(a) = u(b) = 0$. Assume that $f(u)$ is a reasonably well behaved function, that $\frac{\partial}{\partial u} f(u)$ exists and can be computed for functions u in some neighborhood of the solutions to (1), and let $\bar{u}$ be in this neighborhood. Then

$$\left(L_2 - \frac{\partial f}{\partial \bar{u}}(\bar{u})\right) u = f(u) - \frac{\partial f}{\partial \bar{u}}(\bar{u})u,$$

or

$$\bar{L}_2 u = f(u) - \bar{f}'\, u,$$

where $\bar{L}_2 = L_2 - \bar{f}'$, $\bar{f}' = \frac{\partial f}{\partial \bar{u}}(\bar{u})$. If there now exists a Green's function for $\bar{L}_2$, say $\bar{G}$, then

$$u = -\bar{G}(f(u) - \bar{f}'u).$$

One now looks at the iterative equation

$$u^{n+1} = -\bar{G}^n \left(f(u^n) - \bar{f}'^n u^n\right) \tag{2}$$

where

$$\bar{f}'^n = \frac{\partial f}{\partial \bar{u}}(\bar{u}) \Big|_{\bar{u} = u^n}, \quad \bar{L}_2^{\,n} = L_2 - \bar{f}'^n,$$

and $\bar{G}^n$ is the Green's function for $\bar{L}_2^n$.

The iterative scheme (2) can be shown to converge quadradically. That is, it is always locally stable provided that the indicated entities exist. The system (2) is analogous to the Newton-Raphson method.

While this method has the advantage of being locally stable, it has the disadvantage of requiring the recalculation of $\bar{G}^n$, the Green's function, at every step. Whether this is practical or not needs to be determined by numerical experiments.

A technique similar to (2) above exists if f also depends on the derivatives of u; i.e., if $f = f(u,u')$.

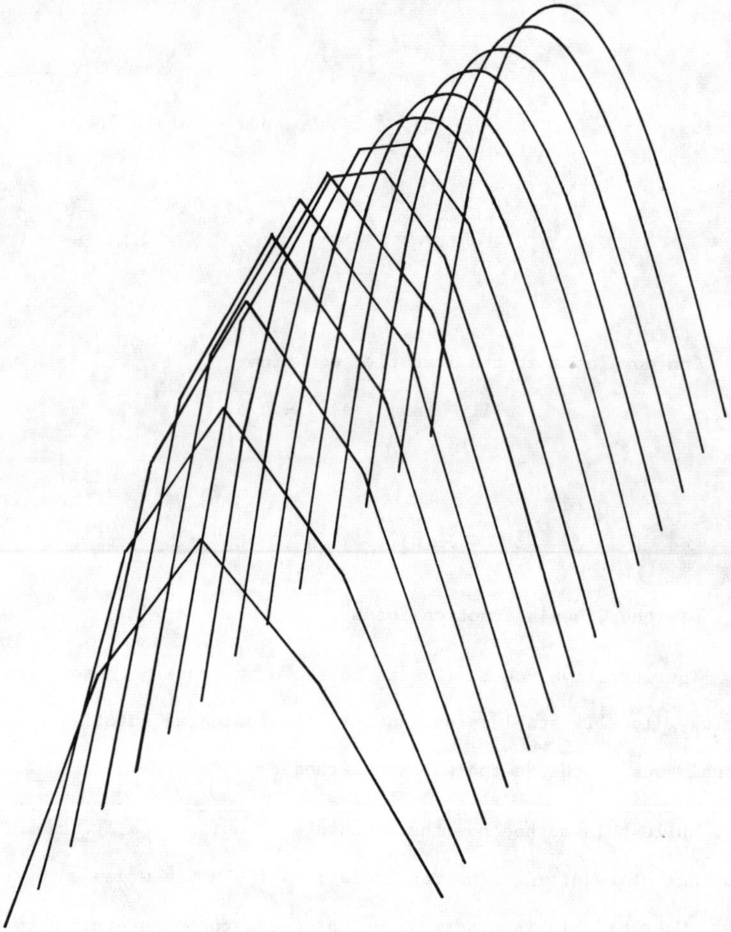

FIGURE 1a

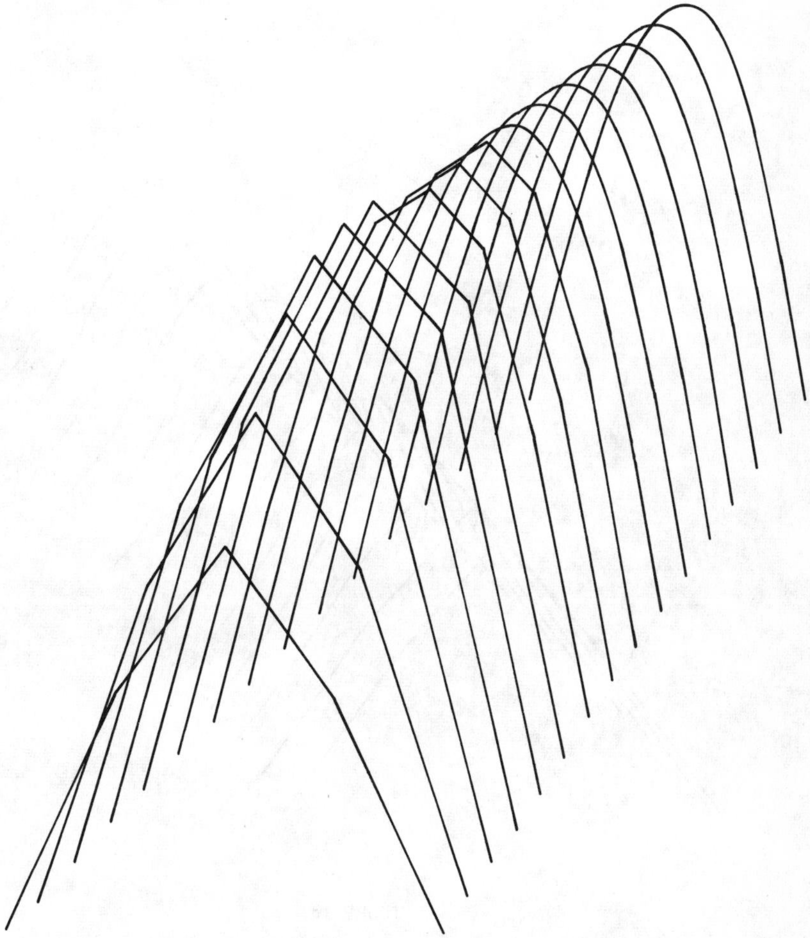

FIGURE 1b

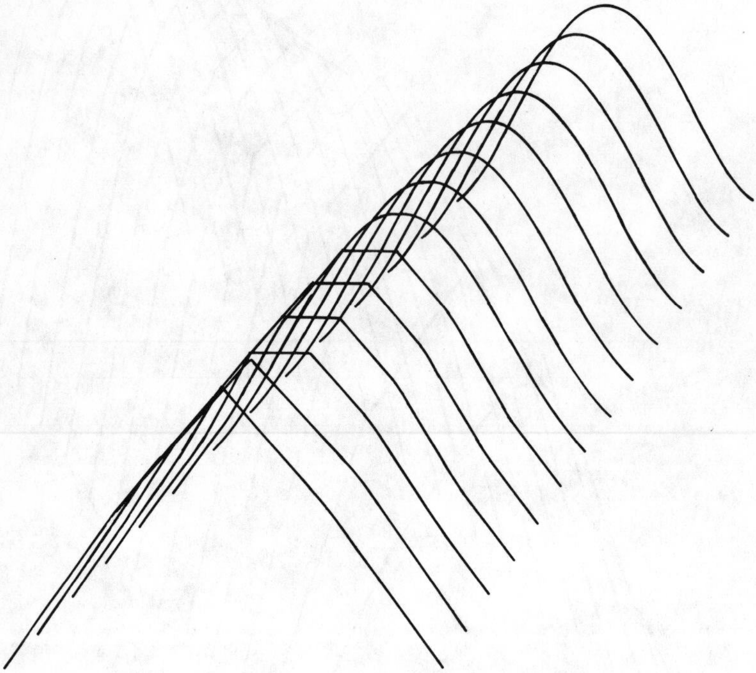

FIGURE 1c

Figure 1 shows the sequence of iterates for solutions of the equations $\frac{d^2}{dx^2}u(x) = f(x,u(x))$; $u(0) = u(1) = 0$.

Figure 1a is for $f(x,u) = -2\pi^2 \sin\ u$.

Figure 1b is for $f(x,u) = -\pi^2\ (2 - \cos \pi x) \sin u$.

Figure 1c is for $f(x,u) = -\pi^2\ ((u - 3u^{-1/3})/4 + 2\sin^{3/2}\pi x)$.

Initially a coarse mesh, linear interpolation and a crude quadrature scheme was used. As convergence approaches, the program switches to progressively finer mesh, more accurate quadrature, and spline-like interpolation. For clarity, successive iterations are displaced with respect to each other and scaled down slightly. The relaxation parameter, $\omega = \frac{1}{2}$, was used.

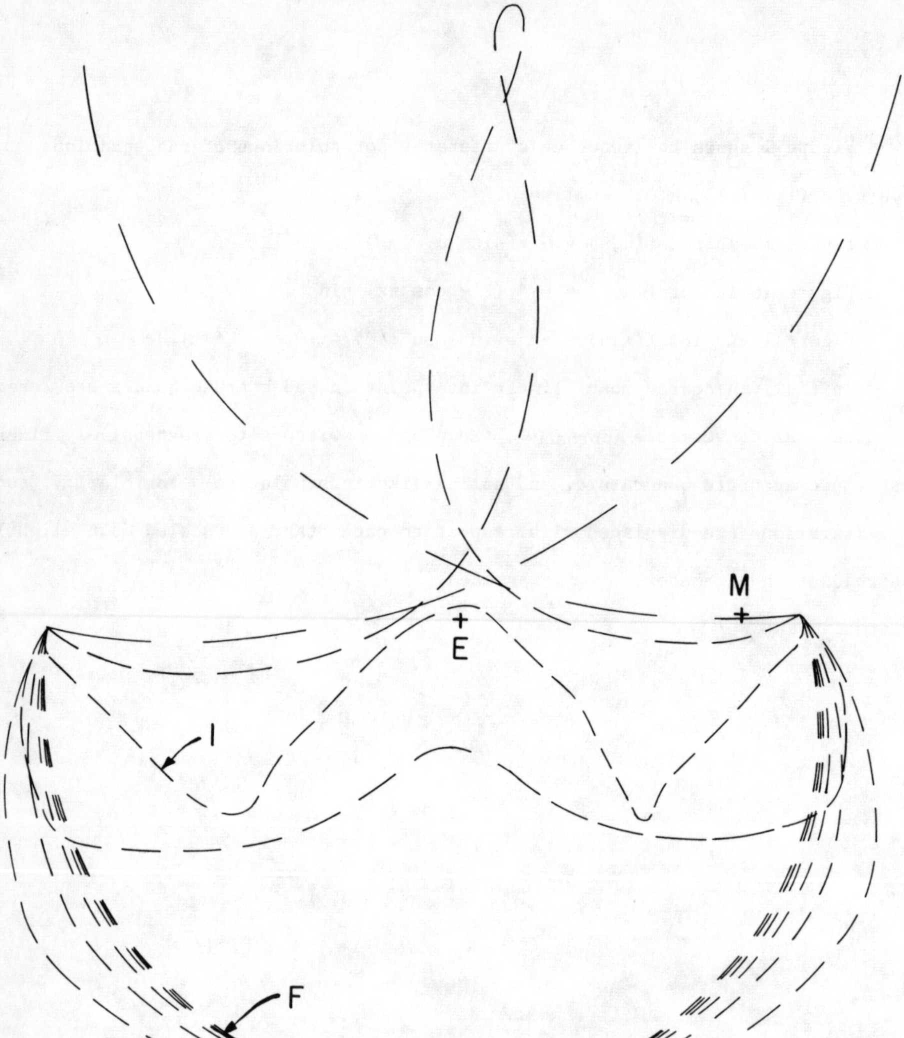

FIGURE 2

Figure 2 shows the convergence of the iteration scheme for a rendezvous type orbit (fixed time between two fixed points) of the resticted three body system.

The three body parameters are chosen so that the earth-moon system is represented ($\mu \approx 0.012$). The orbit is represented in rotating reference frame, normalized to unit angular velocity (period = 2π) and unit earth-moon distance. The length of the dashes is proportional to the speed in that part of the orbit. The time for the orbit is 3.06.

The earth, moon, initial and final orbits are indicated with the letters E, M, I, F respectively.

VI. Bibliography

1. Ahlberg, J. H., E. N. Nilson, and J. L. Walsh, "Convergence Properties of General Splines," Proc. Nat. Acad. Sci. USA 54 (1965) 344-350.

2. Arenstorf, R. F., American Journal of Mathematics, LXXXV (1963) 27.

3. Bailey, P. B., L. F. Shampine, and P. E. Waltman, Nonlinear Two Point Boundary Value Problems, New York: Academic Press, 1968.

4. Bellman, R. E. and R. E. Kalaba, Quasilinearization and Non-linear Boundary Value Problems, New York: American Elsevier Publishing Company, 1965.

5. Ben-Israel, A., and A. Charnes, "Contributions to the Theory of Generalized Inverses," SIAM J. Appl. Math. 11 (1963) 667-699.

6. Bounitzky, E., "Sur la Function de Green des Équations Differentielles Linéaires Ordinaires," J. Math.Pures Appl. (6) 5 (1909) 65-125.

7. Ciarlet, P. G., F. Natterer, and R. S. Varga, "Numerical Methods of High Order Accuracy for Singular Nonlinear Boundary Value Problems," Numerische Mathematik 15 (1970) 87-99.

8. Ciarlet, P. G., M. H. Schultz, and R. S. Varga, "Numerical Methods of High-Order Accuracy for Nonlinear Boundary Value Problems. I. One Dimension Problem," Numer. Math. 9 (1967) 394-430.

9. Ciarlet, P. G., M. H. Schultz, and R. S. Varga, "Numerical Methods of High-Order Accuracy for Nonlinear Boundary Value Problems. V. Monotone Operator Theory," Numer. Math. 13 (1969) 51-77.

10. Ciarlet, P. G., Variational Methods for Non-Linear Boundary-Value Problems, Doctoral Thesis, Case Institute of Technology, 1966.

11. Elliott, W. W., "Generalized Green's Functions for Compatible Differential Systems," Amer. J. Math. 50 (1928) 243-258.

12. Fox, L., The Numerical Solution of Two-Point Boundary Value Problems in Ordinary Differential Equations, London: Oxford University Press, 1957.

13. Henrici, P., Discrete Variable Methods in Ordinary Differential Equations, New York: John Wiley & Sons, 1962.

14. Jamet, P., Numerical Methods and Existence Theorems for Singular Linear Boundary-Value Problems, Doctoral Thesis, University of Wisconsin, 1967.

15. Jamet, P., "On the Convergence of Finite-Difference Approximations to One-Dimensional Singular Boundary-Value Problems," Numer. Math. 14 (1970) 355-378.

16. Lees, M., "Discrete Methods for Nonlinear Two-Point Boundary Value Problems," Numerical Solution of Partial Differential Equations, ed. by J. H. Bramble, New York: Academic Press, 1966. 59-72.

17. Loud, W. S., "Generalized Inverses and Generalized Green's Functions," SIAM J. Appl. Math. 14 (1966) 342-369.

18. Loud, W. S., "Some Examples of Generalized Green's Functions and Generalized Green's Matrices," SIAM Review 12, 2 (1970) 194-210.

19. Mikhlin, S. G., "Variational Methods of Solving Linear and Non-Linear Boundary Value Problems," Differential Equations and Their Applications, ed. by I. Babuska, New York: Academic Press, 1963 77-92.

20. Perrin, F. M., H. S. Price, and R. S. Varga, "On Higher-Order Numerical Methods for Nonlinear Two-Point Boundary Value Problems," Numer. Math. 13 (1969) 180-198.

21. Phillips, G. M., "Analysis of Numerical Iterative Methods for Solving Integral and Integradifferential Equations," Computer J. 13 (1970) 297-300.

22. Reid, W. T., "Generalized Green's Matrices for Compatible Systems of Differential Equations," Amer. J. Math. 53 (1931) 443-459.

23. Reid, W. T., "Generalized Green's Matrices for Two-Point Boundary Problems," SIAM J. Appl. Math. 15 (1967) 856-873.

24. Reid, W. T., "Generalized Inverses of Differential and Integral Operators," Theory and Application of Generalized Inverses of Matrices: Symposium Proceedings, Texas Technological College Mathematics Series, No. 4, Lubbock, Texas 1968, 1-25.

25. Schoenberg, I. J., On Spline Functions, Math. Research Center MRC Tech. Summary Report No. 625, University of Wisconsin (May 1966).

26. Schoenberg, I. J., "On Interpolation by Spline Functions and Its Minimal Properties," Proceedings Conference on Approximation Theory (Oberwohlfach, Germany, 1963), Basel: Birkhauser, 1964, 109-219.

27. Sharma, A., and A. Meir, "Degree of Approximation of Spline Interpolation," J. Math. Mech. 15 (1966) 759-767.

28. Varga, R. S., "Hermite Interpolation-Type Ritz Methods for Two-Point Boundary Problems," Numerical Solution of Partial Differential Equations, ed. by J. H. Bramble, New York: Academic Press, 1966. 365-373.

29. Vašakmadze, T. S., "On the Numerical Solution of Boundary-Value Problems," Ž. Vyčisl. Mat. i Mat. Fiz. 4 (1964) 623-637.

SHOOTING-SPLITTING METHOD FOR SENSITIVE TWO-POINT BOUNDARY VALUE PROBLEMS

by

P. J. Firnett, Informatics, Inc.

and

B. A. Troesch, University of Southern California

A method for the solution of sensitive two-point boundary value problems is described. The method is conceptually simple, easily set up, and requires only the availability of any standard integration routine for systems of ordinary differential equations. The approach is suitable for relatively low order systems as they appear in fluid dynamics, heat transfer and plasma physics.

1. Introduction

Methods for the solution of nonlinear two-point boundary value problems for systems of ordinary differential equations have been investigated extensively. In this paper we will describe a method which has been proven quite successful for problems in fluid dynamics and magneto-hydrodynamics. The method is very closely related to the classical shooting method. But it is well known (cf., for instance, the work of Holt [5], Morrison et al. [9], Sylvester and Meyer [15], and the books by Keller [6], Roberts and Shipman [11], Stoer and Bulirsch [14]) that for some problems the shooting method fails, even if the guesses for the missing initial data are chosen quite accurately. The integration encounters a singularity of the initial value problem before the endpoint of the interval is reached. Problems of this type are called "sensitive" (cf. [11]).

In contrast to the characteristic properties of stiff systems of differential equations, the difficulty is analytical rather than numerical in nature. Also, the shooting process leads to a true singularity and not just to very rapidly growing solutions, although exponential growth already poses serious difficulties (cf. the approach by Conte in [2]).

In general, the values of the parameters which appear in the problem statement do not seem to be excessively small or large, and therefore do not in themselves indicate the observed difficulty. Furthermore, the rapid deviation from the true solution does not have any physical meaning, as the boundary value problem itself is physically perfectly stable. Often the solution actually represents the steady state of a mathematically well-posed problem for a system of partial differential equations.

The shooting-splitting method is well suited for sensitive two-point boundary value problems of moderately low order, although the deciding factor is really the number of missing initial data. (Compare the concluding remark in section 3 on this point.) Three unknown initial values (as encountered by Sloat in [13]) constitute the practical upper limit. In the examples discussed below, only one initial value is missing.

Since the implementation of the method is so easy and only a standard subroutine for systems of differential equations is required, the machine program is best set up for each problem individually. (At some computing centers there are complete two-point boundary value packages available which also use a more powerful general approach.) It seems to be a desirable feature of the present method that, except for the missing initial data, no guesses are required, neither for the lengths of the subintervals of integration nor for the expected shape of the solution, as in the multiple shooting method described by Morrison et al. [9], or in the finite difference method used by Holt [5]. Incidentally, the method remains the same for finite or infinite intervals, and also for parameter values which lead to sensitive or to non-sensitive problems.

The conceptual simplicity and the modest tools necessary to implement the shooting-splitting method should make it attractive to the engineer, who does not wish to get involved with more sophisticated, albeit more powerful, computing techniques (cf. in this connection the work of Glenn [4], Oida [10], and Sloat [13]). The original problem statement is all that is needed as a starting point. Since there are no finite difference formulas and matrices to set up nor variational equations to derive, the possibility of introducing errors is considerably reduced. This consideration seems particularly important for very involved equations (cf. [4], [13]).

Besides the description of the method, the following sections will also discuss some general information about the properties of the solutions, which can be obtained by analytical means. Among the results are physically interesting a priori bounds for initial guesses, relevant combinations of the problem parameters, or the behavior of the solution at infinity. Clearly, each individual problem will be amenable to a different kind of analysis, but the points brought up below indicate what type of investigation might be feasible and fruitful.

The organization of the paper is as follows: The shooting-splitting method is explained in section 3, together with remarks on the underlying assumptions about the system of differential equations and on some details of implementation. However, we first present in section 2 the statement of a typical sensitive problem from plasma physics and its solution, which will serve as the main example in the paper. Section 4 describes some appropriate analytical results for this example. The reason for the difficulty encountered in the solution of sensitive problems is explored in section 5 for a drastically simplified version of the example. The section 6 discusses briefly a very simple sensitive problem, where only limited physical information about the solution happens to be required. Here the analytical investigation proves to be helpful too. Whereas the first two examples have arisen in real physical situations, the problem in section 7, for which the lengths of the subintervals in the shooting-splitting method shrink as the integration progresses, describes an artificial situation.

2. A Typical Sensitive Problem, Example 1.

As the main example for a sensitive problem we choose a system of ordinary differential equations which describes a situation in magnetohydrodynamics, namely the behavior of an electron sheet entering a plasma, as formulated by Lundgren and Chang [7]. The electromagnetic field, E, B, the the electric potential φ , and the electron number density N are governed by the first order system

$$B' = -2\, t_o N \tag{2.1a}$$

$$N' = t_o N\, (B-E) \tag{2.1b}$$

$$E' = -\frac{2}{\beta}\, (N + 2N_o \sinh \varphi) \tag{2.1c}$$

$$\varphi' = -\, E, \tag{2.1d}$$

valid for the independent variable ranging over $0 \leq x < \infty$. The boundary conditions are

$$N(0) = 1$$
$$B(0) = 0$$
$$E(0) = 0$$
$$\varphi(\infty) = 0.$$

In addition, all dependent variables must remain uniformly bounded, and $N(x)$, as a particle density, must remain positive everywhere. The given parameters, t_o, β, N_o, are all positive, and we will consider, in particular, the representative set of values

$$t_o = .1\,,\ \beta = .5,\ N_o = 1,$$

although N_o may be as large as $N_o = 100$.

The discussion will be mainly based on the system (2.1), but the system can be put into equivalent forms, one of which exhibits more clearly the connection with problems that are known to be sensitive (cf. the example in Roberts and Shipman[11, p. 158]). By introducing

$$x_4 = \exp(\varphi)$$
$$x_5 = \exp(-\varphi)$$

and renaming the dependent variables in an obvious way we obtain the system

$$\begin{aligned} x_1' &= -\,2\, t_o x_2 \\ x_2' &= t_o x_2 (x_1 - x_3) \\ x_3' &= -\frac{2}{\beta}\,(x_2 + N_o x_4 - N_o x_5) \\ x_4' &= -\, x_3 x_4 \\ x_5' &= x_3 x_5 \end{aligned} \tag{2.2}$$

and the boundary conditions

$$x_1(0) = x_3(0) = 0$$

$$x_2(0) = x_4(\infty) = x_5(\infty) = 1.$$

An alternate more compact and symmetric form emphasizes the appearance of the exponential functions. If we set

$$N = \exp(\psi)$$

and hence

$$\psi' = t_o(B-E),$$

we obtain the second order system

$$\psi'' = t_o\left(\frac{2}{\beta} - 2t_o\right)\exp(\psi) + \frac{4}{\beta} t_o N_o \sinh \varphi$$

$$\varphi'' = \frac{2}{\beta} \exp(\psi) + \frac{4}{\beta} N_o \sinh \varphi$$

with the boundary conditions

$$\psi(0) = \psi'(0) = \varphi(\infty) = \varphi'(0) = 0.$$

These equations show clearly that the integration in the backward direction, which is sometimes helpful, will exhibit the same difficulties as the forward integration. The reason for the difficulty will be discussed in section 5 below for a simplified problem. From the second equation it follows immediately (see also section 4.1) that $\varphi(x)$, and in particular $\varphi(0)$, must be negative for a uniformly bounded solution.

In Table 2.1 and Figs. 2.1 the solution of the system (2.1) is presented, including an example of an upward and downward diverging solution. The upward and downward divergence is most pronounced in the variable E, and hence the values of the two branches E_u and E_d are the only ones listed separately in the table. For all other functions the values differ insignificantly at the end of the subintervals.

The results were obtained by the shooting-splitting method, which is described in general terms in the next section. We notice that the subintervals Δx are of about equal length, which seems to be typical for autonomous systems. A different situation will be briefly discussed in section 7 below.

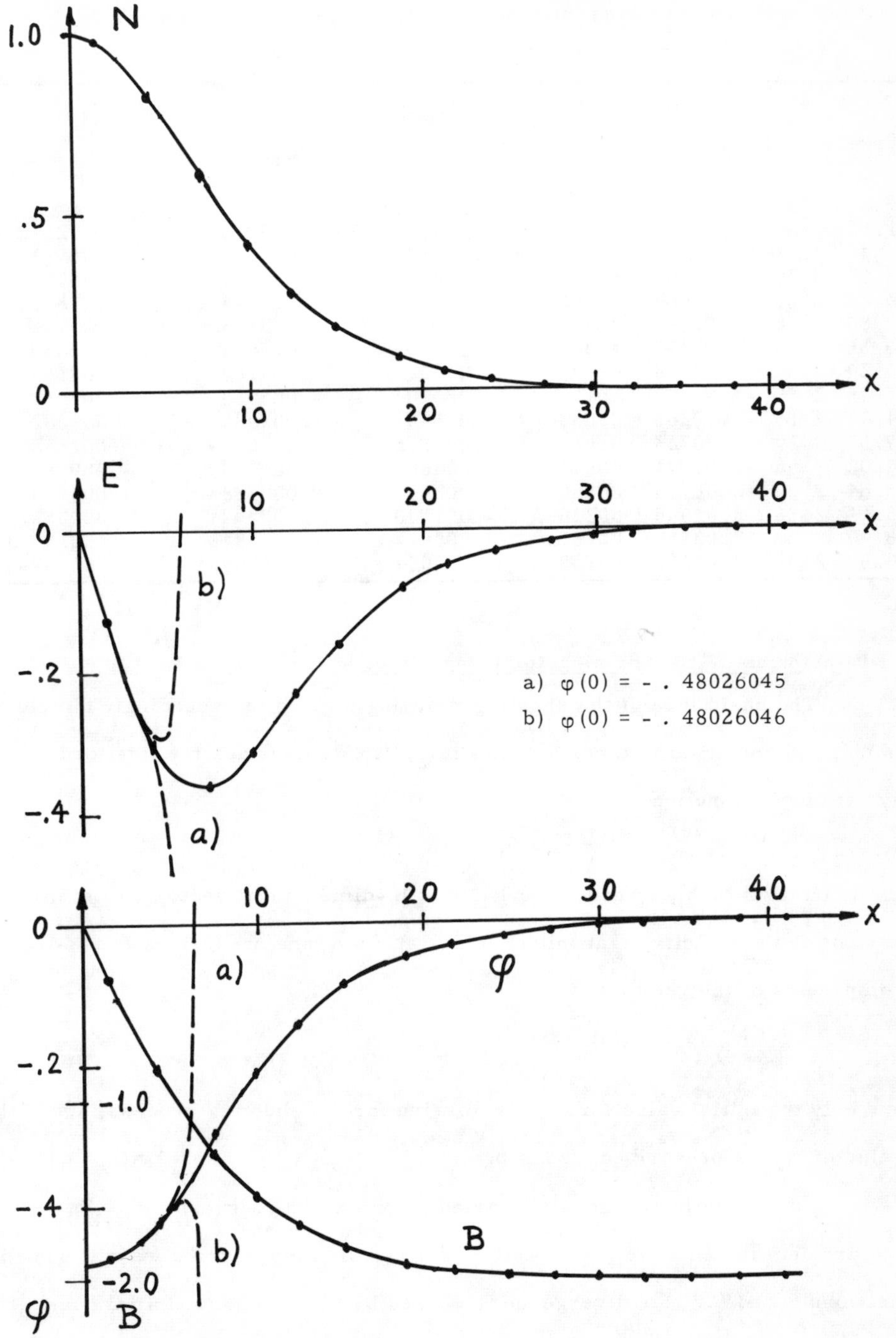

Fig. 2.1 Solutions of the system (2.1) for $t_o = .1$, $\beta = .5$, $N_o = 1$, with an example of upward and downward diverging solutions. Dots mark the end-points of the subintervals.

TABLE 2.1. Results of the integration of eqs. (2.1) with $t_o = .1, \beta = .5, N_o = 1$ at the endpoints of the subintervals of length Δx.

x	Δx	$B_u = B_d$	$N_u = N_d$	$\varphi_u = \varphi_d$	E_u	E_d
0		0	1.0	-.4802605	0	0
1.60	1.60	-0.31741	.97587	-.46952	-.013252	-.013251
4.48	2.88	-0.84261	.82954	-.40325	-.030916	-.030913
7.68	3.20	-1.29907	.59302	-.29242	-.035776	-.035766
10.24	2.56	-1.55564	.41426	-.20602	-.030874	-.030866
12.80	2.56	-1.72989	.27351	-.13671	-.023138	-.023135
15.36	2.56	-1.84277	.17380	-.08711	-.015854	-.015850
19.20	3.84	-1.93808	.08429	-.04232	-.008160	-.008156
21.76	2.56	-1.97208	.05117	-.02570	-.005053	-.005052
24.32	2.56	-1.99265	.03083	-.01549	-.003081	-.003082
27.52	3.20	-2.00723	.01626	-.00872	-.0016389	-.0016386
30.08	2.56	-2.01374	.00972	-.004886	-.0009835	-.0009833
32.64	2.56	-2.01763	.00580	-.002917	-.0005884	-.0005883
35.20	2.56	-2.01995	.00346	-.001740	-.0003514	-.0003513
38.40	3.20	-2.02158	.00181	-.000911	-.00018425	-.00018423
40.96	2.56	-2.02231	.00108	-.000054	-.00010984	-.00010983

3. Description of the Shooting-Splitting Method.

The basic idea of the shooting-splitting method is exceedingly simple and is actually contained in a nutshell in Fig. 3.1. We consider the first order system for k functions

$$\underline{y}' = \underline{f}\ (x, \underline{y})$$

in the interval [a, b], where $\underline{y}$ and $\underline{f}$ are k-dimensional vectors. Let us now assume for simplicity, that initial values at x = a are prescribed for (k-1) components of the vector $\underline{y}$

$$y_j(a) = y_j^{(0)}, \qquad j = 1, 2, \ldots, k-1,$$

and only the initial value for y_k is missing for the shooting process, i.e., the value of y_k is prescribed at x = b.

The shooting process is started at x = a with a guess for $y_k(a)$, and if the problem is sensitive, the solution will rapidly diverge. As we will show in sections 5 and 6.2, this divergence is caused by the erroneous initial guess; the numerical result accurately reflects the behavior of the true solution, so that improved or alternate numerical integration subroutines will not improve the

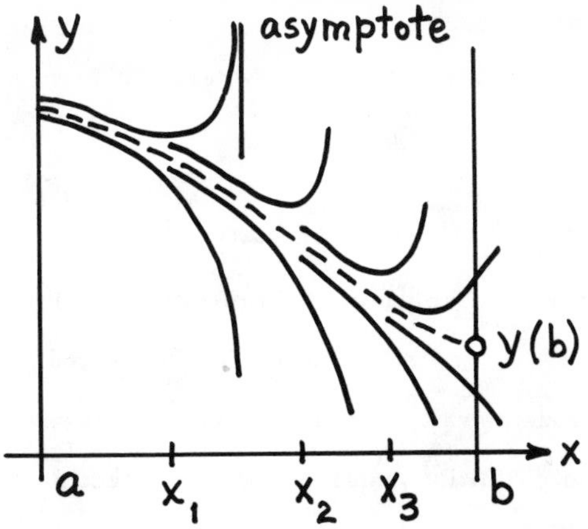

Fig. 3.1 The principle of the shooting- splitting method.

situation, and therefore the user may as well apply his favorite library subroutine. Very likely, not all components of the solution $\underline{y}$ will diverge equally rapidly, and furthermore, some will diverge upwards whereas others will sharply turn downwards (cf. Fig. 2.1). Let us denote the component with the most pronounced divergence behavior by y, without a subscript. (It causes no difficulty, if in a later phase of the computation its role is taken over by another component of the solution vector). Assume now that for the first guess $y_k(a)$ the function y(x) diverges upwards and denote this solution by $y_u(x)$. If a second guess $y_k(a)$ is found for which the solution y(x) turns downwards, let us call it $y_d(x)$, then we are ready to start the shooting-splitting method. A new initial guess is obtained by taking the average of the two previous guesses and, depending on the divergence behavior, the appropriate old guess and the corresponding solution is discarded. This bisection process is continued until two guesses with an upward diverging solution $y_u(x)$ and a downward diverging solution $y_d(x)$ differ by less than a small value δ_1. The choice of this constant depends on the wordlength used in the computation, but must still be large enough, so that systematic (non-random) solution curves are obtained. Let us assume that the accuracy requirement for the final solution is δ_2, typically 10^3 to 10^4 times larger than δ_1. The integration will reach a point $x = x_1$, where the branches y_u and y_d differ by δ_2. The average

$$\underline{y}(x) = (\underline{y}_u(x) + \underline{y}_d(x))/2$$

is then accepted as the final solution of the problem in the interval $[0, x_1]$.

At the point $x = x_1$ the same process is repeated. Since the results $\underline{y}_u(x_1)$ and $\underline{y}_d(x_1)$ are known to lead to upward and downward diverging curves, these are appropriate as initial guesses. Using bisection and observing the divergence pattern, the guesses at x_1 are refined, until they again differ by δ_1 at most. This furnishes the solution in the interval $[x_1, x_2]$, where at x_2 the two branches differ by δ_2. The process is continued through as many subintervals $[x_i, x_{i+1}]$ as required, let us say n, to reach the endpoint b. In the last segment $[x_{n-1}, x_n]$ the standard shooting process is applied to satisfy the boundary condition for $y_k(b)$. This then completes the solution.

The shooting-splitting method shows a somewhat surprising feature of sensitive problems: the specific boundary condition at $x = b$ influences only the solution in that last segment, whereas the solution in $[0, x_{n-1}]$ is solely determined by the requirement that the solution remain bounded.

From the description of the shooting-splitting method it is clear that an important assumption has been tacitly made. The true solution is assumed to lie always between the upward turning and the downward turning curves. In other words, the solution depends monotonically on the missing initial condition and the trajectories sweep out the strip $-\infty < y(x) < +\infty$, $a \leq x \leq b$ without intersecting. Although this requirement is quite restrictive, it seems to be a common feature of a wide class of sensitive problems. In some simple problems (cf. sections 4.1 and 6) it may be feasible to prove this property in general or at least outside some interval of the initial guess (cf. eq. (4.1)). On the other hand, the numerical results for sensitive problems show such a strong tendency to diverge that it seems unlikely that the solution would reverse itself and turn downwards from a very large positive value. Of course, the condition stated above cannot hold globally for problems with multiple solutions, since it ensures existence and uniqueness of the boundary value problem.

Even if the true solution were not straddled by branches diverging in opposite directions, but all approximations diverged in the same direction, the general idea of the shooting-splitting method could still be used. The additional difficulty would be similar to the difficulty of finding the root of a nonnegative function.

There are some details that might be added with regard to the implementation of the shooting-splitting method. Under certain conditions it might be feasible to replace the simple bisection of the guesses by a more refined and faster converging procedure. A guess with slower divergence can be given more weight in the averaging process.

Normally, the integration of a curve through a subinterval is carried out with variable stepsize, but in order to determine properly, when two branches differ by δ_2, the final two branches, $\underline{y}_u$ and $\underline{y}_d$, should be computed in parallel, i.e., with the identical (variable) stepsize.

Experience has also shown that, on occasion, a solution started with a guess for a supposedly upward diverging curve, let us say $y_u(x_i)$, turns downwards. This is caused by the fact that the stepsize at the beginning of the interval $[x_i, x_{i+1}]$ differs from the final stepsize in $[x_{i-1}, x_i]$. Under these conditions it has been necessary to increase the initial guess somewhat, and $\delta_2/4$ has in general proved adequate. In the corresponding case of a downward curve turning upwards, of course, we would take $y_d(x_i) - \delta_2/4$ as a new starting guess.

It should be noted that the search for the proper initial guess remains a one-dimensional search at all starting points x_i, $i = 1, 2, \ldots, n-1$, regardless of the order k of the system of differential equations. For instance, at x_1 the endpoint of the solution vector $\underline{y}(x_1)$ depends only on one parameter, namely on the guess $y_k(a)$. Therefore, the guesses for the next subinterval are chosen to lie on that particular curve in the k-dimensional space, or, as a good approximation, on the chord between $\underline{y}_u(x_1)$ and $\underline{y}_d(x_1)$. From this consideration, it appears that the shooting-splitting method might actually be advantageous for systems of large order k if only one or two initial data are missing. However, the method has not been tried on problems of this kind.

4. Analytical Properties of the Solution of Example 1.

4.1 Bounds for the unknown initial data.

It is a typical property of sensitive problems that an incorrect guess for the missing initial data will be detected after integrating over only a short interval. Nevertheless, it is worthwhile to determine a priori bounds for the initial guess, if this is feasible, first of all to save computing time by avoiding wrong guesses, but also, and this is possibly even more important, to gain an insight into the physical problem, which might be difficult to obtain from a list of numerical results.

For the Example 1, a priori bounds for $\varphi(0)$ can be found from general properties of the equations. (It is always assumed that only solutions with $N(x)>0$ are physically meaningful). If $\varphi(0) > 0$, then the solution $\varphi(x)$ diverges upwards and will never satisfy the boundary condition at infinity. This follows from $\varphi'(0) = 0$ and

$$\varphi'' = \frac{2}{\beta}(N + 2 N_o \sinh\varphi),$$

since the curvature always remains positive.

On the other hand, if $\sinh \varphi(0) < - 1/2 N_o$, then the solution $\varphi(x)$ diverges downwards. Since $\varphi''(0) < 0$ in this case and $\varphi(\infty) = 0$, there must be a smallest value x_o where $\varphi'(x_o) = 0$ and $\varphi''(x_o) \geq 0$. However, for $0 \leq x \leq x_o$, we have $\varphi'(x) \leq 0$, i.e., $E(x) \geq 0$, and since $B(x) \leq 0$ holds everywhere, we conclude that $N'(x) \leq 0$ in that interval. Therefore $N(x_o) \leq 1$, and $\varphi''(x_o) < 0$, which leads to a contradiction. The initial guess must therefore satisfy the inequalities

$$- 1/2 \, N_o \leq \sinh \varphi(0) \leq 0 \qquad (4.1)$$

A similar, but somewhat more involved consideration shows that $N''(0)$ cannot be positive. In the cases where $t_o\beta \leq 1$ this leads to the sharper upper bound

$$\sinh \varphi(0) \leq -1/2\, N_o + t_o \beta/2\, N_o .$$

For the Example 1 this restricts the range of the initial guess quite narrowly to

$$-.5 \leq \sinh \varphi(0) \leq -.475$$

which is in agreement with the numerical value in Table 2.1.

4.2 The limits of E(x) and N(x) at infinity.

There is only one boundary condition given at infinity, namely $\varphi(\infty) = 0$, but it follows from the differential equations (2.1) that

$$E(\infty) = -\varphi'(\infty) = 0$$

$$N(\infty) = 0$$

must also hold. The first relation is established by observing that $\varphi''(x)$ is uniformly bounded (cf. Rudin [12, p. 101]):

$$\varphi'' = \frac{2}{\beta}\,(N + 2\, N_o \sinh \varphi)$$

and both N and φ are uniformly bounded by assumption.

To prove the second relation, we simply show that N(x) stays below a decaying exponential function. The function B(x) is negative and monotonically decreasing, so that from a certain point x_o on, because $E(\infty) = 0$,

$$t_o(B - E) < -a^2$$

for some nonvanishing constant a. Now

$$N(x) = N(x_o)\ \exp \int_{x_o}^{x} t_o\,(B-E)\, d\xi < Ce^{-a^2 x}$$

for all $x > x_o$, which proves the assertion.

4.3 The behavior of the solution for large argument.

From the Table 2.1 it is apparent that the subintervals remain of about the same length throughout; in other words, the basic difficulty persists, even when the functions N, E and φ have become small and B has nearly reached its asymptotic value B_*. However, for small functions the system should behave much better. It is therefore worthwhile to take advantage of this fact and investigate the linearized system

$$b' = -2\,t_o N$$
$$N' = t_o\, B_* N$$
$$\varphi'' = \frac{2}{\beta}\,(N + 2\,N_o \varphi)$$

where $b(x) = B(x) - B_*$ is a small function.

The bounded linearized solutions for N and φ are then

$$N(x) = A \exp\,(t_o\, B_* x)$$
$$\varphi(x) = C \exp\,(-2\sqrt{N_o/\beta}\; x) + D \exp\,(t_o\, B_* x)$$

where the coefficient A is undetermined and

$$D = 2A\,/(\beta\; t_o^2\, B_*^2 - 4\,N_o).$$

There are two decaying exponential functions appearing in $\varphi(x)$, but for large x only the term which decays more slowly needs to be retained. For the parameter values used for the results in Table 2.1 $2\sqrt{N_o/\beta} > -t_o B_*$, and hence the first term in $\varphi(x)$ is dropped. For different parameter values t_o, β, and N_o the situation might be reversed.

The functions E and b then behave for large x like

$$E(x) = -\,t_o B_* D \exp\,(t_o B_* x)$$

and

$$b(x) = -\,(2A/B_*) \exp\,(t_o B_* x).$$

The solution should therefore tend to the constant ratios

$N(x) : \varphi(x) : E(x) : b(x) = 1 : (-.5025) : (-1.017) : (.9884)$,

and this is indeed confirmed by the numerical results in Table 2.1 to within 1% for $x > 30$. The coefficient A turns out to be $A = 4.29$ in this case.

4.4 A first integral of the system.

A first integral of a system of differential equations expresses a conservation law, and therefore its form is often suggested by the physical interpretation of the problem. Because of the simplicity of Example 1, a first integral can, in this case, be obtained by inspection. From eq. (2.1a) we have

$$NB = -\frac{1}{2t_o} BB'$$

and from eqs. (2.1c) and (2.1d)

$$NE = -\frac{\beta}{2} EE' + 2N_o \varphi' \sinh \varphi ,$$

so that eq. (2.1b) contains derivative terms only

$$N' = t_o \left(-\frac{1}{4t_o} B^2 + \frac{\beta}{4} E^2 - 2 N_o \cosh \varphi \right)' .$$

By integrating and taking into account the conditions at infinity, we obtain the desired result

$$4N(x) + B^2(x) - \beta t_o E^2 (x) + 8t_o N_o \cosh \varphi (x) = B_*^2 + 8 t_o N_o .$$

This permits us to establish a relation between the final value B_* and the missing initial condition $\varphi (0)$, namely

$$B_*^2 = 4 + 8 t_o N_o (\cosh \varphi (0) - 1).$$

First of all, we observe that the inequality $B_* \leq -2$ always holds. Furthermore, from the numerical results (cf. Fig. 2.1) we conclude that the numerical integration needs only to be carried out over a small interval ($x < 6$, let us say) in order to determine $\varphi (0)$, and therefore also B_*, very accurately. In the approximations discussed in section 4.3, all the quantities are then known, except for the coefficient A, which must be determined from an interval where the numerical solution and the approximation overlap.

4.5 Expansion in terms of a suitable parameter.

Sensitive problems must often be solved for many sets of parameters. For extreme parameter values it might then be possible to obtain approximate solutions by an expansion with respect to a suitable combination of these values. This possibility exists, for instance, in the case of eqs. (2.1), if N_o is assumed to be large. Without going into details, it can be shown that the proper combination turns out to be $\beta t_o / 4N_o$, which is quite small, namely 1/800, in the example in section 2.

After some straightforward calculations, one obtains the first two terms of the result, for instance for B,

$$B(x) = -2 \tanh t_o x - (t_o/4N_o)\{\tanh t_o x + (t_o x - 2 \tanh t_o x) \cosh^{-2} t_o x\}$$

Even the first term alone furnishes an accuracy of about 1% throughout the entire range $[0, \infty)$, and the two terms an accuracy of about 1/10%.

5. Explanation of the Difficulty in the Shooting Method

Nonlinear boundary value problems are often so complicated that the cause for the difficulty in the shooting method is hard to pinpoint. But sometimes the problem can be drastically simplified and still show the typical behavior of a sensitive problem. This is indeed the case for the system (2.1): If we set $N(x) = 0$, then we obtain for $\varphi(x)$ an equation of the form

$$y_{xx} = a \sinh by \qquad (5.1)$$

Incidentally, this equation also appears in simplified versions of other sensitive problems, for instance in a problem investigated by Weibel [17].

First, we will determine the type of singularity that the solution of this equation possesses, and then also estimate the position of the singularity. As initial conditions we choose, for simplicity, $y(0) = 0$ and vary the slope at the origin $y_x(0) = s$. For $s = 0$ the solution is $y(x) = 0$ and has no singularity. But for reasonably large values of a and b even a very small slope s will lead to a singularity close to the origin, beyond which the standard shooting process cannot be carried out.

If eq. (5.1) is multiplied by y_x and integrated, we obtain

$$y_x = ((2a/b)(\cosh by - 1) + s^2)^{1/2}$$

or

$$x = \int_0^y ((2a/b)(\cosh b\eta - 1) + s^2)^{-1/2} \, d\eta$$

In order to simplify the analysis we perform a change of variables

$$z = \exp(by/2),$$

and hence

$$x = (2/b) \int_1^z ((a/b)(\zeta^4 - 2\zeta^2 + 1) + s^2\zeta^2)^{-1/2}\, d\zeta \qquad (5.2)$$

The singularity occurs at $x = x_s$ where z becomes infinite

$$x_s = (2/b) \int_1^\infty ((a/b)(\zeta^4 - 2\zeta^2 + 1) + s^2\zeta^2)^{-1/2}\, d\zeta .$$

The character of the singularity, i.e., the behavior of z (or y) near the singular point, is then obtained from

$$x_s - x \doteq (4/ab)^{1/2} \int_z^\infty \zeta^{-2}\, d\zeta ,$$

where only the leading term in the integrand for large z has been retained. It follows therefore that

$$z \doteq \frac{2}{\sqrt{ab}\,(x_s - x)}$$

and

$$y \doteq \frac{1}{b} \log \frac{4}{ab\,(x_s - x)^2} .$$

This shows that in Example 1 (cf. eqs. (2.1 c, d)) we may expect $\varphi(x)$ to have a logarithmic singularity,

$$\varphi(x) \doteq \log \frac{4}{N_o\,(x_s - x)^2} .$$

A change of variables obviously also changes the type of the singularity; for instance the variable $x_4 = \exp(\varphi)$ in the system (2.2) then has a second order pole, namely

$$x_4 \doteq \frac{\beta}{N_o (x_s - x)^2} .$$

For the position x_s of the singularity we try to determine an upper bound (or a good approximation to it) as the point beyond which the numerical integration cannot proceed. In order to accomplish this, we increase the integrand (i.e., we decrease the expression in parentheses) and obtain for the upper bound x_s^*

$$x_s < x_s^* = (2/b) \int_1^\infty (\zeta + 1)^{-1} ((a/b)(\zeta - 1)^2 + s^2/4)^{-1/2} \, d\zeta$$

$$= (2/b)(4a/b + s^2/4)^{-1/2} \left(\sinh^{-1}\left(\frac{4\sqrt{a/b}}{s} \right) + \sinh^{-1}\left(\frac{s}{4\sqrt{a/b}} \right) \right)$$

Without loss of generality, we have restricted ourselves to the case $s > 0$. If the initial slope s is small, then a good approximation for x_s, although no longer an upper bound, is therefore

$$x_s^* \doteq \frac{1}{\sqrt{ab}} \sinh^{-1}\left(\frac{4\sqrt{a/b}}{s} \right) \doteq \frac{1}{\sqrt{ab}} \log \left(\frac{8\sqrt{a/b}}{s} \right)$$

(cf. the exercise in Keller's book [6, p. 70]). Since the regular shooting method requires the singularity to lie outside the interval of integration, this may impose severe restrictions on the permissible range of the initial guess s

$$s = y_x(0) < 8 \sqrt{a/b} \exp(-\sqrt{ab}\, x_s).$$

Although these results do not apply directly to Example 1 because of the different boundary conditions, nevertheless with $a = 4N_o/\beta$, $b = 1$, and $s = 5.10^{-8}$ the formula gives

$$x_s^* \doteq 7,$$

which is surprisingly close to the numerically computed position of the singularity (cf. Fig. 2.1). In another problem in plasma physics by Weibel [17] physically realistic numbers are $\sqrt{ab} = 70$ for an integration interval $[0,1]$, so that an accuracy of

$$s \sim \exp(-70)$$

would be required in order to integrate across the interval.

The problem (5.1) is more fully discussed by Stoer and Bulirsch [14], where the closed form solution in terms of elliptic functions is given. It can be observed from eq. (5.2) that a solution of this type should indeed be possible. The reference [14] also contains further illuminating details about the problem.

6. The Second Example

6.1 The problem statement.

The example to be considered in this section is based on a problem in steady-state one-dimensional heat conduction with radiation. The original engineering problem was put into the mathematical form of a two point boundary value problem by Moore [8]:

$$y'' = \beta^2 (y^4 - \tau^4), \qquad 0 \leq x \leq 1 \tag{6.1}$$

with the boundary conditions

$$y(0) = 1, \quad y(1) = \tau,$$

where the constant τ lies between 0 and 1. Since y is a temperature it must remain positive for a physically meaningful solution. For large values of β (values as large as $\beta = 30$ can occur in a physically realistic situation) the problem turns out to be sensitive, so that the regular shooting method is not feasible. However, it so happens that the only result desired is essentially the initial slope, namely $\rho = y'(0)/\beta$. As we will see, this feature makes the problem particularly well suited for the shooting-splitting method.

The form of the differential equation (6.1) suggests that, for the numerical computation, the parameter β be absorbed into the independent variable by introducing

$$\xi = \beta x,$$

so that the problem simplifies to

$$\frac{d^2y}{d\xi^2} = y^4 - \tau^4, \qquad 0 \leq \xi \leq \beta \tag{6.2}$$

$$y(0) = 1, \quad y(\beta) = \tau.$$

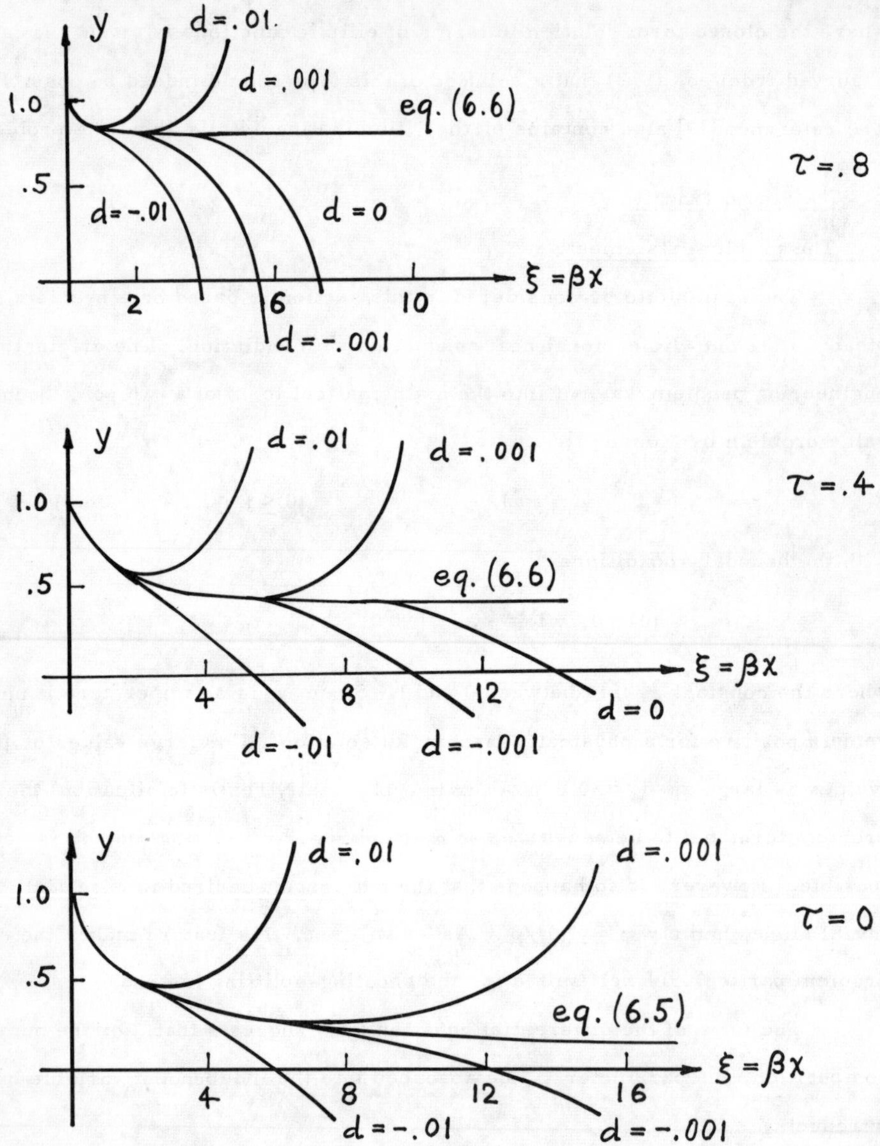

Fig. 6.1. Numerical integration of eq. (6.2) with $y'(0) = \rho_\infty + d$, for $\tau = .8, .4, 0$.

The required result (the heat flow into the system) is then

$$\rho = y'(0),$$

where the dash denotes differentiation with respect to ξ. This is just the missing initial condition for the shooting process. If the differential equation is integrated with a guess ρ which is even slightly in error, then a strongly diverging solution is obtained. This behavior is clearly shown in Figs. 6.1. The figures also give an indication for what values of β the endpoint $\xi = \beta$ can be reached without encountering the singularity, so that the standard shooting method would be applicable.

It is feasible in this simple example to analyze the problem a little more closely. The differential equation (6.2) is reduced to a first order equation

$$y' = \pm \left(\frac{2}{5} y^5 - 2\tau^4 y + c \right)^{1/2},$$

and in an interval where the curve $y(\xi)$ is decreasing

$$\xi(y) = - \int_1^y \left(\frac{2}{5}\eta^5 - 2\tau^4 \eta + c \right)^{-1/2} d\eta .$$

The constant c is determined from the initial condition

$$\rho^2 = \frac{2}{5} - 2\tau^4 + c,$$

and hence

$$y' = \pm \left(\frac{2}{5}(y^5 - 1) - 2\tau^4 (y-1) + \rho^2 \right)^{1/2} \qquad (6.3)$$

Since the problem becomes more difficult to solve as β gets larger, we determine the solution as β tends to infinity, i.e., the solution over an infinite interval with $y(\infty) = \tau$. This solution has a horizontal asymptote $y'(\infty) = 0$, which furnishes the initial slope, denoted by ρ_∞, as

$$\rho_\infty^2 = \frac{8}{5}\tau^5 - 2\tau^4 + \frac{2}{5} .$$

Eq. (6.3) then becomes

$$y' = -\left(\frac{2}{5}(y^5 - \tau^5) - 2\tau^4 (y - \tau)\right)^{1/2} \qquad (6.4)$$

6.2 Location and type of the singularity.

If the initial slope is chosen slightly larger than ρ_∞, then the solution diverges to plus infinity. The behavior near the singular point is determined from eq. (6.4), but in this context the positive square root must be taken. It is assumed that y is so large that only the dominant term needs to be retained:

$$y' \doteq \left(\frac{2}{5} y^5\right)^{1/2}$$

The singular behavior therefore does not depend on τ and is readily found to be

$$y(\xi) \doteq (a_o(\xi_s - \xi))^{-2/3},$$

with $a_o = 3/\sqrt{10}$. This behavior has been confirmed by the numerical computations.

The location of the singularity is more difficult to determine beforehand, although it is of greater interest than the type of the singularity, because it indicates where the integration will break down. Let us restrict ourselves to the simple case $\tau = 1$, i.e.,

$$y'' = y^4 - 1$$

where the exact solution is $y(\xi) = 1$ with $y'(0) = 0$. If the initial slope $y'(0) = d$ is positive, then eq. (6.3) becomes

$$y' = \left(\frac{2}{5} y^5 - 2y + \frac{8}{5} + d^2\right)^{1/2},$$

or with the change of variable $z = y-1$,

$$z' = \left(4z^2 + 4z^3 + 2z^4 + \frac{2}{5} z^5 + d^2\right)^{1/2}$$

$$z(0) = 0.$$

Thus, the position ξ_s of the singularity is

$$\xi_s = \int_0^\infty \left(4z^2 + 4z^3 + 2z^4 + \frac{2}{5} z^5 + d^2\right)^{-1/2} dz.$$

Since different terms of the integrand are dominant in different intervals, we divide the integral into two parts. At the same time, we replace the integrand by larger functions and obtain

$$\xi_s < \int_0^{z_0} (4z^2 + d^2)^{-1/2}\,dz + \int_{z_0}^{\infty} \left(\frac{2}{5} z^5\right)^{-1/2}\,dz.$$

A more refined analysis, although possible, is not justified for this exploratory investigation. The bound for ξ_s now depends on z_0 and takes on its minimum value approximately when $4z_0^2 = (2/5)z_0^5$ or $z_0^3 = 10$.

A straightforward calculation then leads to

$$\xi_s < \frac{1}{3} + \frac{1}{2}\log\left(2z_0 + (4z_0^2 + d^2)^{1/2}\right) - \frac{1}{2}\log d,$$

and the results in Table 6.1 show that the singularity indeed moves toward the origin rather rapidly.

TABLE 6.1 An upper bound for the position of the singularity.

d	.1	.01	.001	10^{-6}	0
$\xi_s <$	2.56	3.72	6.02	8.32	∞

From the computed results in Fig. 6.1, we conclude that the situation fortunately improves for smaller τ values.

6.3 Additional remarks.

In the case where β tends to infinity, additional information about the solution can be readily obtained. For $\tau = 0$ the analytical solution follows from eq. (6.4) and turns out to be

$$y(\xi) = (a_0\,\xi + 1)^{-2/3} \tag{6.5}$$

with $a_0 = 3/\sqrt{10}$.

Although for $\tau \neq 0$ a solution in terms of elementary functions does not exist, a linearization around $y(\infty) = \tau$ furnishes

$$y(\xi) \doteq \tau + A\,\exp(-2\tau^{3/2}\,\xi) \tag{6.6}$$

valid for large ξ, where the undetermined coefficient A has to be matched to the numerical solution.

Let us mention here that eq. (6.1) does not satisfy the condition we have stipulated for the shooting-splitting method in section 3. If the missing initial condition ρ varies from $-\infty$ to $+\infty$, the solutions do not sweep over the half-plane $x \geq 0$ with non-intersecting trajectories. Indeed, the solution becomes oscillatory around the x-axis for sufficiently large negative ρ values. The basic assumption is however met, if eq. (6.1) is replaced by

$$y'' = \beta^2 (|y| \; y^3 - \tau^4).$$

Incidentally, the original problem statement by Moore [8] considers also the alternate boundary condition

$$y(1) = 0,$$

but the basic difficulty of the problem remains the same.

7. An Example with Shrinking Intervals.

Experience with sensitive problems indicates that in those cases where the singularity is movable and could in principle be pushed out to infinity (with the proper analytical initial data) all the subintervals are of about the same length (see Table 2.1). However, it may happen that in a physical problem there is a fixed singularity at the endpoint of the integration interval. Although an expansion in the neighborhood of the singularity is then often feasible, its range of validity may in practice be quite restricted. If the shooting-splitting method is applied to such a problem, the subintervals must necessarily become shorter and shorter.

Let us consider a very simple problem of this type, namely

$$xy' + 6y = 8x^2, \quad -1 \leq x \leq 0 \qquad (7.1)$$

$$y(0) = 0.$$

We assume that we attempt to shoot from $x = -1$, where no data is prescribed. The analytical solution is, of course,

$$y(x) = x^2 + c\,x^{-6},$$

and the correct initial guess is $y(-1) = 1$. Let us now carry out the shooting-splitting method under the following (artificial) assumption: we have started the integration at $x = -1$ with arbitrary values for $y(-1)$ and have narrowed down the choice to $y(-1) = 1+c$ and $y(-1) = 1-c$ by observing the positive and negative divergence. These two branches are integrated to x_1, where the solutions differ by a preset amount d, let us say, for simplicity, $d = 128c$. It follows from the analytical solution that this will happen at

$$x_1 = -(2c/d)^{1/6} = -1/2 .$$

Splitting the initial condition at x_1 in such a way that the divergent solutions are again $\pm c$ above and below the true solution, one finds readily that the new branches differ by the above value of d at the abscissa $x_2 = -(2c/d)^{2/6} = -\frac{1}{4}$, and similarly $x_n = -1/2^n$. The intervals become, as expected, progressively shorter (cf. Fig. 7.1), but despite this difficulty, the solution can still be obtained correctly over most of the full interval. Hopefully, it can then be matched with an expansion around the singularity. A nonlinear sensitive problem of this type has been successfully solved by Glenn [4].

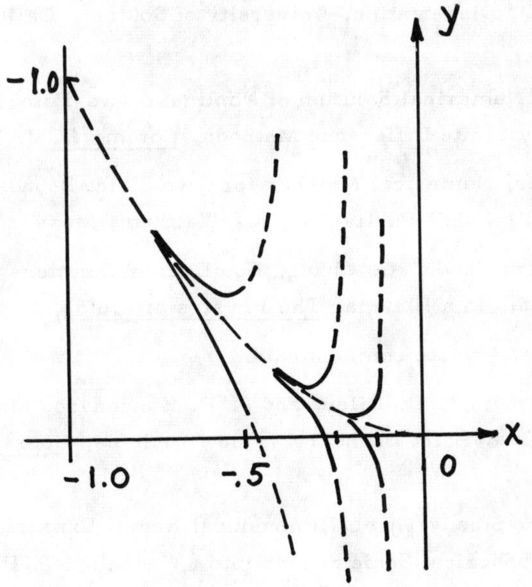

Fig. 7.1. Case of shrinking subintervals, eq. (7.1).

Acknowledgment. The authors wish to thank Professor R. H. Edwards, Dr. P.D. Weidman and Dr. F. T. Krogh for their interest and valuable discussions, and also T.M. Troesch for the programming of the second problem. They also express their gratitude to Professor R. Bulirsch, who made his unpublished results available.

References

[1] R. Bulirsch and J. Stoer, Numerical Treatment of Ordinary Differential Equations by Extrapolation Methods, Numerische Mathematik, 8, 1-13 (1966).

[2] S.D. Conte, The Solution of Linear Boundary Value Problems, SIAM Review, 8, 309-321 (1966).

[3] P. J. Firnett, A Shooting-Pinching Method for a Nonlinear Two-Point Boundary Value Problem, Aerospace Corp., Internal Memorandum TN-3 (1962).

[4] L. A. Glenn, Stationary Radial Source Flow of Liquid Particles into Vacuum, AIAA Journal, 7, 443-450 (1969), (for more details see L.A. Glenn, Ph.D. Dissertation, University of Southern California, January 1968).

[5] J. F. Holt, Numerical Solution of Nonlinear Two-Point Boundary Value Problems by Finite Difference Methods, Comm. ACM, 7, 363-373. (1964).

[6] H. B. Keller, Numerical Methods for Two-Point Boundary-Value Problems, Blaisdell Publishing Co., Waltham, Mass., 1968.

[7] T. S. Lundgren and C.C. Chang, Equilibrium Solutions of a Plane Electron Beam in a Plasma, The Physics of Fluids, 5, 807-814 (1962).

[8] J. K. Moore, private communication.

[9] D.D. Morrison, J. D. Riley, and J. F. Zancanaro, Multiple Shooting Method for Two-Point Boundary Value Problems, Comm. ACM, 5, 613-614 (1962).

[10] Y. Oida, The Steady Three-Dimensional Radial Expansion of a Viscous Gas from a Spherical Sonic Source into a Vacuum, Ph.D. Dissertation, University of Southern California, June 1971.

[11] S.M. Roberts and J. S. Shipman, Two-Point Boundary Value Problems: Shooting Methods, American Elsevier, New York, 1972.

[12] W. Rudin, Principles of Mathematical Analysis, Second Edition, McGraw-Hill Book Company, New York, 1964.

[13] T.N. Sloat, Investigation of Mass Transfer between Two Parallel Walls at Different Temperatures by a Moment Method, Ph.D. Dissertation, University of Southern California, January 1971.

[14] J. Stoer and R. Bulirsch, Einführung in die Numerische Mathematik II, Heidelberger Taschenbuch 114, Chapter 7, Springer-Verlag, Berlin, 1973.

[15] R. J. Sylvester and F. Meyer, Two Point Boundary Value Problems by Quasilinearization, Journal SIAM, 13, 586-602 (1965).

[16] B.A. Troesch, Intrinsic Difficulties in the Numerical Solution of a Boundary Value Problem, Space Techn. Labs., Internal Memorandum NN-142, (1960).

[17] E.S. Weibel, On the Confinement of a Plasma by Magnetostatic Fields, The Physics of Fluids, 2, 52-56 (1959) (cf. also Ref. 6 therein).

ON THE CONVERGENCE AND ERROR OF THE BUBNOV-GALERKIN METHOD

by

Frank Stenger

University of Utah

Salt Lake City, Utah

1. Introduction

Let H be a Hilbert space, and let A be a non-singular linear operator such that A: $D(A) \to H$, where D(A) is a dense linear subspace of H. Let N be a non-linear operator defined on a bounded subset S(N) of D(A), such that both operators, $A^{-1}NS(N)$ and $NA^{-1}C$ are compact, where C is any bounded subset of AS(N). We assume that A has a discrete spectrum, that is,

$$(1.1) \qquad Av_k = \lambda_k v_k \ , \ \lambda_k \neq \lambda_\ell \quad \text{if } k \neq \ell \ ,$$

where the sequence $\{v_k\}_{k=1}^{\infty}$ is a complete orthonormal sequence in H, and where $0 < |\lambda_1| \leq |\lambda_2| \leq \cdots$.

We consider the solution of the equation

$$(1.2) \qquad Au + Nu = 0$$

for $u \in S(N)$, by the Bubnov-Galerkin method. We thus choose a sequence $\{\phi_k\}_{k=1}^{\infty}$, where $\phi_k \in D(A)$, and where the sequence $\{\phi_k\}_{k=1}^{\infty}$ is a complete orthonormal sequence in H. Setting

$$(1.3) \qquad u_n = \sum_{k=1}^{n} a_k \phi_k \ ,$$

we determine the constants a_k by solving the system

$$(1.4)\qquad (Au_n + Nu_n\ ,\ \phi_k) = 0, \quad k=1,\ 2,\ \dots,\ n \quad .$$

We assume that the equation (1.2) has at most one solution $u_o \in S(N)$, with non-zero index ([2]) .

Let T be a unitary operator, such that

$$(1.5)\qquad Tv_k = \phi_k \ , \quad k=1,\ 2,\ 3,\ \dots,$$

and define B by

$$(1.6)\qquad B = TAT^{-1}$$

Then $B\phi_k = TAT^{-1}(Tv_k) = TAv_k = T\lambda_k v_k = \lambda_k\phi_k$, and the domain of definition of B is given by

$$(1.7)\qquad D(B) = TD(A) \quad .$$

In this paper we prove the following theorems:

Theorem 1: *If either* $(Au,u) \geq c_1 \|Au\|\ \|u\|$, $c_1 > 0$, *or else if* $(Au,Bu) \geq c_2 \|Au\|\ \|Bu\|$, $c_2 > 0$, *for all* $u \in D(A) \cap D(B)$, *where* B *is defined in* (1.6), *then the equation* (1.4) *has a unique solution* $u_n \in S(N)$ *for all* n *sufficiently large, and* $\|u_n - u_0\| \to 0$ *as* $n \to \infty$.

Theorem 2: *Let the operator* N *have a Fréchet derivative* N_u, *such that*

$$(1.8)\qquad b \equiv \sup_{u \in S(N)} \|(I + A^{-1}N_u)^{-1}\| < \infty \quad .$$

If the solution u(t) *of*

$$(1.9)\qquad Au(t) + Nu(t) = t(Au_n + Nu_n)$$

is in S(N) *for all* $t \in [0,1]$, *and if* $\|A^{-1}B\| < \infty$, *then*

(1.10) $$\|u_n - u_0\| \leq \frac{b\,\|A^{-1}B\|\ \ \|Au_n + Nu_n\|}{|\lambda_{n+1}|}$$

Remark 3: *In practice* B *is often not explicitly known, and hence a bound on* $\|A^{-1}B\|$ *may be difficult to obtain. We note however, that if* $\phi_k = v_k$, $k = 1,2,\ldots,$ *then* $B = A$, *and* $\|A^{-1}B\| = 1$. If $\|A^{-1}B\|$ is not known, then instead of (1.10) we may use

(1.10') $$\|u_n - u_0\| \leq \frac{b\,\|Au_n + Nu_n\|}{|\lambda_1|}$$

Remark 4: *If* S(N) *is taken to be a ball with center* u_0, *then by successively decreasing the radius of* S(N) *it follows that*

(1.11) $$\|u_n - u_0\| \leq \frac{(\,\|(I + A^{-1}N_{u_n})^{-1}\| + E_n)\,\|A^{-1}B\|\ \ \|Au_n + Nu_n\|}{|\lambda_{n+1}|}$$

or

(1.11') $$\|u_n - u_0\| \leq \frac{(\,\|(I + A^{-1}N_{u_n})^{-1}\| + E_n)\,\|Au_n + Nu_n\|}{|\lambda_1|}$$

where $E_n \to 0$ *as* $n \to \infty$.

In Section 2 we prove Theorem 1, in Section 3 we prove Theorem 2, and in Section 4 we apply the Bubnov-Galerkin method to the problem

(1.12) $$u'' + f(u,x) = 0 \ , \quad u(0) = u(1) = 0.$$

In particular, we use the above procedure to obtain approximate solutions and error estimates for these solutions for the problem

$$u'' + u + \frac{1}{2}u^3 - \frac{1}{2}x(1-x) = 0 \ , \ u(0) = u(1) = 0$$

which is similar to the one considered by Cesari [5]. For the case of (1.12), when

D(A) is the space of functions that are analytic in the ellipse $\underline{E}_\rho$ with foci at 0 and 1, and sum of axes equal to ρ, and $\{\phi_k\}_{k=1}^{\infty} = \{x(1-x)\ P_k(x)\}_{k=1}^{\infty}$, where the polynomials $P_k(x)$ are orthonormal over the interval (0,1) with respect to the weight function $x^2(1-x)^2$, we obtain the error estimate $\|u_n - u_0\| = 0\left((\rho-\varepsilon)^{-n}\right)$, where $0 < \varepsilon < \rho-1$. This error estimate is "essentially" best possible with regards to order, in that we cannot replace ε by 0.

In [2] (see also[1] , for a summary) Krasnosel'skii solves the problem

$$w = Mw \tag{1.13}$$

where M is a completely continuous non-linear operator, by solving $w_n = P_n M w_n$. Here $\{P_n\}_{n=1}^{\infty}$ is a sequence of projections with uniformly bounded norm ($\|P_n\| \leqq K$) and P_n is a projection of H onto a finite dimensional subspace L_n of H, where $L_1 \subset L_2 \subset \ldots$. If w_0 denotes an isolated solution of (1.13) with non-zero index, Krasnosel'skii obtains the estimate

$$\|w_n - w_o\| \leqq (1 + E_n) \|P_n w_0 - w_0\| \tag{1.14}$$

where $E_n \to 0$ as $n \to \infty$. The estimate (1.14) was shown to be incorrect by Shirali (3). In (3) Shirali obtains the estimate

$$\|w_n - w_0\| \leqq (1 + K \|(1-M_{w_o})^{-1}\| + E_n) \|P_n w_o - w_o\| \tag{1.15}$$

where M_{w_o} denotes the Fréchet derivative of M at w_o, and where $E_n \to 0$ as $n \to \infty$.

The Bubnov-Galerkin method (1.4) applied to (1.2) is more convenient in practice than that studied by Krasnosel'skii for the equation (1.13), since in order to apply Krasnosel'skii's results it is necessary to know A^{-1} explicitly, and to be able to evaluate $A^{-1}Nw$ for given w in H. Also, in practice one has available the error $\|Au_n + Nu_n\|$ and our error estimates (1.10) and (1.11) are therefore more conveniently applied than the Shirali estimate (1.15).

The error bound (1.10) is slightly sharper than that found by Cesari in (5). However, the main emphasis in (5) is to prove the existence of exact solutions, particularly, in situations where A is singular, and the index of the ensuing solution is not known, while in the present paper we assume that A is nonsingular and that an exact isolated solution of non zero index exists.

Theorems 1 and 2 are extensions of results of Dzhishkariani (4), who obtains similar results for the case when N is linear. This paper also extends the work of Varga and others (6) in that the sequence $\{u_n\}$ defined by (1.3) and (1.4) need not be a minimizing sequence of the Lagrangian corresponding to 1.2 . We mention however, from the point of view of computational considerations, that our subspaces are $S_m \equiv \text{span}\ \{\phi_1, \phi_2, \ldots, \phi_m\}$, and therefore satisfy $S_{m+1} \subset S_m$. That is, since spline subspaces do not in general satisfy this requirement unless the mesh spacing is restrictive, we have extended the work of Varga, et al (6) to splines only insofar as $S_m \subset S_{m+1}$. The procedure (1.3)-(1.4) may also be applied to integral equations (Dolph (7), Amann (8), Ikebe (14)), and to the work of Varga (3), Gordon and Hall (10), and Strang (12)--in the latter three cases, subject again to the restriction that $S_m \subset S_{m+1}$.

2. Proof of Theorem 1

Let us set

$$\text{(2.1)} \qquad \begin{cases} z = TAu \\ z_n = TAu_n \end{cases}$$

where u and u_n are defined as in (1.2) and (1.4), and let us introduce the notation

$$\text{(2.2)} \qquad \begin{cases} \psi_k = \dfrac{TA\,\phi_k}{\lambda_k} \\ g_k = T\,\phi_k \end{cases}$$

The system of functions $\{\psi_k\}_{k=1}^{\infty}$ and $\{g_k\}_{k=1}^{\infty}$ are independent and complete in H. We therefore have

$$(2.3)\qquad z_n = TA \sum_{k=1}^{n} a_k\phi_k = \sum_{k=1}^{n} b_k\psi_k,\ b_k = \lambda_k a_k$$

while (1.4) becomes

$$(2.4)\qquad (z_n + TN(A^{-1}T^{-1}z_n)\ ,\ g_k) = 0,\quad k = 1, 2, \ldots, n.$$

Let P_n and Π_n denote projections on subspaces L_n and M_n defined by the span of $\{g_k\}_{k=1}^{n}$ and $\{\psi_k\}_{k=1}^{n}$ respectively. That is, for any $f \in H$,

$$(2.5)\qquad \begin{aligned} P_n f &= \sum_{k=1}^{n} \alpha_k g_k \quad , \quad \alpha_k = (f, g_k) \quad , \\ \Pi_n f &= \sum_{k=1}^{n} \beta_k \psi_k \quad , \end{aligned}$$

where the β_k are obtained by minimizing $\| f - \sum_{k=1}^{n} \beta_k\psi_k \|$ with respect to the β_k. The system (2.4) and the original system to be solved thus become

$$(2.6)\qquad \begin{cases} P_n[z_n + TN(A^{-1}T^{-1}z_n)] = 0 \\ \\ z + TN(A^{-1}T^{-1}z) = 0 \end{cases}$$

Let us now assume without loss of generality that S(N) is a sphere with center u_o and radius r. Let $\partial S(N)$ denote the boundary of S(N), and let us define N* by

$$(2.7)\qquad N^*(u_o + tw) = \begin{cases} N(u_o + tw) \text{ if } 0 \leqq t \leqq 1 \\ \\ N(u_o + w) \ \text{ if } t > 1 \end{cases}$$

for all $u_o + w \in \partial S(N)$ and $t \in [0,\infty]$. If we replace n by N* in (1.2), then from the definition of N*, the unique invertibility of A and the complete continuity of $A^{-1}N$, we find that if the radius r of S(N) is sufficiently small, the equation $Au + N^*u = 0$ has a unique solution $u_o \in D(A)$.

Instead of the first of (2.6) we examine the system

$$P_n[z_n + TN^*(A^{-1}T^{-1}z_n)] = 0. \tag{2.8}$$

Since the second of (2.6) has a non-zero index at $z = z_o = TAu_o$, (2.8) has a unique solution $z_n \in M_n$ for all n sufficiently large. In view of the complete continuity of $N^* \circ A^{-1}$, the sequence $\{TN^*(A^{-1}T^{-1}z_n)\}$ has a convergent subsequence $\{\xi_{n_k}\}$ such that $\xi_{n_k} \to \delta_o$ as $k \to \infty$. The sequence $\{\delta_{n_k}\} = \{P_{n_k}\xi_{n_k}\}$ also converges to $\delta_o = \lim_{k\to\infty} P_{n_k}\delta_o$. Therefore

$$P_{n_k}z_{n_k} + \delta_o \to 0 \tag{2.9}$$

as $k \to \infty$.

Now we consider the auxiliary system

$$\tilde{z}_{n_k} + \Pi_{n_k}\delta_o = 0 \tag{2.10}$$

where Π_{n_k} is defined as in (2.5). Taking the projection P_{n_k} of (2.10) and subtracting from (2.9), we obtain

$$\|P_{n_k}(z_{n_k} - \tilde{z}_{n_k})\| = \|\delta_o - P_{n_k}\Pi_{n_k}\delta_o\| \to 0 \tag{2.11}$$

as $k \to \infty$

Now it is shown in [4] that if either $(Au,u) \geqq c_1\|Au\| \ \|u\|$, where $c_1 > 0$, or else if $(Au,Bu) \geqq c_2\|Au\| \ \|Bu\|$, where $c_2 > 0$, $u \in D(A) \cap D(B)$, then there is a constant $c > 0$ and independent of n such that $\|P_nu\| \geqq c\|u\|$ for all $u \in M_n$.

Since both z_{n_k} and $\tilde{z}_{n_k}$ are in M_{n_k}, (2.11) implies that

$$\|z_{n_k} - \tilde{z}_{n_k}\| \to 0 \tag{2.12}$$

as $k \to \infty$.

However, by (2.9),

$$\|\tilde{z}_{n_k} + \delta_o\| \to 0 \tag{2.13}$$

and $k \to \infty$, and (2.12) and (2.13) together imply that

$$\|z_{n_k} + \delta_o\| \to 0 \tag{2.14}$$

as $k \to \infty$, that is, $z_{n_k} \to z_o' \equiv -\delta_o$ as $k \to \infty$.

In view of the uniqueness of the solution of $Au + N^*u = 0$ in $D(A)$ and the fact that T and A have unique inverses, we must have $z_o' = z_o = TAu_o$. Hence, since $\{\xi_{n_k}\}$ was an arbitrary convergent subsequence of the sequence $\{TN^*(A^{-1}T^{-1}z_n)\}$, the sequence $\{TN^*(A^{-1}T^{-1}z_n)\}$ must itself converge to δ_o.

Since N and N^* coincide on $S(N)$, the equation (1.4) has a unique solution u_n in $S(N)$ for all n sufficiently large, and moreover $\|u_n - u_o\| \to 0$ as $n \to \infty$.

This completes the proof of Theorem 1.

3. Proof of Theorem 2

Let us set

$$\delta_n = Au_n + Nu_n \tag{3.1}$$

where u_n satisfies (1.4), and let $u(t)$ be the unique solution in $S(N)$ of the equation

$$Au(t) + Nu(t) = t\delta_n \ , \ 0 \leqq t \leqq 1.$$

Upon differentiating (3.2) with respect to t, we get

(3.) $$Au'(t) + N_{u(t)}u'(t) = \delta_n,\ 0 \leqq t \leqq 1$$

where N_u denotes the Fréchet derivative of N. If the operator $A + N_u$ is non-singular, we get

$$\|u_n - u_o\| \quad \|\int_0^1 u'(t)\,dt\| = \|\int_0^1 (I + A^{-1}N_{u(t)}{}^{-1}A^{-1}\delta_n\,dt\|$$

(3.) $$\leqq b\,\|A^{-1}\delta_n\|$$

where b is defined in (1.8) .

Now

(3.) $$\delta_n = \sum_{k=n+1}^{\infty} (\delta_n,\phi_k)\phi_k\ ;\ \|\delta_n\| = \Big(\sum_{k=n+1}^{\infty} |(\delta_n,\phi_k)|^2\Big)^{\frac{1}{2}}$$

so that, since $B^{-1}\phi_k = \lambda_k^{-1}\phi_k$, $k = 1,2,\ldots,$

(3.) $$\|B^{-1}\delta_n\| = \sum_{k=n+1}^{\infty}\left[\frac{|(\delta_n,\phi_k)|^2}{|\lambda_k^2|}\right]^{\frac{1}{2}} \leqq \frac{\|\delta_n\|}{|\lambda_{n+1}|} \; .$$

Since $\|A^{-1}\delta_n\| \leqq \|A^{-1}B\| \quad \|B^{-1}\delta_n\|$, (1.10) follows.

The inequality (1.10') follows from (3.5), since $\|A^{-1}\| = |\lambda_1|^{-1}$.
This completes the proof.

4. <u>Examples</u>

4.1 Consider the differential equation

(4.1) $$u'' + f(u,u',x) = 0,\ 0 < x < 1,$$

which is to be solved subject to the conditions

$$u(0) = u(1) = 0 \tag{4.2}$$

In this case it is convenient to take $H = L^2(0,1)$, to define A by $Au = -u'' = -\left(d^2/(dx)^2\right) u$ and N by $Nu = -f(u,u',x)$. We thus define D(A) to be the subspace of functions u in H that are twice differentiable on $[0,1]$, satisfy (4.2) and $u'' \in H$. Clearly D(A) is then a dense subspace of H. The sequence $\{v_k\}_{k=1}^{\infty} = \{2^{\frac{1}{2}}\sin k\pi x\}_{k=1}^{\infty}$ is then a complete orthonormal sequence in H, for which $Av_k = k^2\pi^2 v_k$. In solving (4.1) by the Bubnov-Galerkin method it is often less convenient to take $\phi_k = v_k$ than to take $\phi_k = x(1-x)P_k$ where P_k is a polynomial of degree k-1, and the sequence $\{P_k\}_{k=1}^{\infty}$ is constructed such that

$$(\phi_k,\phi_\ell) = \int_0^1 x^2(1-x)^2 P_k(x)P_\ell(x)\, dx = \delta_{k\ell} .$$

The sequence $\{\phi_k\}_{k=1}^{\infty}$ is then also a complete orthonormal sequence in H, and D(A) = D(B), where B is defined in (1.6) .

If the pair of equations (4.1) and (4.2) have a unique isolated solution $u_0 \in D(A) \cap S_\delta(N)$, where* $S_\delta(N) = \{u \in H : \|u - u_0\| < \delta\}$, $\delta > 0$, and if $N(u) = -f(u,u',x)$ is in H for all u in $S_\delta(N)$, then $A^{-1}N$ is completely continuous in $S_\delta(N)$, where for $h \in H$, $A^{-1}h$ is given by

$$(A^{-1}h)(x) = \int_0^1 G(x,t)h(t)dt \quad . \tag{4.3}$$

In (4.3) the Green's function G(x,t) is defined by

$$G(x,t) = \begin{cases} t(1 - x) & \text{if } 0 \leqq t \leqq x \\ x(1 - t) & \text{if } x \leqq t \leqq 1 \end{cases} . \tag{4.4}$$

Furthermore, if the functions $f_u(u,u',x)$, $f_{u'}(u,u',x)$ and $f_x(u,u',x)$ are in H for all $u \in S_\delta(N) \cap D(A)$, the operator $N \circ A^{-1}$ is completely continuous on $A\left(S_\delta(N) \cap D(A)\right)$.

*In the notation of the preceding theory, $S(N) = S_\delta(N) \cap D(A)$.

Finally, if the above conditions are satisfied, and if D(A) can be suitably restricted so that

$$(Au,u) = \int_0^1 -u'' \, u \, dx = \int_0^1 (u')^2 dx \geqq c \; \|u''\| \; \|u\|$$

for some constant $c > 0$, then it follows from Theorem 1 that the Bubnov-Galerkin method applied to (4.1) and (4.2), with $\phi_k(x) = x(1-x)P_k(x)$, $k = 1,2,\ldots$, converges to u_0.

If $f(u,u',x)$ is differentiable with respect to u and u', and if we can obtain a bound on the solution v of the equation

$$v(x) = \int_0^1 G(x,t)\left(f_u(u_n,u_n',t)\, v(t) + f_{u'}(u_n,u_n',t)\, v'(t)\right)dt - h(x) \tag{4.5}$$

in the form $\|v\| \leqq \gamma \; \|h\|$, where $h \in H$, and if $u_n = \sum_{k=1}^{n} a_k\phi_k$ denotes the n'th Bubnov-Galerkin approximation, then using (1.10')

$$\|u_n - u_0\| \leqq \frac{\gamma}{\pi^2} \; \|Au_n + Nu_n\| \; \left(1 + o(1)\right) \tag{4.6}$$

as $n \to \infty$.

4.2 Rate of Convergence of the Bubnov-Galerkin Method

We shall here obtain an estimate of the rate of convergence of the Bubnov-Galerkin method for the problem

$$u'' + f(u,x) = 0 \; , \; u(0) = u(1) = 0 \tag{4.7}$$

Let $\underline{E}_\rho$ denote the ellipse of complex numbers with foci at 0 and 1, and sum of axes equal to ρ. Let $f(u,x)$ be real when u and x are real, and let (4.7) have a unique solution $u_0 \in S(N)$, where $S(N) = \{u$ analytic in $\underline{E}_\rho$: $u(0) = u(1) = 0$ and $\|u - u_0\| < \delta\}$, for some $\delta > 0$. We assume also that $f(u,x)$ is an analytic function of x in $\underline{E}_\rho$ for all $u \in S(N)$. Furthermore, we assume that for all real u in S(N) and $x \in \left(0,1\right)$,

(4.8) $$f_u(u,x) \geqq -\gamma^2 > -\pi^2 \quad .$$

Theorem 5: *If the above conditions are satisfied, and if either* $\phi_k(x) = x(1-x)P_k(x)$ $k = 1,2,3,\ldots$, *or else if* $\phi_k(x) = \sqrt{2}\sin k\pi x$, $k = 1,2,3,\ldots$, *then the system of equations*

$$\int_0^1 \left(u_n''(x) + f(u_n(x),x)\right) \phi_k(x)dx = 0 \ , \ k = 1,2,3,\ldots, n$$

has a solution u_n *in* $S(N)$ *for all* n *sufficiently large, such that for any* $\varepsilon \in (0,\rho-1)$

(4.9) $$\|u_n - u_0\| = 0\left((\rho-\varepsilon)^{-n}\right)$$

as $n \to \infty$. *Moreover, we cannot take* $\varepsilon=0$ *in* (4.9).

Proof: We shall give a proof only for the case when $\phi_k(x) = x(1-x)P_k(x)$; the proof for the case of $\phi_k(x) = \sqrt{2}\sin k\pi x$ is similar, and we omit it.

Under the conditions of the theorem, it follows that $u_0(x)$ has the expansion

(4.10) $$u_0(x) = x(1-x)\sum_{k=1}^{\infty} a_k{}^*P_k(x)$$

where

(4.11) $$a_k{}^* = \int_0^1 t^2(1-t)^2\, u_0(t)\, P_k(t)dt,$$

where the P_k are defined as in Section 4.1, and where the series on the right of (4.10) converges for all complex x in $\underline{E}_\rho$. It may be shown that (Freud[11, p. 123]) for every complex x,

(4.12) $$\limsup_{n\to\infty} \frac{|P_n(x)|^{1/n}}{|2x + \sqrt{(2x-1)^2-1}|} = 1$$

and therefore

$$(4.13) \qquad a_n{}^* = 0\left[(\rho-\varepsilon)^{-n}\right]$$

as $n \to \infty$ for every $\varepsilon \in (0,\rho-1)$. Consequently, if we set

$$(4.14) \qquad u_n{}^* = x(1-x) \sum_{k=1}^{n} a_k{}^* P_k(x) \quad ,$$

then

$$(4.15) \qquad \|u_n{}^* - u_0\| = 0\left((\rho-\varepsilon)^{-n}\right) \ , \quad \|u_n{}^{*\prime} - u_0'\| = 0 \ (\rho-\varepsilon)^{-n})$$

as $n \to \infty$.

The condition (4.8) ensures that (Ciarlet, Schultz and Varga [6, p. 396]) for n sufficiently large, u_n minimizes the functional

$$(4.16) \qquad F(u) = \int_0^1 \{\frac{1}{2} u'(x)^2 + \int_0^{u(x)} f(y,x)dy\} \ dx$$

over all real u in S(N), of the form

$$(4.17) \qquad u = x(1-x) \sum_{k=1}^{n} b_k P_k(x)$$

and that $F(u_0)$ is the minimum over all real u in S(N) of F(u). Furthermore, one has the relations

$$(4.18) \qquad \begin{aligned} F(u_n) - F(u_0) &= \int_0^1 \left(\frac{1}{2}(u_n'-u_0')^2 + \int_{u_0}^{u_n} f(y,x)dy\right)dx \\ &\geq \frac{\pi^2-\gamma^2}{2} \int_0^1 (u_n-u_0)^2 dx. \end{aligned}$$

From (4.18) it follows that

$$\|u_n - u_0\|^2 \leqq c\,|F(u_n) - F(u_0)|$$

$$\leqq c\,|F(u_n^*) - F(u_0)| \tag{4.19}$$

$$\leqq \frac{c}{2}\int_0^1 (u_n^{*\prime} - u_0')^2 dx + \frac{K}{2}\int_0^1 (u_n^* - u_0)^2 dx$$

where $c = 2/(\pi^2-\gamma^2)$, and K is the maximum over all real u in S(N) and x in $[0.1]$ of $f_u(u,x)$. That is, using (4.15)

$$\|u_n - u_0\| = 0\left((\rho-\varepsilon)^{-n}\right)$$

as $n\to\infty$. We cannot take $\varepsilon = 0$ in (4.9) as the problem

$$u'' = \sum_{k=1}^{\infty} \frac{k^5}{\rho^k}\left[x(1-x)P_k(x)\right]'' , \quad u(0) = u(1) = 0 \tag{4.20}$$

clearly shows.

4.3 Let us apply the above results to the equation

$$u'' + u + \tfrac{1}{2}u^3 + \tfrac{1}{2}x(1-x) = 0 , \quad u(0) = u(1) = 0 ,$$

which is a variant of the equation considered by Cesari [5]. As above, we take

$$A\,u = -u'' , \quad Nu = -\left[u + \frac{1}{2}u^3 + \frac{1}{2}x(1-x)\right]$$

so that $N_u(v) = -\left[1 + (3/2)u^2\right] v$. Setting $u_1 = \alpha\phi_1(x) = ax(1-x)$, the equation $(Au_1 + Nu_1, \phi_1) = 0$, yields the following equation for a:

$$\frac{1}{2}\frac{\Gamma(5)^2}{\Gamma(10)}\, a^3 - \frac{3}{10}\, a + \frac{1}{2}\frac{\Gamma(3)^2}{\Gamma(6)} = 0$$

which has a solution $a \cong .111$. Given $h \in H$, the equation $v + A^{-1}N_{u_1} v = h$ has a solution $v \in D(A)$, which satisfies

$$\|v\| \cdot \left(1 - \frac{1}{\pi^2} \|Nu_1\|_\infty\right) \leqq \|h\| \quad ,$$

or, since $\|Nu_1\|_\infty = \sup_{(x \in (0,1))} \left([1 + \frac{3}{2} u_1^2(x)]\right) \leqq 1.0006$,

$$\|v\| \leqq \frac{\pi^2}{\pi^2 - 1.0006} \|h\| \cong 1.11 \|h\| \quad .$$

Hence $\|A^{-1}v\| \leqq (1.11/\pi^2) \|h\| \leqq .113 \|h\|$

Next, we find that $\int_0^1 [Au_1 + Nu_1]^2 dx \cong .02$ which, by (4.6) yields the error estimate

$$\|.111\, x(1-x) - u_0\| \leqq 0.016 \tag{4.21}$$

Similarly,

$$\|.0202 \sin \pi x - u_0\| \leqq 0.005 \quad . \tag{4.22}$$

Although these estimates suggest that the approximation .0202 sin πx is more accurate than .111 x(1-x), this may not actually be the case, since in (4.21) we used the bound (1.10'), while in (4.22) we used (1.10) .

References

1. M.A. Krasnosel'skii, "On some problems of non linear analysis," Usp. Math Nauk v.9 No 3 (1959). English translation in Amer. Math. Soc. Translations, Series 2 V. 10, pp. 345-409.

2. M.A. Krasnosel'skii, "Topological Methods in the theory of non-linear integral equations"(Gostekhteoretisdat, Moscow 1956). English translation by Pergamen Press, Oxford 1964.

3. S. Shirali, "A note on Galerkin's method for nonlinear equations," Aeq. Math. 4 (1969) 198-200.

4. A.V. Dzhishkariani, "On the Bubnov-Galerkin method," Zh. vȳchisl. Math.mat. Fiz. 7,6 (1967) 1398-1402. English translation in Comput. Math. and Math. Phys. (1969) 251-257.

5. L. Cesari, "Functional analysis and Galerkin's method,"Mich. Math. Jour. II (1964) 385-414.

6. P.G. Ciarlet, M.H. Schultz, and R.S. Varga, "Numerical methods of high order accuracy for non-linear boundary value problems,"I. One dimensional problem, Numer. Math. 9 (1967) 394-430.

7. C.L. Dolph, "Non-linear integral equations of the Hammerstein type," Trans. Amer. Math. Soc. 66 (1949) 289-307.

8. H. Amann, "Zum Galerkin-Verfarhren fur die Hammersteinsche Gleichung," Arch. Rat. Mech. Anal. 35 (1969) 114-121.

9. P. Davis, "Interpolation and Approximation." Blaisdell, N.Y. (1963)

10. W.J. Gordon and C.A. Hall, "Discretization error bounds for transfinite elements," General Motors Research Report.

11. G. Freud, Orthogonal Polynomials, Pergamen Press, Oxford (1971).

12. G. Strang, "Approximation in the Finite Element Method," Numer. Math. 19 (1972) 81-98.

13. R.S. Varga, "Functional Analysis and Approximation Theory in Numerical Analysis," SIAM Regional Conference Series in Applied Mathematics, No. 3, Philadelphia (1971)

14. Y. Ikebe, "The Galerkin method for the numerical solution of Fredholm integral equations of the second kind," SIAM Review, 14 (1972) 465-9.

NUMERICAL INTEGRATION OF GRAVITATIONAL N-BODY SYSTEMS WITH THE USE OF EXPLICIT TAYLOR SERIES

by

Myron Lecar, Rudolf Loeser, and Jerome R. Cherniack

Smithsonian Institution

Astrophysical Observatory

Cambridge, Massachusetts

Abstract

The Newtonian equations of motion of N bodies are integrated by expanding the positions and velocities in Taylor series in the time. In the method presented here, the coefficients of the Taylor series (i.e., the time derivatives of the accelerations) are given analytically as explicit functions of the current positions and velocities. This method uses no past information and thus is tailored for variable time steps. We compare the efficiency of this explicit scheme with that of a differencing scheme (which also uses variable time steps) for a variety of solar-system configurations. When the time step changes rapidly, as in an elliptic orbit of high eccentricity, the longer time required to compute the explicit coefficients is more than compensated by the ability to take longer time steps.

1. Introduction

We describe a method of integrating the gravitational N-body problem using explicit Taylor series. The equation of motion of the ith body of a system of N gravitationally interacting bodies is

$$\ddot{\underline{r}}_i = G \sum_{j=1}^{N\prime} \frac{m_j \underline{r}_{ij}}{r_{ij}^3} , \tag{1}$$

where $\underline{r}_{ij} = \underline{r}_i - \underline{r}_j$ and $r_{ij} = |\underline{r}_{ij}|$. From the values of the r_i's and v_i's at time t, their values a t + Δ are given by

$$\underline{r}_i(t+\Delta) = \underline{r}_i(t) + \underline{v}_i(t)\cdot\Delta + \sum_{k=0}^{n-2} \underline{a}_i^{(k)}(t)\cdot\frac{\Delta^{k+2}}{(k+2)!}\ , \tag{2}$$

$$\underline{v}_i(t+\Delta) = \underline{v}_i(t) + \sum_{k=0}^{n-2} \underline{a}_i^{(k)}(t)\cdot\frac{\Delta^{k+1}}{(k+1)!} \tag{3}$$

where n is the order of integration.

Expressions for the acceleration and its first two time derivatives are given below.

$$\underline{a}_i^{(n)} = \sum_j \underline{a}_{ij}^{(n)}\ ,$$

$$\underline{a}_{ij}^{(0)} = \frac{Gm_j\,\underline{r}_{ij}}{r_{ij}^3}\ , \tag{4}$$

$$\underline{a}_{ij}^{(1)} = Gm_j\left[\frac{\underline{v}_{ij}}{r_{ij}^3} - \frac{3\underline{r}_{ij}}{r_{ij}^5}\cdot(\underline{r}_{ij}\cdot\underline{v}_{ij})\right]\ ,$$

$$\underline{a}_{ij}^{(2)} = Gm_j\left\{\frac{\underline{A}_{ij}^{(0)}}{r_{ij}^3} - \frac{6\underline{v}_{ij}}{r_{ij}^5}\ (\underline{r}_{ij}\cdot\underline{v}_{ij}) - 3\,\frac{\underline{r}_{ij}}{r_{ij}^5}\cdot\left[\left(\underline{r}_{ij}\cdot\underline{A}_{ij}^{(0)}\right) + (\underline{v}_{ij}\cdot\underline{v}_{ij}) - 5\frac{(\underline{r}_{ij}\cdot\underline{v}_{ij})^2}{r_{ij}^2}\right]\right\}\ ,$$

where $\underline{v}_{ij} = \underline{v}_i - \underline{v}_j$ and $\underline{A}_{ij}^{(0)} = \underline{a}_i^{(0)} - \underline{a}_j^{(0)} \neq \underline{a}_{ij}^{(0)}$. The expressions for the higher derivatives become increasingly complex, and the derivation of the expression for $\underline{a}_{ij}^{(4)}$ exhausted our algebraic skills. To check our algebra and to evaluate $\underline{a}_{ij}^{(5)}$, we appealed to the algebraic manipulation code SPASM (Hall and Cherniack, 1969). SPASM is described in Appendix A, which also contains the expressions for the first five time derivatives of the acceleration.

Our integration code ICARUS chooses individual and variable time steps for each body according to the following algorithm:

(5) $$\Delta_i = \varepsilon\omega_i^{-1} \quad , \quad \omega_i = \min\,(\Omega_i, \nu_i) \quad ,$$

(6) $$\Omega_i = \min_j \left\{ \left[\frac{G(m_i + m_j)}{r_{ij}^3} \right]^{1/2} \right\} \quad ,$$

(7) $$\nu_i = \min_j \left(\frac{\underline{r}_{ij} \cdot \underline{v}_{ij}}{r_{ij}^2} \right) \quad .$$

With this choice of time step, the relative error of the energy using an nth-order integrator is approximately ε^n. As the time step is proportional to ε, the ratio of the time step of an (n + 1)-order integrator to that of an (n)-order integrator is $\varepsilon^{-1/n+1}$ to integrate to the same accuracy. In this study, we used a 6th-order integrator. With a typical value of $\varepsilon = 0.1$, a 7th-order integrator could use time steps about 1/3 longer. An examination of the expression for $\underline{a}_{ij}^{(5)}$ (see Appendix A) suggested that the computation of that term would consume that extra time.

In general, the bodies are not synchronized, so that at time t_i, when body i is to be integrated to $t_i + \Delta_i$, the coordinates of the other bodies are not available. They are available at t_j, where $t_j < t_i < t_j + \Delta_j$. They are extrapolated to t_i by using the Taylor coefficients at t_j. The equations for extrapolating $\underline{r}$ and $\underline{v}$ are the same as those for integration (Eqs. 2 and 3). In addition, the $\underline{a}_i^{(k)}$ for $k \leq n-4$ are extrapolated by using

$$\underline{a}_i^{(k)}(t + \Delta) = \sum_{\ell=0}^{n-(k+1)} \underline{a}_i^{(\ell+k)}(t)\frac{\Delta^\ell}{\ell!} \quad .$$

However, in extrapolation, the $\underline{a}_i^{(k)}$ are not recomputed, so extrapolation is approximately N-times faster than integration.

The lack of synchronization and the consequent need for a decision whether to integrate or extrapolate require ICARUS to do a fair amount of bookkeeping. The logic and flow of ICARUS are described in Appendix B.

Because the calculation of each derivative of the acceleration requires reference to the remaining N - 1 bodies, while in a difference scheme this is true only for the acceleration itself, we would expect that the explicit scheme would become relatively less efficient as N increases. On the other hand, because the explicit scheme uses only current information, we would expect that the derivatives are calculated with higher accuracy. For example, if the force increases rapidly (e.g., a hard-sphere collision where the velocity is discontinuous), past information degrades rapidly and differencing becomes correspondingly less accurate. As variable time-step procedures are designed for just such problems, we would expect that for small N, the explicit scheme would be the more efficient.

This study is aimed at determining the crossover value of N --i.e., that value of N below which the explicit scheme is more efficient and above which the difference scheme is more efficient. We compare the explicit scheme (Gonzales and Lecar, 1967, and this paper) with the difference scheme developed by R. Wielen and S. Aarseth (Aarseth, 1970 and references therein). The comparison was done for simulations of configurations of solar-system type and for various numbers of bodies from 2 to 80. In these configurations, a central massive body dominates the gravitational field and determines the physical time scale (i.e., the average period of an orbiting body). Bodies close to the central one are integrated most often, and as N increases, extrapolations are 10 to 20 times more frequent than integrations. The result is that for the same physical time, the computing time increases as N rather than as N^2, as would be true for a stellar configuration. Therefore, in a stellar configuration, the difference scheme would probably become more efficient at a smaller value of N.

2. A Comparison between the Explicit and Difference Schemes for a Two-Body Problem

We integrated two-body elliptic orbits of very high eccentricity with both the explicit and the difference schemes. The eccentricities were $1 - 10^{-4}$ and $1 - 10^{-6}$, so that the ratios of the time steps at apocenter and pericenter were of the order of 10^6 and 10^9. As expected, in this case of small N and rapidly varying time steps, the explicit scheme was the more efficient of the two.

The orbits were integrated for six periods in the forward time direction; the velocities were reversed, and the orbits were integrated for six periods backward in time. In Table I, for the two methods of integration, we present the value of the time-step control (ε), the relative error in the energy ($\delta E/E$), the relative error in the phase ($\delta\theta$), and the central-processor computing time (t_c). The computations were performed on a CDC 6400.

3. Comparisons between the Explicit and the Differencing Schemes for N-Body Configurations

In the first of two N-body comparisons, a massive central body (i.e., the Sun) had a mass of 0.99, a planet (i.e., Jupiter) had a mass of 0.001 and orbited the central body at a radius of 1, and the remaining bodies had equal masses whose total was 0.001, and whose radii varied from 2 to 100. These bodies were distributed out of the Sun-Jupiter plane to a distance of 10. In all cases, the problem was integrated for 100 revolutions of Jupiter. In Table II, we present the results of this comparison. We give the number of bodies (N), the time-step control (ε), the number of integration (Nint), the number of extrapolations (Next), the relative energy error ($\delta E/E$), the central-processor computing time (t_c), and the number of integrations per second (Nint/t_c).

In the second N-body comparison, the central massive body had a mass of 1.0 and each of the other bodies had a mass of 0.001. In this case, all the bodies were confined to a plane. Note that the cases with different N were not run to the same physical time. The results of this comparison are presented in Table III.

4. Conclusions

This numerical comparison confirmed our qualitative predictions and provided the necessary quantitative data to evaluate the two methods. As we expected, the explicit scheme can take longer time steps but each step takes longer to compute. And in one of the comparisons, the difference scheme did become more efficient as N increased beyond about 10.

For solar-system configurations, we prefer the explicit scheme. Typically, these problems require long, accurate integrations, and the longer time step of the explicit scheme slows the buildup of roundoff error. However, as the difference in the efficiencies of the two methods is not striking, we use the difference method when the force law is so complicated that evaluation of the explicit coefficients becomes tedious.

TABLE I

Two-Body Elliptic Orbit

	Eccentricity	ε	$\frac{\delta E}{E} \times 10^7$	$\delta\Theta \times 10^9$	t_c (sec)
Explicit	$1 - 10^{-4}$	0.01	4	6	110
Difference	$1 - 10^{-4}$	0.01	500	800	84
Difference	$1 - 10^{-4}$	0.002	9	10	435
Explicit	$1 - 10^{-6}$	0.006	300	60	255
Difference	$1 - 10^{-6}$	0.006	5,000	840	197
Difference	$1 - 10^{-6}$	0.004	1,500	250	300

TABLE II

N-Body Solar-System Configuration

	N	ε	Nint x 10^{-3}	Next x 10^{-3}	$\frac{E}{E}$ x 10^{-3}	t_c (sec)	Nint/t_c
Explicit	20	0.6	4.5	33.9	6.3	310	14.5
Difference	20	0.163	15.3	120.2	8.2	395	38.9
Explicit	40	0.62	4.9	74.7	6.3	685	7.1
Difference	40	0.163	17.2	276.3	7.7	795	21.6
Explicit	60	0.62	3.5	66.1	3.8	1125	3.1
Difference	60	0.163	11.8	274.0	3.8	1280	9.2
Explicit	80	0.6	3.3	73.9	3.2	1620	2.1
Difference	80	0.163	11.5	281.0	3.4	1820	6.3

TABLE III

N-Body Configuration

	N	ε	Nint x 10^{-3}	Next x 10^{-3}	$\frac{\delta E}{E} \times 10^3$	t_c (sec)	Nint/t
Explicit	9	0.455	1.9	6.1	2.4	60	32.2
Difference	9	0.2	4.5	14.2	2.1	63	72.0
Explicit	27	0.322	4.3	10.9	48.0	390	11.1
Difference	27	0.1	45.6	107.6	47.0	320	34.1
Explicit	54	0.643	1.4	3.4	7.5	240	6.0
Difference	54	0.2	29.7	70.0	7.2	190	18.1
Explicit	81	0.563	1.0	4.7	1.7	190	5.4
Difference	81	0.3	2.4	11.9	3.6	130	18.3

APPENDIX A: Computing the Coefficients

Given $\underline{S} = \underline{r} + \underline{v}t + \underline{a}^{(0)}t^2/2 + \underline{a}_1\ \underline{t}^3/3! + \ldots$, we seek consecutive coefficients of t in the truncated expansion of

$$\underline{Q} = \frac{\underline{S}}{(\underline{S}\cdot\underline{S})^{3/2}}$$

(In the notation of the introduction, $\underline{Q} = \underline{a}_{ij}^{(0)}/Gm_j$.)

Our tool is SPASM, a list-based, Fortran-accessible, expression-manipulating system. Representable expressions include functions, and our algorithm is simplified by taking advantage of this generality. Expressions in SPASM are nested polynomials with rational coefficients and (signed) integer exponents. The variables of these polynomials are atomic variables, names of other expressions, or names of functions. Function names point to argument lists, which are in turn lists of names of expressions. Expressions are kept in a canonical order. When a new subexpression is formed, it is converted to canonical order, compared to all other "live" subexpressions in the program, and given the name of its duplicate if one exists.

One immediate consequence of the canonical ordering and renaming conventions is that when most primitive operations (e.g., addition or multiplication of polynomials) are performed, no descent into the nest is required. Thus, primitive operations in SPASM take no longer than in any other list-based, polynomial-manipulating systems.

The other usual operations (differentiation, substitution of expression for variable, automatic truncation, run-time function definition, etc.) are provided. There also exist operations for traversing and creating nested polynomials and functions.

The only technical difficulty in the computation of $\underline{Q}$ arises from the mixture of scalars and vectors in $\underline{S}$. There are numerous techniques for overcoming this difficulty; we chose the one described below to minimize programing and debugging time.

Using the run-time function definition feature, we defined VEC as a function of one variable. Without this definition, the SPASM input routine would identify VEC(X) as the product of "VEC" and "X." Instead, we want VEC(X) to stand for $\underline{X}$.

We punched

VEC(R) + VEC(V) T + VEC(A0) T ** 2/2
+ VEC(A1) T ** 3/6 + VEC(A2) T ** 4/24 $

on consecutive cards in free format, using $ to terminate the expression.

The program read the above cards from the input stream, set automatic truncation of T to (say) 5, and squared the input. The result is

$$VEC^2(R) + 2\ VEC(R) * VEC(V) * T + \dots \quad .$$

We wanted the result to appear as

$$R^2 + 2\ IP(R,V) * T\dots \quad ,$$

where IP(R,V) means the inner product of R and V. We wrote a one-page program to do this transformation, using the features that transverse and create polynomials and functions. This was the only operation not "off the shelf."

After this transformation, vectors are either converted to scalars (e.g., R^2) or separated from scalars by parenthesis (e.g., IP(R,V)). Thus, the remaining computations were straightforwardly done without concern for the mixture of scalars and vectors:

$$(\overline{S} \cdot \overline{S})^{-3/2} = R^{-3} * (1 + 2\ IP(R,V) * T/R^2 + \dots)^{-3/2}$$

was computed by using the SPASM binomial expansion and multiply routines. The product of $\underline{S}$ and $(\underline{S} \cdot \underline{S})^{-3/2}$ was formed, without ambiguity, by the SPASM multiply routine. After some simple editing, the results were output to form Fig. A1.

Fig. A1 – Time Derivatives of the Acceleration

$$Q = a_{ij}^{(0)} / Gm_j$$

$$= r_i + V_i \, \Delta + \sum_{k=0}^{n-2} a_i^{(k)} \, \Delta^{k+2} / (k+2)!$$

```
r_i        = VEC(R)*(1/R^3)

V_i        = VEC(V)*(1/R^3)+VEC(R)*(-3/R^5*IP(R,V))

2! · a_i^(0) = VEC(A0)*(1/2/R^3)+VEC(V)*(-3/R^5*IP(R,V))+VEC(R)*(15/2/R^7*IP^2(R,V)-3/2/R^5*IP(R,A0)-3/2/R^5*V^2)

3! · a_i^(1) = VEC(A1)*(1/6/R^3)+VEC(A0)*(-3/2/R^5*IP(R,V))+VEC(V)*(15/2/R^7*IP^2(R,V)-3/2/R^5*IP(R,A0)-3/2/R^5*V^2)+VEC(R)*(-35/2/R^9
   *IP^3(R,V)+15/2/R^7*IP(R,V)*IP(R,A0)+15/2/R^7*V^2*IP(R,V)-3/2/R^5*IP(V,A0)-1/2/R^5*IP(R,A1))

4! · a_i^(2) = VEC(A2)*(1/24/R^3)+VEC(A1)*(-1/2/R^5*IP(R,V))+VEC(A0)*(15/4/R^7*IP^2(R,V)-3/4/R^5*IP(R,A0)-3/4/R^5*V^2)+VEC(V)*(-35/2/R^9
   *IP^3(R,V)+15/2/R^7*IP(R,V)*IP(R,A0)+15/2/R^7*V^2*IP(R,V)-3/2/R^5*IP(V,A0)-1/2/R^5*IP(R,A1))+VEC(R)*(315/8/R^11*IP^4(R,V)
   -105/4/R^9*IP^2(R,V)*IP(R,A0)-105/4/R^9*V^2*IP^2(R,V)+15/8/R^7*IP^2(R,A0)+15/2/R^7*IP(R,V)*IP(V,A0)+5/2/R^7*IP(R,A1)*IP(R,
   V)+15/4/R^7*V^2*IP(R,A0)+15/8/R^7*V^4-1/8/R^5*IP(R,A2)-1/2/R^5*IP(V,A1)-3/8*A0^2/R^5)

5! · a_i^(3) = VEC(A2)*(-1/8/R^5*IP(R,V))+VEC(A1)*(5/4/R^7*IP^2(R,V)-1/4/R^5*IP(R,A0)-1/4/R^5*V^2)+VEC(A0)*(-35/4/R^9*IP^3(R,V)+15/4/R^7
   *IP(R,V)*IP(R,A0)+15/4/R^7*V^2*IP(R,V)-3/4/R^5*IP(V,A0)-1/4/R^5*IP(R,A1))+VEC(V)*(315/8/R^11*IP^4(R,V)-105/4/R^9*IP^2(R,V
   )*IP(R,A0)-105/4/R^9*V^2*IP^2(R,V)+15/8/R^7*IP^2(R,A0)+15/2/R^7*IP(R,V)*IP(V,A0)+5/2/R^7*IP(R,A1)*IP(R,V)+15/4/R^7*V^2*IP(
   R,A0)+15/8/R^7*V^4-1/8/R^5*IP(R,A2)-1/2/R^5*IP(V,A1)-3/8*A0^2/R^5)+VEC(R)*(-693/8/R^13*IP^5(R,V)+315/4/R^11*IP^3(R,V)*IP(R,
   A0)+315/4/R^11*V^2*IP^3(R,V)-105/8/R^9*IP(R,V)*IP^2(R,A0)-105/4/R^9*IP^2(R,V)*IP(V,A0)-35/4/R^9*IP(R,A1)*IP^2(R,V)-105/4/R^9
   *V^2*IP(R,V)*IP(R,A0)-105/8/R^9*V^4*IP(R,V)+15/4/R^7*IP(V,A0)*IP(R,A0)+5/8/R^7*IP(R,A2)*IP(R,V)+5/2/R^7*IP(V,A1)*IP(R,V
   )+5/4/R^7*IP(R,A1)*IP(R,A0)+15/4/R^7*V^2*IP(V,A0)+5/4/R^7*V^2*IP(R,A1)-1/4/R^5*IP(A0,A1)-1/8/R^5*IP(V,A2)-1/40/R^5*IP(R,
   A3)+15/8*A0^2/R^7*IP(R,V))
```

Where VEC(A) Means **A**, IP(A,B) Means **A·B**

APPENDIX B: ICARUS

In this appendix, we describe in detail how integration based on an explicit Taylor series with variable time steps is done in our computer program ICARUS. We first discuss the case where all the bodies have mass, and then we show how massless test particles can be handled. For brevity, we use the term "marble" for a massive body and the term "ghost" for a massless test particle.

Let us call the integration time universal time (UT). The position and velocities of all marbles are given at UT = 0. To begin, from Eq. (4) we compute for every marble all the acceleration coefficients, as well as a time step, i.e., the time interval within which the Taylor series converges to the desired accuracy. We obtain a provisional time step Δ_i from Eq. (5), and then the actual time step D_i from

$$D_i = \Delta_0 \cdot f^n \quad , \tag{B1}$$

where

$$\Delta_0 \cdot f^n \leq \Delta_i < \Delta_0 \cdot f^{n+1} \quad ;$$

Δ_0 and f are input (typically, f = 2). (The case $\Delta_i < \Delta_0$ is a special situation that can be handled in several ways.) Having computed all D_i values, we select D_{min}, the smallest of them, and use it as the initial value of "marble target time" (MT).

Figure B1A shows the situation at UT = 0 for three marbles. Time increases along the ordinate; the abscissa merely serves to show each marble separately. The filled circles indicate the time for which positions, velocities, and acceleration coefficients are known; the arrows indicate the length of each marble's time step. The shortest arrow defines MT.

From Eqs. (2) and (3), we can now compute new positions and velocities at time MT. We then subtract the interval (MT - UT) from all values of D_i. This leads to $D_1 = 0$, while D_2, $D_3 > 0$. Since $D_1 = 0$, new acceleration coefficients must be computed for it, which in turn requires that acceleration coefficients for marbles 2 and 3 at time MT be computed as well. We obtain the latter by extrapolation from Eq. (8). We then calculate the new acceleration coefficients for marble 1 and a new value of D_1, set UT = MT, select D_{min}, and set MT = UT + D_{min}. This situation is shown in Fig. B1B: All positions, velocities, and acceleration coefficients are now known at the new value of UT, accelerations for marble 1 having been recomputed by considering the gravitational interaction of all marbles, and those for marbles 2 and 3 by simple extrapolation (dotted lines) without reference to other marbles.

This process continues, as shown in Fig. B1C and B1D, until we approach PT, which is that value of UT for which we want a printout of the situation. We obtain synchronization of all marbles at UT = PT by enforcing that always MT $\leq$ PT. After printing, we start afresh by setting UT = 0, RT = RT + PT, and PT = PT + DP, where RT is the "running time" of the integration, and DP is the print interval, an input parameter.

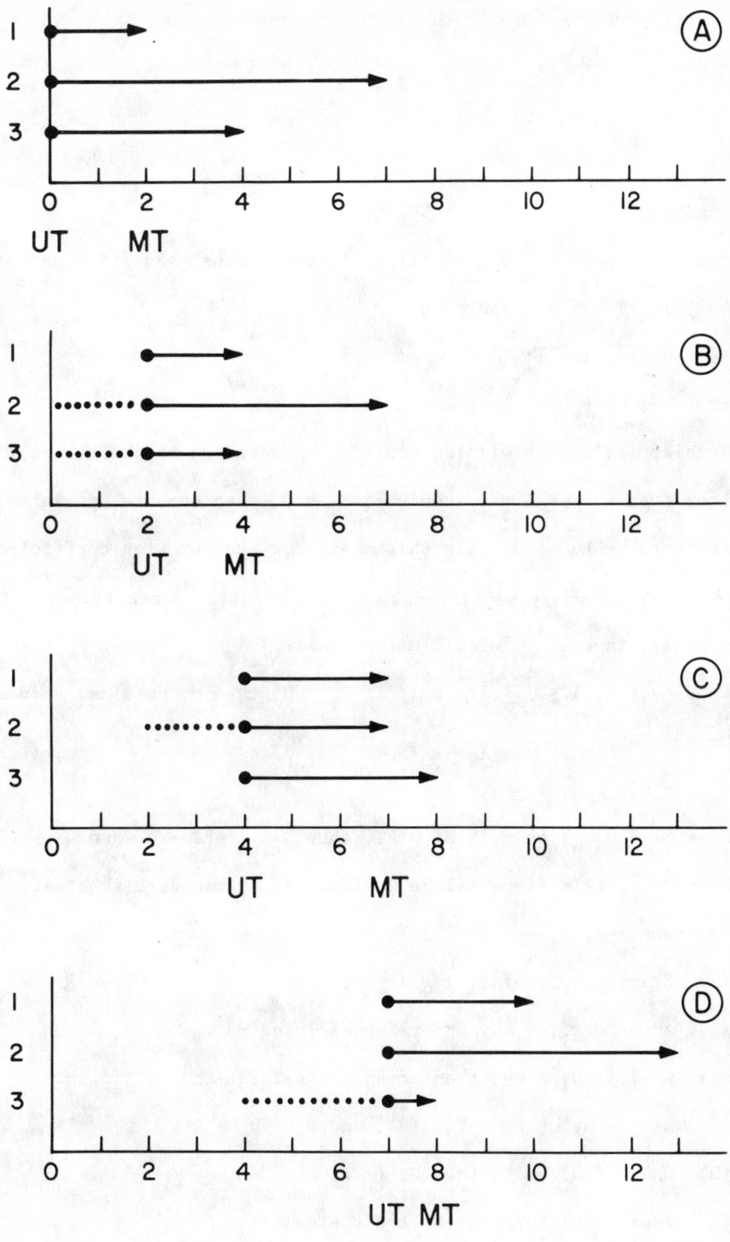

Fig. B1. Schematic representation of sequence of calculations involving marbles only.

Let us now specify precisely the steps of the algorithm. Given mass, initial positions, and velocities for I marbles, and values of ε, Δ_0, f, DP, and n (the total number of print intervals), then

1. RT ← 0; STP ← n · DP.
2. Compute all accelerations by Eq. (4).
3. Print RT, positions, velocities, accelerations, and other data.
4. Only if RT = STP, then stop.
5. $D_i \leftarrow 0$, $1 \leq i \leq I$;
   PT ← DP; UT ← 0
6. Nothing (this step will be modified to accommodate ghosts).
7. Loop over i, $1 \leq i \leq I$: only if $D_i = 0$, then compute acceleration coefficients by Eq. (4). (In this step, new acceleration coefficients are computed for every marble whose D = 0; first $\underline{a}_i^{(0)}$ are computed for all marbles; then $\underline{a}_i^{(1)}$ for all marbles; etc.)
8. Loop over i, $1 \leq i \leq I$: only if $D_i = 0$, then compute new values of D_i by Eq. (9).
9. $D_{min} \leftarrow \min D_i$, $1 \leq i \leq I$.
10. $MT \leftarrow UT + D_{min}$; only if MT > PT, then $D_{min} \leftarrow PT - UT$ and MT ← PT.
11. $D^* \leftarrow D_{min}$ (this step will be modified to accommodate ghosts).
12. Do steps A-E:
    (A) Start loop over i, $1 \leq i \leq I$.
    (B) Advance $\underline{r}_i$, $\underline{v}_i$ by D* by Eqs. (2) and (3).
    (C) Only if $D_i > D^*$, then advance acceleration coefficients of the ith marble by D* from Eq. (8), and $D_i \leftarrow D_i - D^*$.
    (D) Only if $D_i = D^*$, then $D_i \leftarrow 0$.
    (E) Repeat (B) through (D) for the next i.
13. UT ← UT + D*; only if UT < PT, then return to step 7.
14. RT ← RT + UT; return to step 3.

This algorithm can be modified quite easily to handle ghosts: New ghost positions, velocities and time steps are computed as needed before marble positions and velocities are computed, so that we can arrange to compute marble data at the times when ghost acceleration coefficients must be recomputed. We associate with each ghost a value of T, the time for which the ghost's position and velocity are known. The smallest of the T_k is selected as the "ghost target time" (GT). Initially, all $T_k = 0$, and GT = 0.

We begin as in Fig. B2A. Marble positions, time steps, UT, and MT are shown as before. The open circles indicate the time for which ghost data are known, analogous to the filled circles for marbles.

For each ghost, we then compute acceleration coefficients from Eq. (4) and time steps from Eq. (9) and compute new positions and velocities from Eqs. (2) and (3), at new times obtained from the full time step of each ghost. We add the time steps to T_k and select GT, the smallest T_k. This is shown in Fig. B2B.

We then compute marble data for time GT; since GT < MT, all marble acceleration coefficients can be extrapolated (dotted lines) by Eq. (8) without reference to other marbles. The marble time steps are reduced. We put UT = GT, see Fig. B2C.

We can now compute new acceleration coefficients and time step for ghost 1, use them to compute new values of position, velocity, and T_1, and obtain a new value of GT. This is shown in Fig. B2D.

Since now GT > MT, we first compute marble data for time MT. The acceleration coefficients for marble 2 are extrapolated and its time step is reduced; but the accelerations for marble 1 and its new time step must be recomputed explicitly. We put UT = MT. (Note that UT is incremented only as marble time advances.) This is shown in Fig. B2E.

This process continues until print time. Just like marbles, so also ghosts are not allowed to advance beyond PT.

Let us now specify precisely how the above algorithm for marbles is modified to handle ghosts. Given initial positions and velocities for K ghosts, replace step 6 by step 6´, and step 11 by 11´:

6′. $T_k \leftarrow 0,\ 1 \leq k \leq K$.

11 . Do steps (A) through (C):

(A) $GT \leftarrow \min [T_k],\ 1 \leq k \leq K$.

(B) Only if GT = UT, then do steps a through g:

(a) Begin loop over k, $1 \leq k \leq K$.

(b) Only if $T_k > UT$, then skip to step (f).

(c) Compute accelerations from Eq. (4) and time step D_k from Eq. (9) for k^{th} ghost.

(d) Only if $T_k + D_k > PT$, then $D_k \leftarrow PT - T_k$.

(e) Advance $\underline{r}_k$ and $\underline{v}_k$ by D_k from Eqs. (2) and (3); $T_k \leftarrow T_k + D_k$.

(f) Repeat from (b) through (e) for next k.

(g) Return to (A).

(C) Only if GT > UT, then do steps (a) and (b).

(a) Only if $GT \geq MT$, then $D^* \leftarrow D_{min}$; go on to step 12.

(b) Only if GT < MT, then $D^* \leftarrow GT - UT$; go on to step 12.

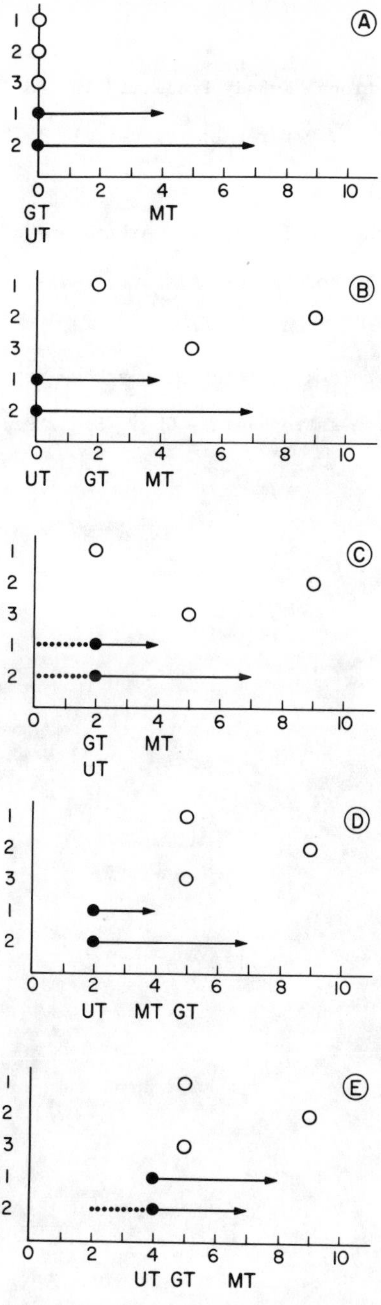

Fig. B2. Schematic representation of sequence of calculations involving marbles and ghosts.

References

1. S.J. Aarseth, "Gravitational N-Body Problem," in The Gravitational N-Body Problem, M. Lecar, Editor, Dordrecht: D. Reidel, 1972, p. 373

2. C. Gonzalez and M. Lecar, "Colloque sur le Probleme des N Corps," Editions du Centre National de la Recherche Scientifique, Paris, 1968

3. N.M. Hall and J.R. Cherniack, Smithsonian package for algebra and symbolic mathematics, Smithsonian Astrophysical Obs. Spec. Rept. 291, 1969

MULTIREVOLUTION METHODS FOR ORBIT INTEGRATION

by

Otis Graf

University of Texas Austin, Texas

1. Introduction

Accurate long term solutions to certain satellite orbit integration problems may be obtained by revolution skipping methods. When many revolutions are similar, the trajectory can be extrapolated several revolutions ahead, without significant loss of accuracy.

These methods require that the change in the elements over one revolution be computed using a reliable orbit integration method. This finite change, due to secular and long period perturbations, is used to extrapolate the orbit M revolutions ahead. Another single revolution integration is performed and the process is repeated. In this way only one (or in some cases two) out of M revolutions are actually integrated. Since the computational work in the extrapolation formula is small, much computer time can be saved. The two-stage algorithm of external extrapolation and internal (single revolution) integration is called "multirevolution integration." The extrapolation algorithm is similar to that for the classical multistep methods for differential equations. In fact, the multirevolution methods are a generalization of the classical methods.

The finite change in the elements over one revolution defines a first order nonlinear difference equation

$$(1.1) \qquad \Delta\bar{y}_\rho = \bar{f}(\bar{y}_\rho, \rho)$$

where $\bar{y}_\rho$ is a vector of elements that completely defines the orbit and

(1.2) $$\Delta\bar{y}_\rho = \bar{y}_{\rho+1} - \bar{y}_\rho \quad .$$

The independent variable is the revolution number, denoted by ρ. The revolutions are counted from any reference point on the orbit, such as apogee, pergee or node. The orbit integration problem has now been transformed from that of solving a system of first order nonlinear differential equations to that of solving a system of first order nonlinear difference equations. There is the disadvantage that the right hand sides of the difference equations cannot be written down in terms of elementary functions. This prevents any attempt at an analytical solution and suggests a numerical solution such as that provided by the multirevolution methods. The solution of Eq. (1.1) provides the osculating orbital elements at any reference point, i.e., $\bar{y}_\rho$, $\rho = 0,1,2,\ldots$.

The earliest investigators of the multirevolution methods were Taratynova (1961), Mace and Thomas (1960), and Cohen and Hubbard (1960). The predictor formula given by Mace and Thomas is a generalization of the Adams-Bashsforth method, and the corrector formula given by Cohen and Hubbard is a generalization of the Adams-Moulton method. Other derivations of the predictor-corrector formulae are given by Boggs (1968) and Velez (1970). Boggs also gives a suitable method for starting the multirevolution integration and discusses modifications of the basic algorithm.

2. Multirevolution Predictor-Corrector Formulae

Without loss of generality, scalar difference equations may be considered. Then Eqs. (1.1) and (1.2) become

(2.1) $$\Delta y_\rho = f(y_\rho, \rho) \quad ,$$

(2.2) $$\Delta y_\rho = y_{\rho+1} - y_\rho \quad ,$$

and ρ is an integer. Let M be the number of revolutions skipped. Then the dependent variable y will be defined on a series of large and small grid, the large grid being separated by M small grid. Let n be the large grid number, then f is given at the large grid by

$$y_{nM+1} - y_{nM} = f(y_{nM}, nM) \quad .$$

Notice that the double subscript in y_{nM} is a product giving the revolution number.

The multirevolution formula is given by

$$(2.3) \qquad \gamma_1 y_{(n+1)M} + \gamma_2 y_{nM} + \gamma_3 y_{(n-1)M} = M \sum_{j=0}^{N} \alpha_j \nabla^{*j} f_{nM} \quad .$$

The difference table is shown in Table 1. Differences with an asterix denote that the difference is taken over the large grid. Equation (2.3) will be a predictor formula or a corrector formula, depending on how the γ_i are chosen.

Table 1. Difference Table for Generalized Multistep Methods

ρ	f	$\nabla^* f$	$\nabla^{*2} f$	$\nabla^{*3} f$	$\nabla^{*4} f$
0	f_0				
		$\nabla^* f_M$			
M	f_M		$\nabla^{*2} f_{2M}$		
		$\nabla^* f_{2M}$		$\nabla^{*3} f_{3M}$	
2M	f_{2M}		$\nabla^{*2} f_{3M}$		$\nabla^{*4} f_{4M}$
		$\nabla^* f_{3M}$		$\nabla^{*3} f_{4M}$	
3M	f_{3M}		$\nabla^{*2} f_{4M}$		$\nabla^{*4} f_{5M}$
		$\nabla^* f_{4M}$		$\nabla^{*3} f_{5M}$	
4M	f_{4M}		$\nabla^{*2} f_{5M}$		
		$\nabla^* f_{5M}$			
5M	f_{5M}				

$$f_{nM} = f(y_{nM}, nM)$$

$$\nabla^* f_{nM} = f_{nM} - f_{(n-1)M}$$

$$\nabla^{*2} f_{nM} = \nabla^* f_{nM} - \nabla^* f_{(n-1)M}$$

$$\nabla^{*3} f_{nM} = \nabla^{*2} f_{nM} - \nabla^{*2} f_{(n-1)M}$$

$$\vdots$$

Since M is constant, the following subscript notation is adopted:

i) upper and lower case roman letter subscripts not containing a product refer to values at the large grid,

ii) lower case greek letters refer to values at the small grid.

With this notation, Eq. (2.3) becomes

$$\gamma_1 y_{n+1} + \gamma_2 y_n + \gamma_3 y_{n-1} = M \sum_{j=0}^{N} \alpha_j \nabla^{*j} f_n \quad . \tag{2.4}$$

The corrector formula is given from Eq. (2.4) with $\gamma_1 = 0$, $\gamma_2 = 1$, $\gamma_3 = -1$, and $\alpha_0 = 1$,

$$y_n - y_{n-1} = M \left[f_n + \alpha_1 \nabla^* f_n + \alpha_2 \nabla^{*2} f_n + \ldots + \alpha_N \nabla^{*N} f_n \right] \quad . \tag{2.5}$$

To determine the coefficients α_i , consider the exponential function

$$f_\rho = Z^{\rho/M} \quad ,$$

where Z is any constant parameter. The difference equation is

$$y_{\rho+1} - y_\rho = Z^{\rho/M} \quad ,$$

with the solution

$$y_\rho = \left(\frac{1}{Z^{1/M} + 1} \right) Z^{\rho/M} \quad ,$$

At the large grid the solution is

$$y_n = \left(\frac{1}{Z^{1/M} + 1} \right) Z^n \quad , \tag{2.6}$$

and the left hand side of Eq. (2.5) is

$$y_n - y_{n-1} = Z^n \left[\frac{1 - 1/Z}{Z^{1/M} - 1} \right] \quad .$$

The differences on the right hand side of Eq. (2.5) are given by

$$\nabla^{*j} f_n = Z^n (1 - 1/Z)^j \quad , \quad j = 0,1,2,3,\ldots \quad .$$

Then Eq. (2.5) becomes

$$Z^n\left[\frac{1 - 1/Z}{Z^{1/M} - 1}\right] = MZ^n\left[1 + \alpha_1(1 - 1/Z) + \alpha_2(1 - 1/Z)^2 + \ldots + \alpha_N(1 - 1/Z)^N\right] .$$

Dividing through by Z^n and letting $\eta = 1 - 1/Z$,

$$\text{(2.7)} \qquad \frac{\eta}{(1 - \eta)^{-1/M} - 1} = M(1 + \alpha_1\eta + \alpha_2\eta^2 + \ldots + \alpha_N\eta^N) .$$

From the binomial theorem

$$(1 - \eta)^{-1/M} - 1 = 1/M\left[\eta + (1 + 1/M)\eta^2/2! + (1 + 1/M)(2 + 1/M)\eta^3/3! + \ldots\right] ,$$

which gives for Eq. (2.7)

$$1 = (1 + \alpha_1\eta + \alpha_2\eta^2 + \ldots)\left[1 + (1/M + 1)\eta/2 + (1/M + 1)(1/M + 2)\eta^2/3! + \ldots + (1/M + 1)(1/M + 2)\ldots(1/M + k)\frac{\eta^k}{(k + 1)!} + \ldots\right] .$$

When the two series are multiplied term by term, the coefficients of powers of η in the resulting series should be zero, giving the following recurrence relations for the α_i ,

$$\alpha_1 = -(1/M + 1)1/2! ,$$

$$\alpha_2 = -\alpha_1(1/M + 1)1/2! - (1/M + 1)(1/M + 2)\,1/3! ,$$

$$\alpha_3 = -\alpha_2(1/M + 1)1/2! - \alpha_1(1/M + 1)(1/M + 2)1/3! - (1/M + 1)(1/M + 2)(1/M + 3)1/4!$$

$$\alpha_4 = -\alpha_3(1/M + 1)1/2! - \alpha_2(1/M + 1)(1/M + 2)1/3! - \alpha_1(1/M + 1)(1/M + 2)(1/M + 3)1/4! - (1/M + 1)(1/M + 2)(1/M + 3)(1/M + 4)1/5! ,$$

and in general

$$\alpha_k = -\alpha_{k-1}\ (1/M + 1)1/2! - \alpha_{k-2}(1/M + 1)(1/M + 2)1/3! - \ldots$$
$$- (1/M + 1)(1/M + 2)\ldots(1/M + k)\ \frac{1}{(k + 1)!}\quad .$$

Explicit expressions for the first few α_i are given in Table 2. In order to facilitate the programming of the formula on an automatic computer, it is usually more convenient to write Eq. (2.5) in Lagrangian form,

$$(2.8)\qquad y_{N+n} - y_{N+n-1} = M\left[\beta_N f_{N+n} + \beta_{N-1} f_{N+n-1} + \ldots + \beta_0 f_n\right] .$$

Instead of computing N differences at each step, N+1 successive values of f_n are retained. The β_i are given in terms of the α_i by

$$\beta_{N-j} = (-1)^j\left[\binom{j}{j}\alpha_j + \binom{j+1}{j}\alpha_{j+1} + \ldots + \binom{N}{j}\alpha_N\right] ,$$

$$j = 0,1,2,\ldots,N \quad , \quad \alpha_0 = 1 \quad .$$

Equation (2.9) is derived by making use of the following property of the backward difference operator (Henrici, 1962, p. 190)

$$\nabla^q U_p = \sum_{m=0}^{q} (-1)^m \binom{q}{m} U_{p-m} \quad , \quad q = 0,1,2,\ldots \quad .$$

The predictor formula is given by Eq. (2.4) with $\gamma_1 = 1$, $\gamma_2 = -1$, $\gamma_3 = 0$ and $\alpha_0^* = 1$,

$$(2.10)\qquad y_{n+1} - y_n = M\left[f_n + \alpha_1^*\ \nabla^* f_n + \alpha_2^*\ \nabla^{*2} f_n + \ldots + \alpha_N^* \nabla^{*N} f_n\right].$$

Expansion of Eq. (2.10) in powers of η provides the following recurrence relations for the α_i^* ,

$$\alpha_1^* = (1 - 1/M)1/2! \quad ,$$

$$\alpha_2^* = \alpha_1^*(1 - 1/M)1/2! + (1 - 1/M)(1 + 1/M)1/3! \quad ,$$

$$\alpha_3 = \alpha_2^*(1 - 1/M)1/2! + \alpha_1^*(1-1/M)(1+1/M)1/3! + (1-1/M)(1+1/M)(2+1/M)1/4!$$

$$\alpha_4 = \alpha_3^*(1-1/M)1/2! + \alpha_2^*(1-1/M)(1+1/M)1/3! + \alpha_1^*(1-1/M)(1+1/M)(2+1/M)1/4!$$
$$+ (1-1/M)(1+1/M)(2+1/M)(3+1/M)1/5! \quad ,$$

and in general

$$\alpha_k^* = \alpha_{k-1}^*(1-1/M)1/2! + \alpha_{k-2}^*(1-1/M)(1+1/M)1/3! + \ldots$$
$$+ (1-1/M)(1+1/M)\ldots(K-1+1/M)\,\frac{1}{(k+1)!} \quad .$$

Explicit expressions for the first few α_i^* are given in Table 3. The Lagrangian form of Eq. (2.10) is

$$y_{N+n+1} - y_{N+n} = M\,[\beta_N^* f_{N+n} + \beta_{N-1}^* f_{N+n-1} + \ldots + \beta_0^* f_n] \quad ,$$

where

$$\beta_{N-j}^* = (-1)^j \left[\binom{j}{j}\alpha_j^* + \binom{j+1}{j}\alpha_{j+1}^* + \ldots + \binom{N}{j}\alpha_N^*\right] \quad ,$$

$$j = 0,1,2,\ldots,N \quad , \quad \alpha_0^* = 1 \quad .$$

The predictor-corrector formulae derived by Taratynova (1961), and Cohen and Hubbard (1960), Mace and Thomas (1960), Boggs (1969), and Velez (1970) all reduce to one or both of the formulae given in this section. The derivations given by the different authors vary in their complexity.

Table 2. Coefficients for the Corrector Formula

$\alpha_0 = 1$

$\alpha_1 = -1/2 - 1/2\ 1/M$

$\alpha_2 = -\ 1/12 + 1/12\ 1/M^2$

$\alpha_3 = -1/24 + 1/24\ 1/M^2$

$\alpha_4 = -19/720 + 1/36\ 1/M^2 - 1/720\ 1/M^4$

$\alpha_5 = -3/160 + 1/48\ 1/M^2 - 1/480\ 1/M^4$

$\alpha_6 = -\frac{863}{60480} + 1/60\ 1/M^2 - \frac{7}{2880}\ 1/M^4 + \frac{1}{30240}\ 1/M^6$

$\alpha_7 = -\frac{275}{24192} + 1/72\ 1/M^2 - 1/384\ 1/M^4 + \frac{1}{12096}\ 1/M^6$

$\alpha_8 = -\frac{33953}{36\ 28800} + 1/84\ 1/M^2 - \frac{29}{10800}\ 1/M^4 + \frac{5}{36288}\ 1/M^6 - \frac{1}{12\ 09600}\ 1/M^8$

$\alpha_9 = -\frac{8183}{10\ 36800} + 1/96\ 1/M^2 - \frac{469}{1\ 72800}\ 1/M^4 + \frac{1}{5184}\ 1/M^6 - \frac{1}{3\ 45600}\ 1/M^8$

$\alpha_{10} = -\frac{32\ 50433}{4790\ 01600} + 1/108\ 1/M^2 - \frac{29531}{108\ 86400}\ 1/M^4 + \frac{1069}{43\ 54560}\ 1/M^6 -$

$-\frac{13}{20\ 73600}\ 1/M^8 + \frac{1}{479\ 00160}\ 1/M^{10}$

Table 3. Coefficients for the Predictor Formula

$\alpha^*_0 = 1$

$\alpha^*_1 = 1/2 - 1/2\ 1/M$

$\alpha^*_2 = 5/12 - 1/2\ 1/M + 1/12\ 1/M^2$

$\alpha^*_3 = 3/8 - 1/2\ 1/M + 1/8\ 1/M^2$

$\alpha^*_4 = \frac{251}{720} - 1/2\ 1/M + 11/72\ 1/M^2 - \frac{1}{720}\ 1/M^4$

$\alpha^*_5 = \frac{95}{288} - 1/2\ 1/M + \frac{25}{144}\ 1/M^2 - \frac{1}{288}\ 1/M^4$

$\alpha^*_6 = \frac{19087}{60480} - 1/2\ 1/M + \frac{137}{720}\ 1/M^2 - \frac{17}{2880}\ 1/M^4 + \frac{1}{30240}\ 1/M^6$

$\alpha^*_7 = \frac{5257}{17280} - 1/2\ 1/M + \frac{49}{240}\ 1/M^2 - \frac{49}{5760}\ 1/M^4 + \frac{1}{8640}\ 1/M^6$

$\alpha^*_8 = \frac{10\ 70017}{36\ 28800} - 1/2\ 1/M + \frac{121}{560}\ 1/M^2 - \frac{967}{86400}\ 1/M^4 + \frac{23}{90720}\ 1/M^6 - \frac{1}{12\ 09600}\ 1/M^8$

$\alpha^*_9 = \frac{25713}{89600} - 1/2\ 1/M + \frac{761}{3360}\ 1/M^2 - \frac{89}{6400}\ 1/M^4 + \frac{1}{2240}\ 1/M^6 - \frac{1}{2\ 68800}\ 1/M^8$

$\alpha^*_{10} = \frac{268\ 42253}{958\ 00320} - 1/2\ 1/M + \frac{7129}{30240}\ 1/M^2 - \frac{4523}{2\ 72160}\ 1/M^4 + \frac{3013}{43\ 54560}\ 1/M^6 -$

$-\frac{29}{29\ 03040}\ 1/M^8 + \frac{1}{479\ 00160}\ 1/M^{10}$

3. Calculation of Intermediate Revolutions

For orbit integration problems, it is usually necessary to determine the state at some desired epoch. If the orbit is computed by the multirevolution methods of Section 2, the state will be known only at epochs separated by many revolutions, and then only at the orbital reference points. The state could be found at epochs in between the large grid by a numerical integration of the differential equations. However, in some cases the number of revolutions skipped will be large (20, 50, or more) and some of the efficiency of the revolution skipping procedure would be lost. It is desirable, therefore, to develop a method which can give the state at any intermediate revolution. The formula is a generalization of the corrector equation (2.5). Using the subscript notation of Section 2, define

$$(3.1)\qquad y_{(n-1)M+P} - y_{(n-1)M} = P\left[f_{nM} + \gamma_1\nabla^* f_{nM} + \gamma_2\nabla^{*2} f_{nM} + \ldots + \gamma_N\nabla^{*N} f_{nM}\right] ,$$

$$0 \le P \le M \quad .$$

The subscripts specify the small grid. Thus, $y_{(n-1)M+P}$ and $y_{(n-1)M}$ are separated by P small grid. The differences on the right hand side of Eq. (3.1) are defined over M small grid as shown in Table 1. Therefore, the terms in brackets of Eq. (3.1) should reduce to the terms in brackets of Eq. (2.5), when P = M. Once the right hand side is determined, Eq. (3.1) will give the value of y_ρ at any small grid between $\rho = (n-1)M$ and $\rho = nM$. The recurrence relations for the γ_i are

$$\gamma_1 = -1/2!(1/M+1) - 1/2!(1-P/M) \quad ,$$

$$\gamma_2 = -\gamma_1\, 1/2!(1/M+1) - 1/3!(1/M+1)(1/M+2) - 1/3!(1-P/M)(P/M+1) \quad ,$$

$$\gamma_3 = -\gamma_2\, 1/2!(1/M+1) - \gamma_1\, 1/3!(1/M+1)(1/M+2)$$
$$-1/4!(1/M+1)(1/M+2)(1/M+3) - 1/4!(1-P/M)(P/M+1)(P/M+2) \quad ,$$

and in general

$$\gamma_k = -\gamma_{k-1}\,1/2!(1/M+1) - \gamma_{k-2}\,1/3!(1/M+1)(1/M+2) - \ldots$$

$$- (1/M+1)(1/M+2)\ldots(1/M+k)\,\frac{1}{(k+1)!} - (1-P/M)(P/M+1)(P/M+2)\ldots(P/M+k-1)\,\frac{1}{(k+1)!}$$

It is immediately seen that the γ_i reduce to the α_i of Eq. (2.5) when P = M. For other special values of P ,

i) $P = 0$: $y_{(n-1)M} - y_{(n-1)M} = 0$

ii) $P = 1$: $\gamma_1 = -1$, $\gamma_i = 0$, $i = 2,3,4,\ldots,$

$$y_{(n-1)M+1} - y_{(n-1)M} = f_{nM} - \nabla^* f_{nM} \;,$$

$$y_{(n-1)M+1} - y_{(n-1)M} = f_{(n-1)M} \;.$$

4. Modified Methods

The multirevolution integration formulae discussed in Section 2 will integrate exactly (excluding roundoff error) difference equations which are polynomial functions of the independent variable. When the difference equations for the elements

$$\bar{y}_{\rho+1} - \bar{y}_\rho = \bar{f}(\bar{y}\;,\rho)$$

can be approximated by a polynomial in ρ (the revolution number) over a substantial range of ρ, these methods will give good results. However, there may be some problems of interest for which the differences of the elements cannot be accurately approximated by polynomials. In addition, the perturbations of the elements in almost all problems are composed of terms that are periodic. It is of interest, therefore, to modify the previous methods so that difference equations which are periodic in ρ can be integrated exactly.

The multirevolution formula can be written as

$$(4.1)\quad y_n - y_{n-1} = M\left[f_{n-p} + \alpha_1\nabla^* f_{n-p} + \alpha_2\nabla^{*2} f_{n-p} + \ldots + \alpha_N\nabla^{*N} f_{n-p}\right] \;,$$

and will be a corrector for p = 0 and a predictor for p = 1. When the α_i are given by Tables 2 or 3, Eq. (4.1) will integrate exactly a difference equation that is a polynomial of degree N. The coefficients listed in Tables 2 and 3 will be called the "standard" coefficients.

If $\alpha_1, \ldots, \alpha_{N-2}$ in Eq. (4.1) have their standard values, α_{N-1} and α_N can be modified so that

$$f_\rho = \sum_{i=0}^{N-2} A_i \rho^i + B \sin w\rho + C \cos w\rho$$

is integrated exactly. Modified multistep methods for ordinary differential equations which integrate exactly the equation

$$dZ/dt = \sum_{i=0}^{N-2} a_i t^i + b \sin wt + c \cos wt \quad .$$

are given by Bettis (Stiefel and Scheifele, 1971, pp. 127-134). Bettis' derivation can be generalized to the case of difference equations.

For simplicity, introduce the notation

$$f_\rho = Z^{\rho/M} \quad , \quad Z = e^{2i\sigma} \quad , \quad \sigma = \frac{wM}{2} \; .$$

At the large grid

$$f_n = Z^n \quad ,$$

and from Eq. (2.6)

$$y_n - y_{n-1} = \frac{Z^n(1-1/Z)}{(Z^{1/M} - 1)} \quad .$$

The differences are

$$\nabla^{*i} f_{n-p} = Z^{n-p}(1-1/Z)^i \quad ,$$

so that Eq. (4.1) becomes

$$\text{(4.2)} \qquad \frac{Z^p(1-1/Z)}{(Z^{1/M}-1)} = M\left[1 + \alpha_1(1-1/Z) + \alpha_2(1-1/Z)^2 + \ldots + \alpha_N(1-1/Z)^N\right] \quad .$$

Let

$$\eta = 1 - 1/Z \quad ,$$

and

$$L = \frac{Z^p(1-1/Z)}{M(Z^{1/M}-1)}$$

Then Eq. (4.2) becomes

$$(4.3) \qquad L = 1 + \alpha_1\eta + \alpha_2\eta^2 + \ldots + \alpha_N\eta^N \quad ,$$

and

$$L(\sigma) = \lambda e^{i\sigma(2p-1-1/M)} \quad ,$$

where

$$\lambda(\sigma) = \frac{\sin\sigma}{M \sin\sigma/M}$$

Since it is desired that the real functions sin $w\rho$ and cos $w\rho$ be integrated exactly, Eq. (4.3) must also be satisfied if w is replaced by -w. This leads to a second equation which is the complex conjugate of Eq. (4.3),

$$(4.4) \qquad \bar{L} = 1 + \alpha_1\bar{\eta} + \alpha_2\bar{\eta}^2 + \ldots + \alpha_N\bar{\eta}^N \quad , \quad \bar{\eta} = 1 - e^{2i\sigma} \quad .$$

The coefficients α_{N-1} and α_N will be given in terms of σ, 1/M, and α_i (i = 1,2,..., N-2). To eliminate α_N, multiply Eq. (4.3) by $\bar{\eta}^N$, and Eq. (4.4) by η^N, and subtract to obtain

$$(4.5) \qquad \bar{\eta}^N L - \eta^N \bar{L} = \sum_{k=0}^{N-2} \alpha_k(\eta^k\bar{\eta}^N - \bar{\eta}^k\eta^N) + \alpha_{N-1}(\eta^{N-1}\bar{\eta}^N - \eta^{N-1}\bar{\eta}^N) \quad .$$

It can be shown that (Stiefel and Scheifele, 1971, p. 131)

$$(4.6) \qquad \eta^k\bar{\eta}^j - \bar{\eta}^k\eta^j = 2i \sin 2\sigma \, u^{j-1} S_{k-j} \quad ,$$

where

$$u = 4 \sin^2\sigma$$

The two sequences of rational functions $S_m(u)$ and $R_m(u)$ are defined by the recurrence relations

$$S_{m+1} = u/2(R_m + S_m) \quad , \quad S_0 = 0 \quad ,$$

(4.7)

$$R_{m+1} = S_{m+1} - 2S_m \quad , \quad R_0 = 2 \quad .$$

Using Eq. (4.6), the right hand side of Eq. (4.5) is

$$(4.8) \qquad 2i \sin 2\sigma \; u^{N-1} \left[\sum_{k=0}^{N-2} \alpha_k S_{k-N} + \alpha_{N-1} S_{-1} \right] .$$

Since

$$\eta^m = 2^m (i)^m e^{-im\sigma} \sin^m \sigma \quad ,$$

the left hand side of Eq. (4.5) is

$$(4.9) \qquad \lambda 2^N \left[(-i)^N e^{i\sigma(N+2p-1-1/M)} - (i)^N e^{-i\sigma(N+2p-1-1/M)} \right] \sin^N \sigma .$$

For N even,

$$(-i)^N = (i)^N = (-1)^{N/2} \quad .$$

For N odd,

$$(-i)^N = -(i)^N \quad , \quad (i)^N = i(-1)^{N-1/2} \; .$$

Thus, expression (4.9) becomes

$$i\lambda 2^{N+1} (-1)^{N/2} \sin^N \sigma \sin[\sigma(N+2p-1-1/M)] \quad , \qquad (N \text{ even}),$$

(4.10)

$$i\lambda 2^{N+1} (-1)^{N+1/2} \sin^N \sigma \cos[\sigma(N+2p-1-1/M)], \qquad (N \text{ odd}) .$$

Combining (4.8) and (4.10) and using $S_{-1} = -1$,

$$(4.11) \qquad \alpha_{N-1} = \sum_{k=0}^{N-2} \alpha_k S_{k-N} - \frac{\lambda V}{(2 \sin\sigma)^{N-1} \cos \sigma} \quad ,$$

where

$$V(\sigma) = \begin{cases} (-1)^{N/2} \sin[\sigma(N+2p-1-1/M)] \quad , \quad (N \text{ even}) \\ (-1)^{N+1/2} \cos[\sigma(N+2p-1-1/M)], \ ,(N \text{ odd}) \quad . \end{cases}$$

To find α_N, eliminate α_{N-1} from Eq. (4.3). Multiply Eq. (4.3) by $\bar{\eta}^{N-1}$ and and Eq. (4.4) by η^{N-1} and subtract to obtain

$$\text{(4.12)} \qquad \bar{\eta}^{N-1}L - \eta^{N-1}\bar{L} = \sum_{k=0}^{N-2} \alpha_k(\eta^k \bar{\eta}^{N-1} - \bar{\eta}^k\eta^{N-1}) + \alpha_N(\eta^N \bar{\eta}^{N-1} - \bar{\eta}^N\eta^{N-1}) \quad .$$

The left hand side of (4.12) becomes

$$\text{(4.13)} \qquad \begin{matrix} i\lambda 2^N \ (-1)^{N/2} \sin^{N-1}\sigma \cos[\sigma(N+2p-2-1/M)] \quad , \quad (N \text{ even}) \\ i\lambda 2^N \ (-1)^{N+1/2} \sin^{N-1}\sigma \sin[\sigma(N+2p-2-1/M)] \ , \quad (N \text{ odd}) \end{matrix}$$

and the right hand side is

$$\text{(4.14)} \qquad 2iu^{N-2} \sin 2\sigma \left[\sum_{k=0}^{N-2} \alpha_k S_{k-N+1} + \alpha_N S_1\right] \quad .$$

Combining (4.13) and (4.14) and using $S_1 = u$,

$$\text{(4.15)} \qquad \alpha_N = 1/u\left\{-\sum_{k=0}^{N-2} \alpha_k S_{k-N+1} + \frac{\lambda\ V}{(2 \sin \sigma)^{N-2} \cos \sigma}\right\}$$

where

$$V(\sigma) = \begin{cases} (-1)^{N/2} \cos[\sigma(N+2p-2-1/M)] \quad , \quad (N \text{ even}) \\ (-1)^{N+1/2} \sin[\sigma(N+2p-2-1/M)] \quad , \quad (N \text{ odd}) \end{cases}$$

The functions $S_k(u)$ are given for negative k by the recurrence relations

$$S_m = 1/2(S_{m+1} - R_{m+1}) \ , \quad R_m = 2/u\, S_{m+1} - S_m \ ,$$

$$S_{-1} = -1 \ , \quad S_{-2} = -1 \ , \quad S_{-3} = -1 + 1/u \ ,$$

$$R_{-1} = 1 \ , \quad R_{-2} = 1 - 2/u \ , \quad R_{-3} = 1 - 3/u \quad .$$

For N = 2, Eqs. (4.11) and (4.15) provide

$$\alpha_1 = \frac{\lambda}{\sin 2\sigma} \sin[\sigma(2p+1-1/M)] - 1 \quad ,$$

$$\alpha_2 = \frac{1}{4 \sin^2\sigma} \{1 - \lambda/\cos\sigma \cos[\sigma(2p-1/M)]\} \quad .$$

By expanding α_1 and α_2 in a power series in σ , it can be shown that

$$\lim_{w \to 0} \alpha_1 = 1/2(2p-1-1/M) \quad ,$$

$$\lim_{w \to 0} \alpha_2 = 1/4[1/2(2p-1/M)^2 - 1/6\ 1/M^2 - 1/3] \quad ,$$

which are the coefficients in Table 2 or Table 3 for p = 0 or p = 1, respectively. In general, α_{N-1} and α_N reduce to the standard coefficients when $w \to 0$.

For other values of N,

(i) N = 3

$$\alpha_2 = \frac{1}{4 \sin^2\sigma} - (1+\alpha_1) - \frac{\lambda \cos[\sigma(2+2p-1/M)]}{4 \sin^2\sigma \cos\sigma} \quad ,$$

$$\alpha_3 = \frac{1}{4 \sin^2\sigma} \left\{ 1 + \alpha_1 + \frac{\lambda \sin[\sigma(2p+1-1/M)]}{2 \sin\sigma \cos\sigma} \right\}$$

(ii) N = 4

$$\alpha_3 = \frac{1}{4 \sin^2\sigma} (2+\alpha_1) - (1+\alpha_1+\alpha_2) - \frac{\lambda \sin[\sigma(2p+3-1/M)]}{8 \sin^3\sigma \cos\sigma} \quad ,$$

$$\alpha_4 = \frac{1}{4 \sin^2\sigma} \left\{ 1+\alpha_1+\alpha_2 - \frac{1}{4 \sin^2\sigma} + \frac{\lambda \cos[\sigma(2p+2-1/M)]}{4 \sin^2\sigma \cos\sigma} \right\} \quad .$$

The expressions for α_{N-1} and α_N are given explicitly by Eqs. (4.11) and (4.22) in terms of σ , 1/M, and α_i (i=1,2,3,...,N-2). However, these expressions contain singularities at $\sigma = \frac{m\pi}{2}$, m = 0,1,2,3,... . If w is small, the singularity at σ=0 can cause numerical difficulties. In the applications, there is another singularity which is more severe. Since M is an integer, the equations will be singular for $w=m\pi$ (m = 1,2,3,...), whatever the value of M. These singularities correspond to the periods (in variable ρ) of T=2/m (m = 1,2,3,...). In orbit integration problems, the periods of interest will be longer than unity and hence $w=\pi$ is the only case where an unavoidable singularity exists.

5. Numerical Examples

The multirevolution predictor-corrector formulae (Eqs. (2.8) and (2.11)) with the standard coefficients were applied to two problems of satellite orbit integration: a 24-hour earth satellite and a lunar satellite. The equations of motion for the internal integration were

$$\ddot{x}_1 = \partial U/\partial x_1 \quad , \quad \ddot{x}_2 = \partial U/\partial x_2 \quad , \quad \ddot{x}_3 = \partial U/\partial x_3 \quad ,$$

$$U = K^2 M_c / r - V \quad ,$$

$$r = (x_1^2 + x_2^2 + x_3^2)^{1/2} \quad .$$

The perturbing potential is

$$V = K^2 M_c / r \left[\frac{J_{20} R_0^2}{2r^2} (3 \sin^2\phi - 1) + 3J_{22} \frac{R_0^2}{r^2} \cos^2\phi \cos 2(\lambda - \lambda_{22}) \right.$$
$$+ \frac{J_{30} R_0^3}{2r^3} (5 \sin^3\phi - 3 \sin\phi)$$
$$\left. + \frac{J_{31} R_0^3}{2r^3} \cos \phi \; (15 \sin^2\phi - 3) \cos (\lambda - \lambda_{31}) \right] ,$$

where

ϕ = latitude referred to the true equator of the central body,

λ = longitude referred to the true equator and zero meridian of the central body,

R_0 = mean radius of the central body,

λ_{nm} = gravity constants referred to the true equator and zero meridian of the central body,

K = the Gaussian constant,

M_c = the mass of the central body.

For the earth, the gravity constants are (Wagner, 1964):

$$J_{20} = 1082.48 \times 10^{-6} ,$$

$$J_{30} = -2.56 \times 10^{-6} ,$$

$$J_{22} = -1.2 \times 10^{-6} , \quad \lambda_{22} = -26.4^{\circ} ,$$

$$J_{31} = -1.9 \times 10^{-6} , \quad \lambda_{31} = 4.6^{\circ} .$$

For the moon, the gravity constants are (Melbourne, et al, 1968):

$$J_{20} = 1.9076 \times 10^{-4} ,$$

$$J_{30} = 0.2058 \times 10^{-4} ,$$

$$J_{22} = -0.1411 \times 10^{-4} , \quad \lambda_{22} = -25.32^{\circ} ,$$

$$J_{31} = -0.3554 \times 10^{-4} , \quad \lambda_{31} = 17.11^{\circ} .$$

The equations of motion were integrated with an eighth-order Runge-Kutta-Fehlberg formula (RK78) with error control (Fehlberg, 1968).

The elements used in the extrapolation were: semi-major axis (a), eccentricity (e), inclination (I), argument of perigee (g) and longitude of ascending node (Ω). A sixth element ν was used for the time computation (Boggs, 1968),

$$\nu_{\rho} = t_{\rho} - t_0 - \rho P ,$$

where t_{ρ} is the time at the ρth orbital reference point and P is the average period between reference points. Then,

$$\Delta\nu_{nM} = t_{nM+1} - t_{nM} - P ,$$

and the time at each large grid is given by

$$t_{(n+1)M} = t_{nM} + \nu_{(n+1)M} - \nu_{nM} + MP .$$

The orbital reference point (pericenter) was found by iterating with RK78. The starting values for the multirevolution integration were found by the method proposed by Boggs (1968). A reference solution was computed using a double precision (28 decimal digits) integration program. The reference integration was required to maintain an accuracy of 17 decimal digits at each step.

The initial orbital elements for the 24-hour earth satellite problem were

Semi-major axis	6.63 ER
Period	1442 min
Eccentricity	0.1
Inclination	30°
Argument of perigee	0°
Long. of ascending node	0°
Height of perigee	31679 km
Height of apogee	40136 km

The solutions were compared, at the 1088th revolution, with the reference orbit and the results are shown in the table below.

Method	N	M	$\|\Delta X_1\|^*$	$\|\Delta X_2\|$	$\|\Delta X_3\|$	Revs. Compt. with RK78	Ratio
Multirevolution	10	16	1.5×10^{-9}	1.5×10^{-8}	6.8×10^{-9}	197	.183
"	10	24	1.8×10^{-7}	5.9×10^{-8}	1.4×10^{-8}	163	.151
"	10	32	1.1×10^{-5}	7.4×10^{-6}	3.4×10^{-6}	149	.138
	10	64	1.7×10^{-3}	2.4×10^{-4}	8.0×10^{-5}	135	.125
RK78	_	_	1.8×10^{-9}	7.7×10^{-9}	4.1×10^{-9}	1088	1.000

The multirevolution method with M = 16 provides a solution that is as accurate as RK78 integration, but only 18 percent as many revolutions are computed. Since the extrapolation formulae require only a negligible amount of computational work, this percentage indicates the reduction in computer time that is possible.

---

* Units are Earth Radii, 1 ER = 6378 km.

The initial orbital elements for the lunar satellite problem were:

Semi-major axis	1.08 LR
Period	121 min
Eccentricity	0.05
Inclination	30°
Argument of periapsis	0°
Long. of ascending node	0°
Height of periapsis	45 km
Height of apoapsis	233 km

The solutions were compared at the 500th revolution and the results are shown below.

Method	N	M	$\|\Delta X_1\|$*	$\|\Delta X_2\|$	$\|\Delta X_3\|$	Revs. Compt. with RK78	Ratio
Multirevolution	8	8	3.1×10^{-9}	2.0×10^{-9}	7.5×10^{-11}	159	.318
"	8	12	1.8×10^{-7}	$1.5 \times 10^{-}$	3.7×10^{-8}	125	.250
"	8	16	1.6×10^{-6}	8.9×10^{-7}	7.4×10^{-7}	113	.226
RK78	-	-	3.2×10^{-9}	2.3×10^{-9}	6.6×10^{-11}	500	1.000

The multirevolution integration with M = 8 has the accuracy of the RK78 integration and requires only 32 percent as many revolutions to be computed.

6. Conclusions

Multirevolution integration methods, when applied to certain problems of artificial satellite orbit computation, offer large savings in computer time while providing accuracy comparable to that obtained by classical numerical integration methods. They are most efficient when applied to long term orbit predictions. Modified methods are available which allow periodic functions to be used in the multirevolution integration process.

---

* Units are Lunar Radii, 1 LR = 1738 km.

References

1. D. Boggs, "An Algorithm for Integrating Lifetime Orbits in Multirevolution Steps," AAS Paper No. 68-142, presented at the AAS/AIAA Astrodynamics Specialist Conference, Jackson, Wyoming, Sept. 3-5, 1968.

2. C.J. Cohen and E.C. Hubbard, "An Algorithm Applicable to Numerical Integration of Orbits in Multirevolution Steps," Astron. J., Vol. 65, pp. 454-456, 1960.

3. E. Fehlberg, "Classical Fifth-, Sixth-, Seventh-, and Eighth-Order Runge-Kutta Formulas with Stepsize Control," NASA Technical Report R-287, 1968.

4. P. Henrici, Discrete Variable Methods in Ordinary Differential Equations, John Wiley & Sons, New York, 1962.

5. D. Mace and L.H. Thomas, "An Extrapolation Formula for Stepping the Calculation of the Orbit of an Artificial Satellite Several Revolutions Ahead at a Time," Astron. J., Vol. 65, pp. 300-303, 1960.

6. W.G. Melbourne, J.D.Mulholland, W.L. Sjogren, F.M. Sturms, "Constants and Related Information for Astrodynamics Calculations," JPL Technical Report 32-1306, pp. 23-24, 1968.

7. E.L. Stiefel and G. Scheifele, Linear and Regular Celestial Mechanics, Springer-Verlag, New York, 1971.

8. G.P. Taratynova, "Numerical Solution of Equations of Finite Differences and Their Application to the Calculation of Orbits of Artificial Earth Satellites," ARS Journal, Vol. 31, pp. 976-988, 1961.

9. C.E. Velez, "Numerical Integration of Orbits in Multirevolution Steps," NASA Technical Note D-5915, 1970.

10. C.A. Wagner, "The Drift of a 24-Hour Equatorial Satellite Due to an Earth Gravity Field Through 4th Order," NASA Technical Note D-2103, 1964.

Vol. 215: P. Antonelli, D. Burghelea and P. J. Kahn, The Concordance-Homotopy Groups of Geometric Automorphism Groups. X, 140 pages. 1971. DM 16,–

Vol. 216: H. Maaß, Siegel's Modular Forms and Dirichlet Series. VII, 328 pages. 1971. DM 20,–

Vol. 217: T. J. Jech, Lectures in Set Theory with Particular Emphasis on the Method of Forcing. V, 137 pages. 1971. DM 16,–

Vol. 218: C. P. Schnorr, Zufälligkeit und Wahrscheinlichkeit. IV, 212 Seiten. 1971. DM 20,–

Vol. 219: N. L. Alling and N. Greenleaf, Foundations of the Theory of Klein Surfaces. IX, 117 pages. 1971. DM 16,–

Vol. 220: W. A. Coppel, Disconjugacy. V, 148 pages. 1971. DM 16,–

Vol. 221: P. Gabriel und F. Ulmer, Lokal präsentierbare Kategorien. V, 200 Seiten. 1971. DM 18,–

Vol. 222: C. Meghea, Compactification des Espaces Harmoniques. III, 108 pages. 1971. DM 16,–

Vol. 223: U. Felgner, Models of ZF-Set Theory. VI, 173 pages. 1971. DM 16,–

Vol. 224: Revètements Etales et Groupe Fondamental. (SGA 1). Dirigé par A. Grothendieck XXII, 447 pages. 1971. DM 30,–

Vol. 225: Théorie des Intersections et Théorème de Riemann-Roch. (SGA 6). Dirigé par P. Berthelot, A. Grothendieck et L. Illusie. XII, 700 pages. 1971. DM 40,–

Vol. 226: Seminar on Potential Theory, II. Edited by H. Bauer. IV, 170 pages. 1971. DM 18,–

Vol. 227: H. L. Montgomery, Topics in Multiplicative Number Theory. IX, 178 pages. 1971. DM 18,–

Vol. 228: Conference on Applications of Numerical Analysis. Edited by J. Ll. Morris. X, 358 pages. 1971. DM 26,–

Vol. 229: J. Väisälä, Lectures on n-Dimensional Quasiconformal Mappings. XIV, 144 pages. 1971. DM 16,–

Vol. 230: L. Waelbroeck, Topological Vector Spaces and Algebras. VII, 158 pages. 1971. DM 16,–

Vol. 231: H. Reiter, L^1-Algebras and Segal Algebras. XI, 113 pages. 1971. DM 16,–

Vol. 232: T. H. Ganelius, Tauberian Remainder Theorems. VI, 75 pages. 1971. DM 16,–

Vol. 233: C. P. Tsokos and W. J. Padgett. Random Integral Equations with Applications to stochastic Systems. VII, 174 pages. 1971. DM 18,–

Vol. 234: A. Andreotti and W. Stoll. Analytic and Algebraic Dependence of Meromorphic Functions. III, 390 pages. 1971. DM 26,–

Vol. 235: Global Differentiable Dynamics. Edited by O. Hájek, A. J. Lohwater, and R. McCann. X, 140 pages. 1971. DM 16,–

Vol. 236: M. Barr, P. A. Grillet, and D. H. van Osdol. Exact Categories and Categories of Sheaves. VII, 239 pages. 1971. DM 20,–

Vol. 237: B. Stenström, Rings and Modules of Quotients. VII, 136 pages. 1971. DM 16,–

Vol. 238: Der kanonische Modul eines Cohen-Macaulay-Rings. Herausgegeben von Jürgen Herzog und Ernst Kunz. VI, 103 Seiten. 1971. DM 16,–

Vol. 239: L. Illusie, Complexe Cotangent et Déformations I. XV, 355 pages. 1971. DM 26,–

Vol. 240: A. Kerber, Representations of Permutation Groups I. VII, 192 pages. 1971. DM 18,–

Vol. 241: S. Kaneyuki, Homogeneous Bounded Domains and Siegel Domains. V, 89 pages. 1971. DM 16,–

Vol. 242: R. R. Coifman et G. Weiss, Analyse Harmonique Non-Commutative sur Certains Espaces. V, 160 pages. 1971. DM 16,–

Vol. 243: Japan-United States Seminar on Ordinary Differential and Functional Equations. Edited by M. Urabe. VIII, 332 pages. 1971. DM 26,–

Vol. 244: Séminaire Bourbaki – vol. 1970/71. Exposés 382–399. IV, 356 pages. 1971. DM 26,–

Vol. 245: D. E. Cohen, Groups of Cohomological Dimension One. V, 99 pages. 1972. DM 16,–

Vol. 246: Lectures on Rings and Modules. Tulane University Ring and Operator Theory Year, 1970–1971. Volume I. X, 661 pages. 1972. DM 40,–

Vol. 247: Lectures on Operator Algebras. Tulane University Ring and Operator Theory Year, 1970–1971. Volume II. XI, 786 pages. 1972. DM 40,–

Vol. 248: Lectures on the Applications of Sheaves to Ring Theory. Tulane University Ring and Operator Theory Year, 1970–1971. Volume III. VIII, 315 pages. 1971. DM 26,–

Vol. 249: Symposium on Algebraic Topology. Edited by P. J. Hilton. VII, 111 pages. 1971. DM 16,–

Vol. 250: B. Jónsson, Topics in Universal Algebra. VI, 220 pages. 1972. DM 20,–

Vol. 251: The Theory of Arithmetic Functions. Edited by A. A. Gioia and D. L. Goldsmith VI, 287 pages. 1972. DM 24,–

Vol. 252: D. A. Stone, Stratified Polyhedra. IX, 193 pages. 1972. DM 18,–

Vol. 253: V. Komkov, Optimal Control Theory for the Damping of Vibrations of Simple Elastic Systems. V, 240 pages. 1972. DM 20,–

Vol. 254: C. U. Jensen, Les Foncteurs Dérivés de $\varprojlim$ et leurs Applications en Théorie des Modules. V, 103 pages. 1972. DM 16,–

Vol. 255: Conference in Mathematical Logic – London '70. Edited by W. Hodges. VIII, 351 pages. 1972. DM 26,–

Vol. 256: C. A. Berenstein and M. A. Dostal, Analytically Uniform Spaces and their Applications to Convolution Equations. VII, 130 pages. 1972. DM 16,–

Vol. 257: R. B. Holmes, A Course on Optimization and Best Approximation. VIII, 233 pages. 1972. DM 20,–

Vol. 258: Séminaire de Probabilités VI. Edited by P. A. Meyer. VI, 253 pages. 1972. DM 22,–

Vol. 259: N. Moulis, Structures de Fredholm sur les Variétés Hilbertiennes. V, 123 pages. 1972. DM 16,–

Vol. 260: R. Godement and H. Jacquet, Zeta Functions of Simple Algebras. IX, 188 pages. 1972. DM 18,–

Vol. 261: A. Guichardet, Symmetric Hilbert Spaces and Related Topics. V, 197 pages. 1972. DM 18,–

Vol. 262: H. G. Zimmer, Computational Problems, Methods, and Results in Algebraic Number Theory. V, 103 pages. 1972. DM 16,–

Vol. 263: T. Parthasarathy, Selection Theorems and their Applications. VII, 101 pages. 1972. DM 16,–

Vol. 264: W. Messing, The Crystals Associated to Barsotti-Tate Groups: With Applications to Abelian Schemes. III, 190 pages. 1972. DM 18,–

Vol. 265: N. Saavedra Rivano, Catégories Tannakiennes. II, 418 pages. 1972. DM 26,–

Vol. 266: Conference on Harmonic Analysis. Edited by D. Gulick and R. L. Lipsman. VI, 323 pages. 1972. DM 24,–

Vol. 267: Numerische Lösung nichtlinearer partieller Differential- und Integro-Differentialgleichungen. Herausgegeben von R. Ansorge und W. Törnig, VI, 339 Seiten. 1972. DM 26,–

Vol. 268: C. G. Simader, On Dirichlet's Boundary Value Problem. IV, 238 pages. 1972. DM 20,–

Vol. 269: Théorie des Topos et Cohomologie Etale des Schémas. (SGA 4). Dirigé par M. Artin, A. Grothendieck et J. L. Verdier. XIX, 525 pages. 1972. DM 50,–

Vol. 270: Théorie des Topos et Cohomologie Etale des Schémas. Tome 2. (SGA 4). Dirigé par M. Artin, A. Grothendieck et J. L. Verdier. V, 418 pages. 1972. DM 50,–

Vol. 271: J. P. May, The Geometry of Iterated Loop Spaces. IX, 175 pages. 1972. DM 18,–

Vol. 272: K. R. Parthasarathy and K. Schmidt, Positive Definite Kernels, Continuous Tensor Products, and Central Limit Theorems of Probability Theory. VI, 107 pages. 1972. DM 16,–

Vol. 273: U. Seip, Kompakt erzeugte Vektorräume und Analysis. IX, 119 Seiten. 1972. DM 16,–

Vol. 274: Toposes, Algebraic Geometry and Logic. Edited by. F. W. Lawvere. VI, 189 pages. 1972. DM 18,–

Vol. 275: Séminaire Pierre Lelong (Analyse) Année 1970–1971. VI, 181 pages. 1972. DM 18,–

Vol. 276: A. Borel, Représentations de Groupes Localement Compacts. V, 98 pages. 1972. DM 16,–

Vol. 277: Séminaire Banach. Edité par C. Houzel. VII, 229 pages. 1972. DM 20,–